CHEMICAL ECOLOGY

CONTRIBUTORS

JOHN TYLER BONNER
RAYMOND B. CLAYTON
V. G. DETHIER
THOMAS EISNER
ARTHUR D. HASLER
JOHN R. RAPER
JOHN B. SIDDALL
F. W. WENT
R. H. WHITTAKER
CARROLL M. WILLIAMS
EDWARD O. WILSON

CHEMICAL ECOLOGY

Edited by
ERNEST SONDHEIMER
and
JOHN B. SIMEONE

State University College of
Forestry at Syracuse
University
Syracuse, New York

ACADEMIC PRESS
NEW YORK LONDON 1970

ACADEMIC PRESS, INC.
111 Fifth Avenue, New York, New York 10003

United Kingdom Edition published by
ACADEMIC PRESS, INC. (LONDON) LTD.
Berkeley Square House, London W1X 6BA

LIBRARY OF CONGRESS CATALOG CARD NUMBER: 78-117113

PRINTED IN THE UNITED STATES OF AMERICA

This book is dedicated to Conrad Schuerch to whose unwavering efforts the chemical ecology program at this college owes its existence.

List of Contributors

Numbers in parentheses indicate the pages on which the authors' contributions begin.

JOHN TYLER BONNER (1), Princeton University, Princeton, New Jersey

RAYMOND B. CLAYTON (235), Department of Psychiatry, Stanford University School of Medicine, Stanford, California

V. G. DETHIER (83), Department of Biology, Princeton University, Princeton, New Jersey

THOMAS EISNER (157), Langmuir Laboratory, Cornell University, Ithaca, New York

ARTHUR D. HASLER (219), Laboratory of Limnology, University of Wisconsin, Madison, Wisconsin

JOHN R. RAPER (21), Harvard University, Cambridge, Massachusetts

JOHN B. SIDDALL (281), Insect Hormone Research, Zoecon Corporation, Palo Alto, California

F. W. WENT (71), Laboratory of Desert Biology, Desert Research Institute, University of Nevada System, Reno, Nevada

R. H. WHITTAKER (43), Section of Ecology and Systematics, Cornell University, Ithaca, New York

CARROLL M. WILLIAMS (103), Bussey Professor of Biology, Harvard University, Cambridge, Massachusetts

EDWARD O. WILSON (133), The Biological Laboratories, Harvard University, Cambridge, Massachusetts

Preface

Great rivers that had independent courses for eons have been known to send out branches, which, when they touched, formed a mighty new stream. This may also happen on occasion in the sciences. Recently, spectacularly successful methods have been developed for the purification and characterization of natural products in microquantities. At the same time, interest in solving some difficult ecological problems has quickened, due no doubt to social pressures. The result of this has been the development of the new discipline: chemical ecology.

Perhaps this term can best be clarified through an example. Many soil nematodes feed on higher plants such as potatoes, tobacco or beets. The eggs, which are usually encysted may remain dormant for periods up to ten years. It is of obvious survival value to have the eggs hatch at a time when there is an adequate food supply. This explains why many nematode eggs require specific root exudates as hatching factors. Work on the isolation and characterization of these substances is underway in several laboratories. These results are eagerly awaited, not only because they will teach us a great deal about ecological mechanisms, but also because they are expected to lead to the selective eradication of costly pests. If the hatching factors can be manufactured cheaply enough, they can be broadcast over the infected soil at a time when the nematode food sources are unavailable. The eggs will be induced to hatch but the emerging larvae will starve. Since one is using a material that at least at times is a normal soil constituent, possible ecological mischief to other organisms is reduced to a minimum.

As man becomes more reluctant to add synthetic substances of broad and lasting effectiveness to the environment, the impact of chemical ecology on industry, agriculture, and forestry may assume revolutionary significance. This volume will acquaint the reader with the salient facts and theories that are presently encompassed by chemical ecology. It is hoped that this treatise will also draw attention to the possible application of fundamental research in this area to pressing problems of ecological importance.

This book resulted from a series of lectures presented in the fall of 1968 at State University of New York College of Forestry at Syracuse University. Each contributor spent two days on campus, presented a formal lecture, and led a discussion group. In addition, a group of students used this material as the basis for a course and met further requirements through readings as well as oral and written presentations. A generous grant from the New York State Science and Technology Foundation made this program possible.

We are greatly indebted to each of the contributors and particularly to Professor C. M. Williams who, through his role as coordinator of this series, was largely responsible for the assembly of this distinguished roster. Finally we would like to thank Mrs. Bruce A. MacCurdy and Miss Susan Sondheimer for their help with this publication.

November, 1969

E. SONDHEIMER

J. B. SIMEONE

Contents

3. THE BIOCHEMICAL ECOLOGY OF HIGHER PLANTS

R. H. WHITTAKER

4. PLANTS AND THE CHEMICAL ENVIRONMENT

F. W. WENT

5. CHEMICAL INTERACTIONS BETWEEN PLANTS AND INSECTS

V. G. DETHIER

6. HORMONAL INTERACTIONS BETWEEN PLANTS AND INSECTS

CARROLL M. WILLIAMS

1

The Chemical Ecology of Cells in the Soil

JOHN TYLER BONNER

I. INTRODUCTION

Cells that live in the soil have both a changing pattern of distribution and a changing pattern of chemical substances surrounding them. Sometimes the cells will be relatively independent of each other, and sometimes they will be closely linked in some type of colony or even a small multicellular organism. In considering the primitive organization of the soil and the role played by the chemical substances, I want to show that this chemical ecology gives a direct insight into the evolutionary origin of multicellularity and its development in the life cycle of an organism.

The reason for choosing the organisms of the soil rather than aquatic forms is simply that the soil has more structure and more pattern; the environment of water is relatively homogeneous and constant, at least from the point of view of a microorganism. The size of the grains of soil, the air cavities, the moisture content, the distribution of organic, edible material will differ from one region to the next, or even in the same region over a period of time. The soil is a truly heterogeneous environment, and the problems of adapting to such conditions are correspondingly diverse.

1

The simplest kinds of organisms that live in the soil are bacteria. They are the basic food of all other organisms, and, were we to compare a small segment of soil to the world of macroorganisms, bacteria play an ecological role comparable with green plants. Their distribution in the soil is far from uniform, but is distinctly patchy.

It is possible to estimate the number of bacteria in soil, as has been done by numerous workers and summarized by Waksman (1952). The values depend greatly upon the type of soil and the degree of manuring, but in unfertilized soil one might expect values to range from 2 to 14 million cells/cm³. Singh (1949) has made a study of the abundance of soil amebae, which are the prime predators of bacteria, and under comparable conditions, they will vary from about 8 to 17 thousand cells/cm³. In other words, there are about 1000 bacteria (prey) to each ameba (predator), this fact being the foundation of an Eltonian food pyramid for soil. The relative abundance of larger soil animals have also been examined by a number of workers (Park *et al.*, 1939; Williams, 1941), and as expected, in general, the larger the animal, the fewer the number.

If we take the mean values for bacteria (8×10^6 cells/cm³) and amebae (12×10^3 cells/cm³), we find that the average distance between bacteria is 50 μ and the average distance between amebae is about 500 μ. But it must be repeated that in general these cells are not uniformly distributed in the soil, but clumped and grouped in various patterns.

If we look for an explanation of these patterns, we come up with two straightforward causes: localized growth of the cells and directed cell movements. These two factors are of paramount importance in the distribution of one-celled organisms in soil, and, as we shall see, they are largely the result of the chemical environment. We shall also see that cell differentiation plays a role too, although in many ways a less obvious one.

II. UNICELLULAR MICROORGANISMS

A. Growth

The first and foremost stimulus for growth is food, and a most common and effective inhibitor of growth is an absence of food. If there is a bit of organic matter in the soil, such as a dead nematode, one bacterium in the region will produce many millions in little time. And when the food is consumed, the growth will stop. There is much we could say about the capabilities of different species of bacteria for utilizing different kinds of nutrients and their adaptive roles within the soil. There are, for instance, some bacteria that have a very limited range of substances that they can utilize as fuel, yet there are others that have a wide range of possible diets. As Pardee (1961) has shown, the generalists have a battery of enzymes available and pay the price by being relatively slow growers, while the specialists grow with extraordinary rapidity once they have found the particular chemical they can utilize. The soil seems to contain both, for there are apparently two separate categories of niches, permitting the two types to exist in harmony, very much as has occurred in certain instances in macroecology. But this fascinating aspect of the chemical ecology of microorganisms brings us away from the straightforward point I wish to make: the distribution of food is the prime factor for the distribution of bacterial cells because the food can directly result in cell multiplication.

If we turn to the carnivores that eat the bacteria, it has recently been demonstrated by Horn (1969) that there is a marked feeding preference for certain species of bacteria by certain species of cellular slime mold amebae. In other words, the presence of particular food will affect which species of bacteria will grow, which in turn will determine the species of ameba present. Again, this is the kind of niche specialization which is common in higher animals and plants in a large complex environment. The only difference is that in the big forms there is greater consistency, greater stability; in the microworld multiplication is more rapid, and violent change more frequent.

The first step in the control of the external chemical environment may be seen in growth inhibitors. We must realize that bacteria compete; they are subject to natural selection. The success of a species depends upon its contribution to subsequent generations, and if this success can be furthered by preventing competing species from taking the limited food, this will obviously be selectively advantageous. The result is the production of antibiotics that prevent other cells from growing. Such growth inhibitors are a common component of the chemical environment of cells in the soil.

There is also another form of growth inhibition. Many bacteria and amebae produce resistant stages, or spores. These apparently serve the purpose of carrying the cell over adverse conditions. The biochemistry of spore germination is quite specialized, and there are particular substances which prevent this germination. They are not always produced by competing organisms, but sometimes by the parent organism. Such self-inhibition of germination is common in the seeds of higher plants, and it is assumed that their value is to prevent germination when the spores are concentrated. Such germination would be a disadvantage in a fruiting body where there is no food, or in an area in the soil where the spores or cysts are concentrated. After all, only one need germinate to make another generation; the rest would do better to remain spores in the event of sudden adverse conditions.

There is one situation in which it is conceivable that there would be an advantage to the mass germination of concentrated spores: if the cells issuing forth from the spores were gametes and needed to find a partner. In one known case where this does occur in the myxomycetes, Smart (1937) showed that the spores contained a germination stimulant, and the more concentrated the spores, the better the percent germination.

B. Differentiation

The primitive differentiation of single cells in the soil into spores or cysts is a reflection of the fact that the physical conditions in which they exist vary tremendously. They have to survive drought, flood, heat, and cold, conditions of the environment that may change with alarming suddenness. To put the matter another way, the sequence of changes of the physical conditions through time will produce a temporal patchiness in the environment. If an organism is to remain in a particular area of soil, it must differentiate resistant stages. Were it to have no such stage, it could only remain if the climate were so constant that no inclement change occurred. This is unlikely, even in the mildest tropical conditions, and therefore the only way a microorganism without spores could remain in any area is by constant reinvasions after each climatic catastrophe. Obviously, spores are of major adaptive value in view of the variability of the physical world.

A general stimulus for spore formation is starvation. For instance, in bacteria, it is known, depending on the species, that either a lack of a carbon source or a nitrogen source will trigger a new set of enzymes and new metabolic steps that lead ultimately to the production of spores. This is another very obvious way in which the chemical environment can have a profound effect on the organism.

C. Movement

From the point of view of natural selection, the movement of cells in the soil serves the important function of bringing the cells toward food and away from noxious regions. This is all part of the mechanism of dispersal which involves the invasion of new regions, and sometimes the escape of old ones which can no longer support them.

The motion itself may be passive or active. In cases where it is passive the cells or spores are carried by some large animal from one spot to another. Often the spores will be specially adapted to adhere to the body of a passing worm or insect that will bring it some distance to a new location.

The more interesting situation from our point of view is the case of active locomotion, where the cells themselves move by ameboid or flagellar motion, or by some other less clear fashion such as the gliding movement of myxobacteria cells. But whatever the means, each cell has its own propulsion device.

These cell movements of soil microorganisms are generally directed and the control is exerted by a gradient of a chemical substance. In this cellular chemotaxis there are presumably more molecules of some key substance on one side of a cell than the other, and the cell is sensitive to this difference. If the cell moves down the gradient toward the lowest concentration of the substance, it is a negative chemotaxis, and movement up the gradient is a positive chemotaxis.

As has been shown by Samuel (1961), a particular species of soil ameba (*Dictyostelium mucoroides*) tends to be repelled by cells of its own kind. For instance, if a group of cells are artificially placed in one spot on an open surface, they will spread out radially, and their movement away from the center will be oriented, not random. This phenomenon first shown by Twitty and Niu (for review, see Twitty, 1949), was demonstrated by the melanoblast cells of certain salamanders grown in tissue culture on which the cells space themselves so that they are roughly equidistant from one another.

In *Dictyostelium* cells it is possible occasionally to observe such an equidistant distribution of the amebae when they are actively feeding in a colony of bacteria. This presumably makes for dispersed and even grazing, a quality highly favored in the black-faced sheep of the Scottish Highlands. In both cases, the grazing efficiency is high and therefore presumed to be adaptively advantageous; the result is the character has become genetically fixed by selection.

Positive chemotaxis, a more striking event in soil organisms, is easy to demonstrate. It is known as a food-seeking mechanism for many amebae. In instances with the cellular slime mold amebae, it is possible to show that they will move toward a small colony of bacteria in a directed fashion (Konijn, 1961; Samuel, 1961). Recently we have been able to show that the active chemical given off by the bacteria that attracts slime mold amebae is cyclic 3',5'-adenosine monophosphate (Konijn, *et al.*, 1967). It has been known for some time that bacteria produce this substance (Makman and Sutherland, 1965; Okabayashi *et al.*, 1963). It is stable and yet can be enzymatically destroyed by an extracellular phosphodiesterase which the *Dictyostelium* amebae secrete (Chang, 1968).

Besides orientation toward food, positive chemotaxis plays an important part in bringing cells together in the same species. The most obvious cause of such mutual attraction would be sexual fusion, and some sex attraction has been clearly demonstrated in a number of fungi and algae. The soil amebae and the myxobacteria are for the most part asexual (at least sexuality has not been demonstrated), yet they show clear cell aggregation by positive chemotaxis. However, these aggregates appear to be a step toward multicellularity, and therefore are considered in Section III.

D. Size Increase

One of the most interesting roles of the chemical environment of cells in the soil has been to produce cell aggregates. But before we examine these aggregations, which are known in myxobacteria and in cellular slime molds, a few general things should be said about the adaptive value of such multicellularity by aggregation.

Any discussion of adaptive significance of this sort is total speculation.

We know that size increase has occurred, we have considerable evidence that natural selection is responsible for evolutionary change, and therefore we assume that the primitive multicellular forms of the soil arose because of the selective virtues in occupying new niches.

The advantage of size in food seeking, which we think of automatically when contemplating the large carnivores of the macroworld, seems unlikely to play a significant role. The slime molds, the myxobacteria, and even the fungi that invade the soil could all eat as effectively if they remained small. In fact, in the myxobacteria and the cellular slime molds, the eating is done at the unicellular, preaggregation stage, and multicellularity only occurs after the food is consumed; starvation appears to be a stimulus for aggregation and fruiting, another significant effect of the chemical environment to which we shall return (Fig. 1). In the myxomycetes, the feeding stage is large, in the form of a multicellular plasmodium, and perhaps there are advantages here in size for feeding, but what it might be is hard to grasp (Fig. 2).

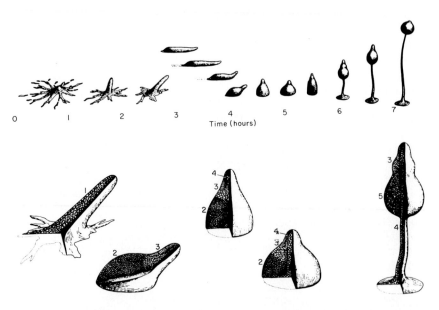

Fig. 1. Part of the life cycle of the cellular slime mold, *Dictyostelium discoideum.* Above: The aggregation, migration, and culmination stages shown in an approximate time scale. Below: Cutaway diagrams to show the cellular structure of different stages. 1, Undifferentiated cells at the end of aggregation; 2, prespore cells; 3, prestalk cells; 4, mature stalk cells; 5, mature spores. Spore germination and the growth of the separate amebae are not shown. (Drawing by J. L. Howard, courtesy of the *Scientific American.*)

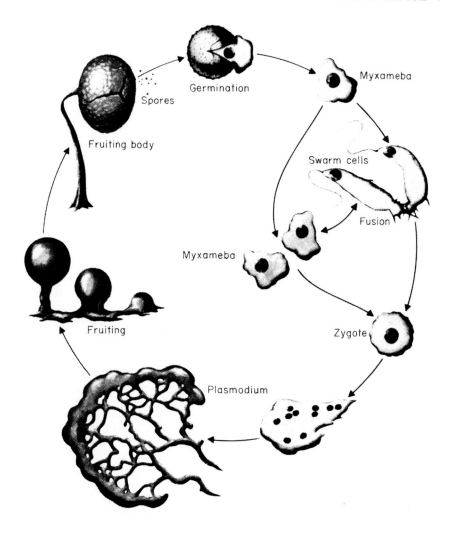

Fig. 2.   Diagram of a generalized life cycle of a true slime mold, a myxomycete. The spore germinates giving rise to a cell which, depending upon the environmental conditions, is either a myxamoba (e.g., in a dry environment) or a flagellated swarm cell (e.g., in a wet environment). After fertilization, the zygote grows into a large multinucleate plasmodium that eventually turns into many spore bearing fruiting bodies. (From C. J. Alexopoulos and J. Koevenig, "Slime Molds and Research," Copyright © 1964 by the Biological Sciences Curriculum Study. Published by D. C. Heath and Company, Lexington, Mass. BSCS pamphlet No. 13.)

It is much more likely that the advantage in size for these soil organisms lies in dispersal. Somehow fruiting bodies which bear a group of
spores rather than a solitary cyst must have advantages in producing
numerous offspring, at least under certain circumstances. One might
imagine that by raising the spores into the air in a small cavity of the soil,
or at the soil surface, the spores might be more resistant to adverse conditions or, more important, placed where they can effectively dispersed
by the wind or by a passing animal.

There is no doubt that with size increase certain physiological properties emerge that are denied to small organisms. For instance, in the cellular slime molds, the multicellular masses are phototactic, presumably
an advantage so that the spore mass will be near the soil surface and
therefore more likely to be dispersed. Since it is possible to produce cell
masses of different sizes merely by controlling the number of cells that
enter the aggregate, we examined the response to light in fruiting bodies
of a wide size range (Bonner and Whitfield, 1965). It was clear that the
small fruiting bodies were quite insensitive to light (as are the individual
cells); only the large ones responded. Since the small fruiting bodies are
normal in all other respects, one might imagine that a certain size is
necessary for the proper amplification of the light response mechanism.
In any event, we presume phototaxis is adaptive for better dispersal, and
this physiological attribute is only possible once a certain size has been
attained.

III. MULTICELLULAR MICROORGANISMS

A. Growth

There are numerous multicellular forms that increase size by growth.
It is as though the daughter cells of frequent divisions failed to separate.
An obvious example are the soil fungi. In this case the hyphae penetrate
by growth into those areas that contain food substances, and are inhibited in their growth in areas that are either without food or contain
growth inhibitors. The process is very simple, for the food is used directly
in the hyphal tip, and the hyphae grow at the apex. There is basically little difference in this example from that of the growth of single-celled
organisms in the soil; the only one being that the cells do not separate
but remain attached. Furthermore, it is well known that the hyphal tips
can grow towards the food, that is, they are capable of chemotropism.
For instance, Stadler (1952, 1953) has demonstrated this for the bread
mold, *Rhizopus*. He showed that the hyphae tend to grow away from one
another by the production of a "staling substance" which caused a negative chemotropism. Certain foods tended to inactivate the staling substance, thereby increasing the gradient with the result that the hyphae
would grow toward the food in a highly oriented fashion.

A number of interesting studies have been done on the pattern of my-celial growth. The hyphae branch at regular intervals, a process gov-erned by some internal control system, and the branches effectively fill the space. They do this because they repel one another by a staling sub-stance and therefore spread evenly, each hypha equidistant from its neighbor (Fig. 3). This is another example of efficient grazing, and also an example (as was the evenly dispersed amebae described previously) of a primitive territorial system, a matter which we will examine in more detail presently.

Many soil fungi have mutual hyphal attraction as well. This appears to be a localized phenomenon, where either two hyphae grow toward each other by chemotropism and anastamose, or a group of growing hyphae will cling together to form a fruiting body (Figs. 3, 4, 5). Obviously, in these cases there are local chemical differences, but the nature of the chemicals is totally unknown.

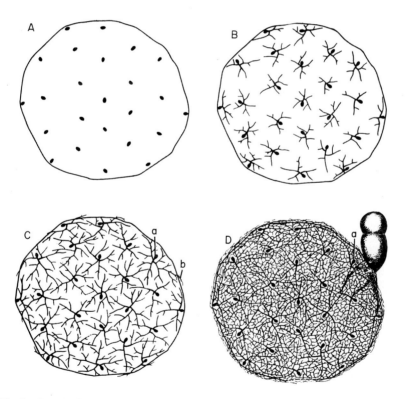

Fig. 3. Growth of the mushroom *Coprinus* on a restricted area of nutrient medium. Note both the even spacing of the hyphae as well as the occasional anastomosis of the hyphae. The final fruit-ing body is produced by the flow of protoplasm from the entire mycelium. A–D indicates time se-quence. (From A. R. H. Buller.)

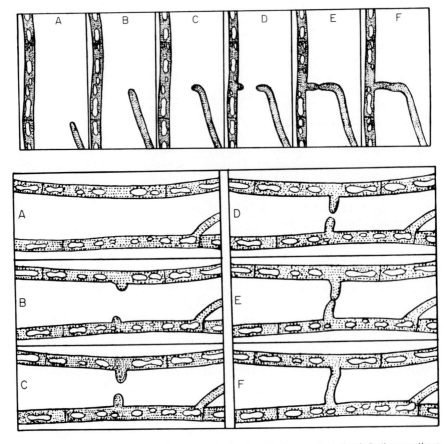

Fig. 4. Two examples of anastamosis between hyphae in a fungus. Note that in both cases there is a mutual attraction over a distance. A–F indicates time sequence. (From A. R. H. Buller.)

As with the single-celled forms, the whole pattern of growth, its stimulus or inhibition, is totally a reflection of the chemical environment in the soil. However, it depends not just on the external substances present by chance, but also on the control chemicals produced by the organism itself, on the removal of food, and on the production of growth inhibitors by competing organisms.

B. Differentiation

For the most part, multicellular forms form spores, often in an organized fruiting structure which consists of a supporting stalk and a mass of spores borne on or near the top. The stimulus for spore formation is starvation. This is true, as we have seen, of many unicellular forms as well, but it is far more striking in the larger species. In many fungi, myxomycetes, and cellular slime molds, once the food is gone, the chemical steps leading to spore formation begin. Again, this seems reasonable from the point of view of selection, for with starvation comes adversity and the need to tide the protoplasm over the lean period in some safe container.

Unlike the situation with bacteria, virtually nothing is known of the biochemical mechanism of this triggering and switching to spore formation in the larger forms. The initial stimulus from the chemical environment is the lack of food; but what comes next is not clear.

In the cellular slime molds, it is known that in the process of spore differentiation there is a stopping place, the so-called prespore cells, and these respond to a further chemical environmental cue to complete the

Fig. 5. Stages in the development of the fruiting body of the basidiomycete *Pterula gracilis*. The fruiting body begins as a single hypha, and as growth and branching proceeds the new hyphae adhere closely to one another. (After Corner.)

final spore differentiation. A very slight lowering of the humidity will induce this change, and by keeping the water vapor content of the air high the process can be deferred for extended periods (Bonner and Shaw, 1957). In another species, *Dictyostelium polycephalum*, Whittingham and Raper (1957) showed that it could only fruit and bear spores at slightly reduced humidities.

This differentiation into spores is the same example of differentiation in time that we saw in unicellular forms: a period of feeding is followed by a period of spore formation and ultimately by a period of rest in the spores. But there has also been a new kind of differentiation in these multicellular organisms. If we call the former temporal, the new one is spatial. For example, when a fruiting body is formed in some species, part of the cells make up the stalk and part of the cells make up the spores. These exist at the same time; it is a division of the labor at one moment in time. In mentioning this kind of differentiation, the discussion becomes removed from the external chemical environment and begins to involve the problem of the internal chemical environment of the cells within a multicellular organism.

C. Movement in Primitive Multicellular Microorganisms

If we look for feeding movement there is only one obvious example, for many of the larger soil microorganisms feed as separate cells before they become multicellular. But in the myxomycetes, the large multinucleate plasmodium is a large feeding structure. It is known that a plasmodium has a chemotactic system and will be attracted to certain foods, for instance, glucose (Coman, 1940). This situation closely parallels the food seeking by growth in the multinucleate hyphae of fungi.

The fruiting movements are of much greater significance. Of these, the primary one is the aggregation of cells into a fruiting body, a phenomenon characteristic of the myxobacteria and the cellular slime molds. In the case of the myxobacteria, the rod-shaped cells glide into collection points that ultimately produce fruiting structures containing spores or cysts. It has, in recent years, been demonstrated without question by McVittie and Zahler (1962) and Fleugel (1963) that this is a chemotactic process, although nothing is known of the chemicals involved.

There has been so much done on the aggregation process in the cellular slime molds that it is difficult to present the matter briefly. I shall confine my remarks to the most recent work and urge those who wish more details to consult reviews (Shaffer, 1962; Bonner, 1967).

The main point is that the amebae are attracted to central collection points by a substance called acrasin. It is also known that there is an enzyme (acrasinase), which is produced by the amebae, that destroys the acrasin. Recently we have found the chemical nature of this substance for one species, *Dictyostelium discoideum*. It is a well-known fact that there is a species specificity to aggregation, a phenomenon clearly designed to keep each species separate. The result is that there are likely to be different acrasins, and the chemical basis of the specificity is not known.

As mentioned previously, we know that cyclic 3',5'-AMP is an attractant produced by bacteria (Konijn *et al.*, 1967, 1968); see Fig. 6. It was then shown by one of us (Chang, 1968) that the acrasinase of *D. discoideum* was a specific phosphodiesterase which converted cyclic 3',5'-AMP to 5'-AMP. This enzyme was secreted into the medium by the amebae and was the reason why we had initially been unable to isolate much from *D. discoideum* for chemical identification. We had previously done this with another species (*Polysphondylium pallidum*), but it was not until recently that we were able to prove conclusively that *D. discoideum* also synthesizes cyclic AMP (Konijn *et al.*, 1969; Barkley, 1969).

We have begun to examine the details of this chemotactic system and have the following information. At the onset of aggregation, some hours after feeding is completed, there is a hundredfold increase in the amount of cyclic AMP synthesized and secreted and a hundredfold increase in the sensitivity of the amebae to cyclic AMP (Fig. 7). The acrasinase is presumably used in both cases as a means of maintaining and even increasing the active gradients (Bonner *et al.*, 1966, 1969).

Recently we have found that there is another substance (of unknown nature) which is produced by bacteria that is especially effective in attracting the feeding amebae. This means that both feeding and the production of multicellularity are the result of external chemical gradients. But once the cells are brought together in masses, acrasin and acrasinase are still produced within the multicellular organism, but their role there is unknown. This sequence is an excellent illustration of the main point of this entire paper: the chemicals which play a role in cells in the soil have, through evolution, become increasingly internal and controlled as the cells group and the size increases. The unpredictable chemical environment becomes a predictable, gene-controlled environment.

There is a good parallel to this aggregation in the fungi. This is the case for the forms that produce fruiting bodies made up of a mass of closely adhering hyphae, the extreme example being the large basidiomycete mushrooms. Here, all the growth is achieved by the mycelium, which removes the nutriment from the soil. When the right environmental conditions appear, the protoplasm from the outlying hyphae migrates to central collection points and makes up the fruiting body. This is the main reason why mushrooms appear with such astounding speed (Fig. 3).

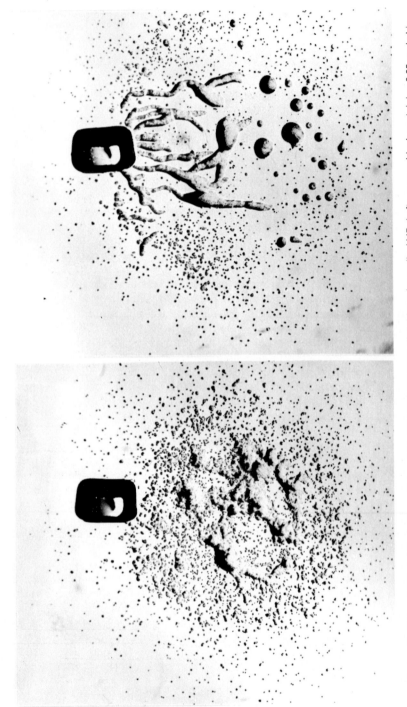

Fig. 6. Attraction of the amebae of the cellular slime mold *Dictyostelium discoideum* by cyclic AMP. Left: an agar block containing 0.05 mg/ml is placed near amebae that are at a sensitive stage. Right: 57 min later.

14

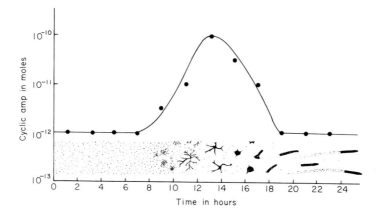

Fig. 7. A graph showing the amount of cyclic AMP given off during successive 2-hr intervals at different stages of development of the cellular slime mold, *Dictyostelium discoideum*. Note there is a hundredfold increase in acrasin given off at the height of aggregation.

Besides aggregation there are well-established cases of inhibition in the formation of fruiting bodies. This may be a widespread phenomenon, but again the cellular slime molds serve as an excellent example where the inhibition has been clearly demonstrated.

It has been shown by a number of workers that once an aggregation center is formed, it inhibits the formation of centers in the close vicinity (Arndt, 1937; Shaffer, 1962; Bonner and Dodd, 1962). It is possible to show, under certain conditions, that the density of the centers is relatively unaffected by the density of the amebae with the result that if there are few amebae, the fruiting bodies are small, and if there are many, they are larger. There are a number of factors that are probably involved in such territory formation, as Kahn (1968) has stressed, but Feit (1969) has recently confirmed the fact that one of the factors is a volatile substance that inhibits neighboring centers from forming. Such inhibition of rival centers is a primitive example of spacing within a species, a phenomenon so common among higher forms. We already encountered it in individual feeding amebae and feeding hyphae, where optimal grazing was achieved. Here it would seem to be that by having aggregation centers spaced, an optimal condition for spore dispersal is achieved.

Another aspect of spore dispersal deserves special attention: it is the orientation of the fruiting body so that the spores can spread most effectively. In the larger mushrooms gravity is a controlling factor, and in this way the gills are oriented so that the spores may stay dry and drop unhindered (Buller, 1909). But the majority of fruiting bodies in or on the surface of the soil are too small for gravity to operate, and other means are necessary.

Orientation to light is one common means. Clearly this will bring the spore mass up to the surface and in a region where dispersal is readily possible. In the cellular slime molds, the light sensitivity of the migratory cell masses is quite remarkable as is their sensitivity to temperature gradients, which would achieve a similar function (Bonner *et al.*, 1950; Francis, 1964).

But many small fruiting bodies are known to rise at right angles from the substratum in conditions of uniform light and heat, regardless of the position of the substratum. Clearly, the bodies are too small to be affected by gravity; the question is why are they perpendicular to the substratum.

The answer to this question for the cellular slime molds is that the fruiting body gives off a volatile substance which has the effect of repelling the rising cell mass (Bonner and Dodd, 1962). If this repellent is given off equally on all sides, clearly, the fruiting body will be exactly upright from the flat substratum. If two fruiting bodies are artificially placed close together, they will lean away from one another; they will repel each other. If a center is placed in a crack as though it were at the bottom of a crevice, the spore mass will be exactly in the center between the two walls of the crevice. This gas-induced orientation is perfectly suited to keep the spores in a position where dispersal is most likely to be effective. It is also of great interest to see that lowly soil organisms make use of volatile substances in the chemical control of their environment, a fact which is also known for various sex hormones in the Mucorales (for review, see Machlis, 1966).

IV. CONCLUSION

We have discussed what we know of the distribution pattern of microorganisms in the soil and would now like to summarize briefly some of the key points in order to use them as a basis for some generalizations.

If one looks at the soil from an ecological point of view, there is a size scale among the organisms, and their relative abundance fits the standard Eltonian pyramid of numbers: the larger the organism, the fewer individuals in a given space.

The soil also has niches for all size levels; this means a great proliferation of species at each level. The situation differs from macroecological conditions in that there is less overall stability. The sudden changes due to heat, cold, dryness, and moisture are especially severe, with the result that spore or cyst formation is ubiquitous, and of all the species present some are bound to be successful in each new incarnation of the environmental conditions.

There is a presumed evolutionary trend toward size increase in the soil, and with this increase one finds the following new features: more effective dispersal mechanisms and more self-produced chemical environment. The latter means more control of an organism's development; it is the first step toward the isolation of the organism from the extreme and fickle variability of the environment. It is of interest to note that, in these small multicellular organisms, the benefit of size increase is not more efficient feeding, but rather more efficient dispersal.

The gamut of size we see in soil organisms is essentially a panoramic view of evolutionary progress. It involves the occupation of new niches that are unavailable to the smaller organisms, and this process has been accompanied by increased chemical control. The most primitive condition is where the organism is entirely dependent upon the chemicals that happen to be in the soil. The next step is where the organism begins to produce chemicals that affect its own species favorably or other chemicals that affect other species adversely. The third step is the internalization of these new chemical control systems so that not only is there a chemical communication system between organisms, but ultimately one within a multicellular organism.

A good example of how the process has occurred comes from the work on cyclic AMP. Bacteria produce it, and while its role is not understood in bacteria, it is thought to be related to some metabolic control mechanism. Cellular slime molds go one step further and also use it as a means of communication between cells so that mutual attraction or aggregation can occur. It has been shown in studies on mammals that it is a second messenger in many hormone reactions (review: Robison *et al.*, 1968). The hormone stimulates the production of cyclic AMP, which in turn stimulates some reaction within the cell. The same substance in this hypothetical evolutionary sequence goes from a link in a metabolism chain to a primary hormone, to a secondary hormone. The moral is that with increased size of organisms there has been an increase in the length of the chain of any biochemical process, and this increase is the mechanism of greater control, both genetic and physiological. There has not only been a compounding of cells in the trend toward size increase, but a compounding of the chemical steps in the development and in the functioning of the organism. Each of these steps is governed in the genome so that there is a complex and stable internal chemical environment. But even though this be true, the multicellular organism still has an external environment replete with chemical information, and instead of ignoring this outside, the multicellular form is, if anything, even more sensitive in its response to the environment.

The chemical ecology of the soil reflects primitive evolutionary progress and at the same time illuminates the origin of the development of organisms and the origin of complex physiological reactions of organisms to their environment. We said that the pattern of distribution of living organisms in the soil is the result of growth, movement, and, indirectly, of differentiation. When the cells have come together to be bound in a simple multicellular organism, such an organism has a life cycle leading from a unicellular stage to the multicellular one. The pattern of the multicellular stage is also achieved by growth and by movement, which leads ultimately to differentiation. The difference between ecological patterns and developmental patterns is simply that in the latter the external chemical processes, with size increase, have become progressively supplemented by self-produced, internal chemical processes. The same is true for the physiological functioning of a multicellular organism; again it is achieved by complex internal chemical systems. And by internal we mean gene controlled, inherited, and therefore subject to natural selection.

Within the framework of evolution and genetics, it is clear from this excursion into primitive ecology that ecology, developmental biology, and physiology are closely allied; the difference is merely where one draws the line between organism and environment.

Acknowledgment

The experimental work described in this paper was supported in part by funds from research Grant No. GB-3332 of the National Science Foundation and by funds from the Hoyt Foundation. We also benefited from the central equipment facilities in the Biology Department, Princeton University, supported by the Whitehall Foundation and the John A. Hartford Foundation.

References

Arndt, A. (1937). Untersuchungen über *Dictyostelium mucoroides* Brefeld. *Arch. Entwicklungsmech. Organ. Wilhelm Roux* 136, 681–747.

Barkley, D. S., (1969). Adenosine-3′,5′-phosphate: identification as acrasin in a species of cellular slime mold. *Science* 165, 1133–1134.

Bonner, J. T. (1967). "The Cellular Slime Molds," 2nd Ed. Princeton Univ. Press, Princeton, New Jersey.

Bonner, J. T., and Dodd, M. R., (1962). Aggregation territories in the cellular slime molds. *Biol. Bull.* 122, 13–24.

Bonner, J. T., and Shaw, M. J. (1957). The role of humidity in the differentiation of the cellular slime molds. *J. Cellular Comp. Physiol.* 50, 145–154.

Bonner, J. T., and Whitfield, F. E. (1965). The relation of sorocarp size to phototaxis in the cellular slime molds *Dictyostelium purpureum. Biol. Bull.* 128, 51–57.

Bonner, J. T., Clarke, W. W., Jr., Neely, C. L., Jr., and Slifkin, M. K. (1950). The orientation to light and the extremely sensitive orientation to temperature gradients in the slime mold *Dictyostelium discoideum. J. Cellular Comp. Physiol.* 36, 149–158.

Bonner, J. T., Kelso, A. P., and Gilmore, R. G. (1966). A new approach to the problem of aggregation in the cellular slime molds. *Biol. Bull.* 130, 28–42.

Bonner, J. T., Barkley, D. S., Hall, E. M., Konijn, T. M., Mason, J. W., O'Keefe, O. G., III, and Wolfe, P. B. (1969). Acrasin, acrasinase, and the sensitivity to acrasin in *Dictyostelium discoideum. Develop. Biol.* 20, 72–87.

Buller, A. H. R. (1909). "Researches on Fungi," Vol. 1. Longmans, Green, New York.

Chang, Y. Y. (1968). Cyclic 3′,5′-adenosine monophosphate phosphodiesterase produced by the slime mold *Dictyostelium discoideum. Science* 160, 57–59.

Coman, D. R. (1940). Additional observations on positive and negative chemotaxis: Experiments with a myxomycete. *A.M.A. Arch. Pathol.* 29, 220–228.

Feit, I. N. (1969). Evidence for the regulation of aggregate density by the production of ammonia in the cellular slime molds. Ph.D. Thesis, Princeton Univ., Princeton, New Jersey.

Fleugel, W. (1963). Fruiting chemotaxis in *Myxococcus fulvus* (Myxobacteria). *Proc. Minn. Acad. Sci.* **30**, 120-123.

Francis, D. W. (1964). Some studies on phototaxis of *Dictyostelium*. *J. Cellular Comp. Physiol.* **64**, 131-138.

Horn, E. G. (1969). Some aspects of food competition among the cellular slime molds. Ph.D. Thesis, Princeton Univ., Princeton, New Jersey. In preparation.

Kahn, A. J. (1968). An analysis of the spacing of aggregation centers in *Polysphondylium pallidum*. *Develop. Biol.* **18**, 149-162.

Konijn, T. M. (1961). Chemotaxis in *Dictyostelium discoideum*. Ph.D. Thesis, Univ. of Wisconsin, Madison, Wisconsin.

Konijn, T. M., van de Meene, J. G. C., Bonner, J. T., and Barkley, D. S. (1967). The acrasin activity of adenosine-3',5'-cyclic phosphate. *Proc. Natl. Acad. Sci. U.S.* **58**, 1152-1154.

Konijn, T. M., Barkley, D. S., Chang, Y. Y., and Bonner, J. T. (1968). Cyclic AMP: A naturally occurring acrasin in the cellular slime molds. *Am. Naturalist* **102**, 225-233.

Konijn, T. M., Chang, Y. Y., and Bonner, J. T. (1969). Cyclic AMP synthesis by *Dictyostelium discoideum* and *Polysphondylium pallidum*. *Nature*, (in press).

Machlis, L. (1966). Sex hormones in fungi. *In* "The Fungi" (G. C. Ainsworth and A. S. Sussman, eds.) Vol. II, pp. 415-433. Academic Press, New York.

McVittie, A., and Zahler, S. A. (1962). Chemotaxis in *Myxococcus*. *Nature* **194**, 1299-1300.

Makman, R. S., and Sutherland, E. W. (1965). Adenosine-3',5'-phosphate in *Escherichia coli*. *J. Biol. Chem.* **240**, 1309-1314.

Okabayashi, T., Yoshimoto, A., and Ide, M. (1963). Occurrence of nucleotides in culture fluids of microorganisms. V. Excretion of adenosine cyclic 3',5'-phosphate by *Brevibacterium liquifaciens* sp. n. *J. Bacteriol.* **86**, 930-936.

Pardee, A. B. (1961). Response of enzyme synthesis and activity to environment. *Symposia Soc. Gen. Microbiol.* **11**, 19-40.

Park, O., Allee, W. C., and Shelford, V. E. (1939). "A Laboratory Introduction to Animal Ecology and Taxonomy." Univ. of Chicago Press, Chicago, Illinois.

Robison, G. A., Butcher, R. W., and Sutherland, E. W. (1968). Cyclic AMP. *Ann. Rev. Biochem.* **37**, 149-174.

Samuel, E. W. (1961). Orientation and rate of locomotion of individual amoebae in the life cycle of the cellular slime mold *Dictyostelium discoideum*. *Develop. Biol.* **3**, 317-335.

Shaffer, B. M. (1962). The Acrasina. *Advan. Morphogenesis* **2**, 109-182.

Singh, B. N. (1949). The effect of artificial fertilizers and dung on the numbers of amoebae on Rothamsted soils. *J. Gen. Microbiol.* **3**, 204-210.

Smart, R. F. (1937). Influence of certain external factors on spore germination in the *Myxomycetes*. *Am. J. Botany* **24**, 145-159.

Stadler, D. R. (1952). Chemotropism in *Rhizopus nigricans*: The staling reaction. *J. Cellular Comp. Physiol.* **39**, 449-474.

Stadler, D. R. (1953). Chemotropism in *Rhizopus nigricans*. II. The action of plant juices. *Biol. Bull.* **104**, 100-108.

Twitty, V. C. (1949). Developmental analysis of amphibian pigmentation. *Growth* **9**, 133-161.

Waksman, S. A. (1952). "Soil Microbiology." Wiley, New York.

Whittingham, W. F., and Raper, K. B. (1957). Environmental factors influencing the growth and fructification of *Dictyostelium polycephalum*. *Am. J. Botany* **44**, 619-627.

Williams, E. C. (1941). An ecological study of the floor fauna of the Panama rain forest. *Bull. Chicago Acad. Sci.* **6**, 63-124.

2

Chemical Ecology among Lower Plants

JOHN R. RAPER

I. INTRODUCTION

In this imperfect world, the basic requirements of life are in short supply, and the first imperative for survival of every organism, be it an individual or a species, is to obtain *more* than its fair share of the scarce essentials of life. Thus the prime basis for biological success is the ability simultaneously to grab the available necessities of life while blocking competitors from the same objectives. The central importance of chemical substances in this continual and deadly struggle among living things was established for soil microorganisms by Professor Bonner (Chapter 1). His survey of numerous microbiological phenomena that are mediated by chemical substances gives some intimation of the awesome complexity of what really goes on in any environment capable of sustaining life.

In respect to the origins and targets of ecologically significant chemical agents, however, two major categories must be recognized. The first category might be termed *extrabiotic* and includes the generalized effects of substances present in the environment, both of biotic and nonbiotic origin, i.e., organic materials that serve as food, growth substances, nonspecific toxicants, etc., as well as inorganic materials of one sort or another. The second or *biotic* category comprises those highly specific effects upon organisms caused by secretions of other organisms. The latter category may be further subdivided into two types of cases that can most conveniently be distinguished as *intraspecific* and *interspecific.* The effects of secretions that elicit specific responses in individuals of the same species are almost invariably regulatory and, in general, increase the adaptiveness of the single species involved. By contrast, the effects of secretions that have been evolved to elicit specific responses in individuals of other species are most commonly antagonistic and increase the adaptiveness of the secreting species at the expense of the target species. Bonner's concept of the evolutionary progression would nicely explain both the prevalence and the variety of intraspecific as well as interspecific chemical effects that are known among eukaryotic microorganisms. Interspecific effects — as exemplified by antibiosis — will be discussed briefly later. Intraspecific effects, however, are less commonly known and deserve further consideration.

Intraspecific chemical ecology no doubt plays an important role in the overall biology of most microorganisms, but the actual cases that can be documented relate to rather few basic activities. These pertain for the most part to the coordination of the activities or of the development of neighboring individuals. Bonner mentioned the interactions among myoamebas of a cellular slime mold that result in their uniform distribution during the vegetative and feeding phase as representative of many such cases of regulated distribution that occur widely among plants and animals.

Although intraspecific chemical ecology finds many expressions among the lower plants, primary attention will be devoted here to one particular category: the coordination and regulation of sexual processes between cells and between individuals.

II. INTRASPECIFIC CHEMICAL ECOLOGY – SEXUALITY

Sexuality, even in its simplest expressions in primitive organisms, is a very complex business, and the mechanisms that have evolved in lower plants to initiate and regulate these activities are surprisingly varied and often display considerable evolutionary sophistication. Most of the lower plants, members of the fungi and algae, are essentially aquatic. Many actually live in an aquatic habitat, but others that occupy niches commonly considered to be terrestrial (such as soil, parasites on land plants, etc.) still pass through stages in their life cycles during which they are critically dependent on water. In such organisms, constantly immersed in an aqueous environment, the evolution and fixation of regulatory systems depending upon diffusible metabolic products are easily rationalized.

Consideration of the regulation of sexual activities can perhaps best be introduced by an examination of several adjustments and functions relating to the sexual process that must be achieved if the sexual progression is to be effective (Raper, 1957). Each of the following six functions can thus be recognized as constituting a distinct prerequisite for the success of the entire sexual venture.

1. Initiation

For any lower plant sexuality demands the channeling of a major fraction of the organism's metabolic activities. In the very simplest lower plants, unicells or simple filaments, the onset of sexual activity often precludes any continuing vegetative growth, but, in any case, sexual activity requires a major adjustment of metabolic processes.

2. Differentiation

Following the initiation of sexual activity, the synthetic processes are usually so altered as to produce specifically differentiated structures that are remarkably unlike the vegetative portions of the same plant.

3. Sequential Regulation

Sexual processes typically involve a number of successive developments, and the temporal or sequential regulation of the various successive stages is obviously of critical importance.

4. Spatial Orientation

Spatial adjustment, to bring together compatible sexual elements, is almost invariably necessary sometime during the sexual process. In forms having motile gametes, juxtaposition of compatible elements is a matter of random or directed motility and is a somewhat different problem from that in many lower plants that lack motile gametes. In these latter cases, other methods of adjustment are necessary.

5. Quantitative Control

If the sexual process is to be efficient in terms of the energy expended, some sort of regulation is necessary to provide ultimately fusing elements of the two compatible types in the appropriate ratio to insure the optimal yield of zygotes.

6. Qualitative Control

Again, if the process is to be effective in terms of its energy requirements, sexual activation and continued sexual activity must be limited to those circumstances that provide a high probability of a successful culmination. Qualitative regulations are necessary in various circumstances, particularly in the restriction of cross-mating and in the determination of species specificities.

The proper execution of these several functions results in the production of sexual organs or sexual cells competent to undergo sexual fusion. Among lower plants, these functions are often achieved through the activities of diffusible metabolic products, or "sexual hormones." These, of course, constitute a single category of those substances, commonly termed pheromones, that elicit specific responses outside the body of the secreting organism. In a very few cases, all six of these obligatory functions have been shown to be effected by chemical substances. In other cases, chemical agents are known to be involved only in certain of the functions, whereas in still other cases, such as the initiation of sexual activity, it is triggered by external stimuli such as light, physiological conditioning, e.g., low nitrate availability, or other means. Critical roles of specific chemical substances in the initiation or coordination of sexual processes have been demonstrated in numerous species of algae and fungi.

Chlamydomonas is a very large genus of unicellular, green algae. In certain species, sexual fusion can occur between unicells of a single clone (i.e., homothallic); in other species, cells of a given clone can mate only with cells of a different, compatible clone (i.e., heterothallic). Each clone thus belongs to one or the other of two mating types, (+) and (−). It was first observed almost a century ago in some heterothallic species of *Chlamydomonas* and certain other algae having isogametes that when cells of compatible clones are mixed, there is rapid agglutination of the cells to form large clumps, and these clumps later break up into individual pairs of copulating cells (de Bary and Strasburger, 1877; but see Machlis and Rawitscher-Kunkel, 1967). Much later, it was found that the filtrate of a clone of either (+) or (−) cells would induce clumping of cells of the opposite mating type (Hartmann, 1932; Moewus, 1933). A little more than 10 years ago, the clumping reaction in one species was shown to be due to two surface-active agents that were specific to the two mating types, and the substances responsible for the interaction proved to be glycoproteins (Förster and Wiese, 1954a,b; Förster *et al.*, 1956; Wiese and Jones, 1963).

The flagella had earlier been shown to be critically involved in the clumping and pairing interactions in a related species of *Chlamydomonas* (Lewin, 1954). It is now generally considered that the surface-active glycoproteins are associated with the flagella, from which they may be liberated into the filtrate to induce clumping (Coleman, 1962).

In most algae in which gametes of both mating types are motile, chemotaxis appears to play no role in sexual behavior (Lewin, 1954; Coleman, 1962). In a single species of *Chlamydomonas*, however, a chemotactic response of (+) cells to secretion(s) of (−) cells has been demonstrated (Tsubo, 1957, 1961). Only gametes were affected, vegetative (+) cells giving no response. Interspecific tests with the active (−) filtrate and several species of *Chlamydomonas* revealed three patterns of response in different species: both (+) and (−) gametes responding, only (+) gametes responding, and neither (+) nor (−) gametes responding.

A colonial form, *Volvox*, a distant relative of *Chlamydomonas*, displays certain refinements in the use of diffusible chemical substances in the regulation of sexual activities. The colony of *Volvox* (Fig. 1) consists of numerous vegetative cells, each structurally very similar to *Chlamy-*

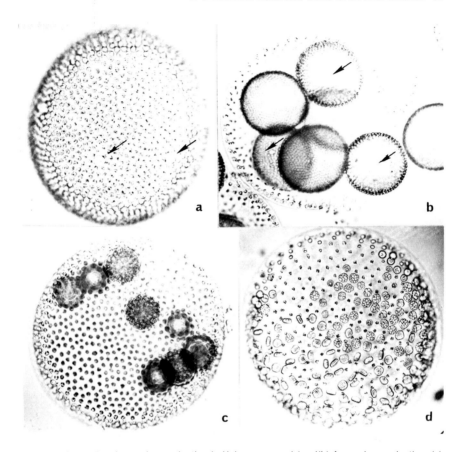

Fig. 1. Asexual and sexual reproduction in *Volvox aureus*. (a)and(b) Asexual reproduction. (a) Young, potentially ♀ colony having a number of gonidia (arrows). In the absence of fertilizing sperm, the gonidia develop asexually into daughter colonies. (b) About half of the asexually derived colonies have gonidia and are potentially ♀ or asexually reproducing colonies (arrows); the remaining colonies lack gonidia and are potentially ♂ or purely vegetative colonies. (c) and (d) Sexual reproduction. (c) The gonidia can also behave as ♀ gametes, i.e., fuse with sperm to form zygotes that develop into daughter colonies. (d) Young colonies lacking gonidia (b) remain vegetative in isolation. In a mixed culture of ♂ and ♀ colonies or in filtrate of ♂ colonies, young gonidia-lacking colonies become differentiated as ♂ colonies, with the production of numerous sperm packets. (From Darden, 1966.)

domonas, arranged in a hollow sphere, and held in position by a gelatinous substance. Fibrils connect the individual cells and apparently serve as channels of communication to coordinate cellular activities. In a typical species (Darden, 1966), the colonies of individual clones are about half male and half female, i.e., sperm- and nonmotile egg-producing, respectively. Differentiation into potential male or potential female colonies occurs early in development. In some colonies, a few cells enlarge and become specialized as reproductive cells, or gonidia. These colonies are potentially female, and the gonidia, if fertilized, serve as eggs; if not fertilized, the gonidia serve as asexual reproductive cells. Gonidia are not formed in about half of the colonies, however, and these colonies, if isolated, remain vegetative. If such vegetative colonies are placed at the proper time in filtrate from a mixed culture consisting of about equal numbers of male and female colonies, they become differentiated as male colonies and produce numerous packets of sperm. This effect was shown to be due to a highly specific agent, proteinaceous in nature, secreted by differentiated male colonies (Darden, 1966). Comparable activities of specific secretions have been demonstrated in several other species of *Volvox*, but the inductive system differs rather markedly from species to species (Starr, 1968).

A considerably more elaborate hormonal coordinating system has been described for an unidentified species of the filamentous green alga, *Oedogonium* (Rawitscher-Kunkel and Machlis, 1962). The life cycle of this and many other species of *Oedogonium* is somewhat complicated by the inclusion of diminutive male plants, the dwarf males, each of which produces a couple of motile sperm (Fig. 2). The species is heterothallic, i.e., ♂ and ♀ sexual organs are produced on different individuals or filaments. When grown in isolation, ♂ filaments produce specialized reproductive cells within which are formed motile androspores, and these may serve as asexual reproductive cells. ♀ filaments in isolation produce enlarged oogonial mother cells that are competent to develop into ♀ sexual cells. In isolation, development proceeds no further than this in either ♂ or ♀. In a mixed culture, however, androspores are actively attracted to the oogonial mother cells, to which a number become attached. The androspores then germinate to produce short filaments that become differentiated into dwarf male plants. As the dwarf males grow and differentiate, however, they become curiously oriented on the ♀ filament. Those attached to the anterior two-thirds or so of the oogonial mother cell bend sharply toward the apical end of the cell, whereas those located posterior to this line bend in the opposite direction. A simple but clever manipulation shows that this orientation as well as further differentiation of ♂ and ♀ cells are both highly specific effects dependent upon actual contact. If the ♀ filament is encased in a thin sheath of agar, androspores are still attracted to the region of the oogonial mother cells and become attached to the agar sheath in the immediate vicinity. There they germinate and develop into vegetative filaments, oriented not in the highly specific way of the dwarf males, but simply perpendicular to the surface to which they are attached.

Actual attachment of differentiated dwarf males to the wall of the oogonial mother cell for a period of no less than 8 hr is requisite to the further development of the ♀ sexual apparatus. Only after the differentiation of dwarf males attached to its wall does the oogonial mother cell

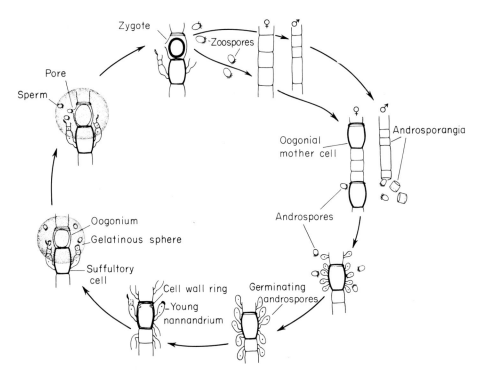

Fig. 2. The life cycle of a nannandrous (dwarf male), heterothallic species of *Oedogonium*. A chemotactic agent, secreted by the oogonial mother cells, attracts the androspores, which become attached to the ♀ cells. Additional secretions that are effective over short distances or that may require actual physical contact between the interacting elements appear to be responsible for (1) differentiation of the dwarf male plants, (2) the orientation of the dwarf males on the oogonial mother cells, and (3) the differentiation of the oogonial mother cell to form the oogonium and the subtending suffultory cell. (From Rawitscher-Kunkel and Machlis, 1962.)

become differentiated. This process involves the division of the oogonial mother cell to form two cells, a basal suffultory cell and an apical oogonium, the contents of which become organized into a single large, nonmotile egg. The differentiated oogonium develops a receptive papilla and secretes about itself a gelatinous sheath that is thought to serve as a sperm trap. In any event, the anteriorly directed dwarf males attached to the suffultory cell discharge their sperm into the gelatinous sheath, and a single sperm then enters the receptive papilla to fertilize the single egg.

That a number of sexual hormones are involved in the sexual progression of *Oedogonium* is quite well established, although the number of specific substances has not yet been determined with certainty, nor is anything known of the chemical nature of the secretions involved.

Comparable systems for the regulation of sexual processes by chemical secretions are known among the various groups of fungi. These again range from apparently simple surface-active interactions to highly complicated systems involving numerous specific secretions. To what extent the apparent simplicity of certain systems might be due to lack of information cannot now be stated; of the complexity of other systems, however, there can be no doubt.

There have been a few reports of the initiation or regulation of sexual activity by diffusible chemical agents in unicellular fungi. True unicellular forms occur only in two very different groups of the fungi: (a) in the yeasts of the Ascomycetes and (b) in the haploid phase of the smuts and a few saprophytic members of the Basidiomycetes.

The first report of the control of sexual activity in a yeast involved interspecific stimulation: the enhancement of conjugation in a homothallic species of *Zygosaccharomyces* by a filtrate of the mold, *Aspergillus niger* (Nickerson and Thimann, 1941, 1943). Addition of a filtrate of the mold to a culture of the yeast approximately tripled the frequency of conjugation, and greater enhancement was achieved with filtrate fractions. Two active components were found to be involved, and these could be replaced by riboflavin and glutaric acid.

A number of other cases of chemical control of sexual processes in unicellular forms clearly involve intraspecific systems. Baker's yeast, *Saccharomyces cerevisiae*, is heterothallic, the two self-sterile, cross-fertile classes being designated a and α. Secretion of cells of either mating type induces a specific swelling response in cells of the compatible mating type (Yanigishima *et al.*, 1968). This response is interpreted as a preliminary stage in sexual activation. One of the two agents, that secreted by cells of the a mating type, has been studied in some detail, and present evidence suggests it to be a steroidlike compound. Also in baker's yeast, cells of a and α mating types in close proximity on an agar surface interact without necessary physical contact. Hyphal processes, copulatory or conjugation tubes, form on a cells and grow to α cells (Levi, 1956). A two-way, a-α, interaction appears to be a requisite for sexual activation. Cells of neither strain gave a "mating reaction" when placed on agar on which the other had grown, but a cells reacted by the formation of conjugation tubes when placed on agar on which a and α cells had interacted. These facts suggest that secretion(s) of the a cells stimulates the α cells to secrete (in the absence of any morphological change) an agent(s) that induces the formation of conjugation tubes on a cells. Such cryptic activating agents are known in other systems (see *Mucor* and *Ascobolus* discussed later).

A comparable case of initiation and induction of sexual activity has been described in the wood-rotting basidiomycete, *Tremella mesenterica*, the haploid vegetative phase of which consists of yeastlike unicells that multiply asexually by budding. The cells of a single clone are incapable of fusing among themselves but they fuse readily with cells of other compatible clones. Filtrates of such vegetative cells induce the formation of conjugation tubes on cells of compatible clones (Bandoni, 1963, 1965). Copulation in this form, as in the yeast, seems to be effected by at least two specific substances, each secreted by cells of one mating type and having a specific inductive effect on cells of the compatible mating type.

In at least one other heterothallic yeast, however, *Hansenula wingei*, conjugation tubes are not formed on cells of either mating type under the influence of filtrate from cells of the other type. Conjugation tubes are only formed when compatible cells are actually in contact. Surface-active agents, however, are active in this species, for, when suspensions of cells of the two mating types are mixed, massive agglutination is the result (Brock, 1958a,b). Two complementary macromolecular substances are postulated as responsible for this agglutination, and such a factor has recently been obtained from one of the reacting strains. This factor comprises about equal amounts of protein and carbohydrate and is demonstrably involved in agglutination (Brock, 1965). Although agglutination has not been obligatorily connected with mating per se, the author of this work points out the obvious facilitation to mating afforded by agglutination.

The agents responsible for agglutination in this yeast are quite different from those involved in induction of conjugation tubes in the baker's yeast and in *Tremella*. In both of these cases, the active agents are obviously small molecules, as they pass readily through dialyzing membranes. Actually, small molecular size and the consequent rapidity of diffusion would seem to be an absolute requirement of substances responsible for the initiation of sexual activity and the coordination of sexual development between individuals.

Spatial orientation was mentioned earlier as a requirement for effective sexual activity, and something of a special case was made of those forms having motile gametes. Mate location, however, appears to be differently managed in various types of organisms having motile gametes. In *Chlamydomonas*, for example, agents capable of inducing chemotactic response in the unicells have, to date, been found in only a single species, and the location of a mate appears in most cases to depend upon active, random movement. Once compatible cells bump into each other, however, surface-active agents provide opportunity for them to become intimately acquainted. A very different situation, however, exists in the water mold, *Allomyces*, because here the ♀ gametes secrete a chemical lure, *sirenin*, which causes in the ♂ gametes a positive chemotactic response and to which the gametes are extremely sensitive (Machlis, 1958a,b,c). Sirenin is of particular significance in that it is the first sexual hormone of any plant to be isolated (Machlis *et al.*, 1966) and to be chemically identified (Machlis *et al.*, 1968). Sirenin is a colorless, optically active, viscous liquid having a molecular weight of 236 and the following chemical structure:

$$\text{HOH}_2\text{C} \diagdown \underset{\text{H}_3\text{C}}{\text{C}} = \underset{\text{H}}{\text{C}} - \underset{\text{H}_2}{\text{C}} - \underset{\text{H}_2}{\text{C}} - \text{C} \underset{\text{CH}_3}{\overset{\text{H H}_2}{\diagup}} \cdots$$

(chemical structure)

The statement that this is the first identified plant sexual hormone requires the following justification. More than a century ago, the chemotactic attraction of sperm to the ripened archegonia of certain mosses was first described, and Pfeffer (1884), almost a century ago, demonstrated the positive chemotactic activity of sodium malate for the sperm of several liverworts, mosses, and ferns. During the next two decades, numerous workers showed an extensive array of simple organic salts (malate, maleate, succinate, fumerate, and tartrate), other organic compounds (sucrose, a few proteins, e.g., hemoglobin), as well as various inorganic ions to elicit chemotactic responses by the sperm of various liverworts, mosses, ferns, and "fern allies." It is interesting to note that in this wide array of chemotactic agents the sperm of most individual species would respond only to one agent or to a very few agents (cf. Machlis and Rawitscher-Kunkel, 1967, for review). The specific substance(s) that are liberated in the maturation of the archegonia in these cases and serve as active sperm lures, however, have never been identified.

No other stage in the sexual progression of *Allomyces* is known to be mediated by secreted chemical substances. This is perhaps not too surprising since the species of *Allomyces* in which sirenin is known to be active is homothallic; i.e., ♂ and ♀ sexual cells or gametangia are produced in juxtaposition on the gametophytic plant, and coordination and regulation of all stages prior to the release of the gametes can be managed internally. Liberated ♂ and ♀ gametes, however, are not identical. The ♀ gametes are rather large and, although flagellated, are essentially sedentary, whereas the ♂ gametes are smaller and extremely active. A more or less sedentary habit of the ♀ gamete permits the establishment of a concentration gradient of sirenin that can very effectively encourage any ♂ gamete in the vicinity.

Beyond these forms in which chemical substances have been shown to regulate single events in the sexual progression, there are a few cases among the fungi in which the entire sexual progression has been shown to be under control of specific secretions. These are *Achlya*, of the biflagellate aquatic Phycomycetes; *Mucor*, of the Mucorales or "black bread molds"; and *Ascobolus*, a coprophilous ascomycete, a member of the "cup fungi." Detailed descriptions of all three of these systems cannot be given here, for the hormonal mechanisms in the three cases are quite as dissimilar as are the life cycles and morphological characteristics of the three species involved. They do, however, have certain basic features in common that would seem to be of considerable ecological significance.

In each of the three cases, (a) successive stages in the progression alternate from one sexual partner to the other, (b) the achievement of each successive stage is dependent upon the proper development of the last preceding stage, and finally, (c) each successive event in the sexual progression can be correlated with a single member in a series of successive secretions.

Achlya is a small inconspicuous water mold that grows on such organic substrates as dead insects, seeds, etc. It is small, the individual plant consisting of a large number of radiating filaments or hyphae and measuring no more than a centimeter or so in diameter. Individual plants in the species of interest here are self-sterile but belong to a rather large number of interfertile strains, most of which can react either as ♂ or as ♀ depending upon the stronger sexual affinity of the mate. Our concern here, however, is between plants that interact, one as ♂ and the other as ♀.

The sexual progression in *Achlya* consists of a number of distinct stages (Fig. 3). When ♂ and ♀ plants are placed together, intense branching is soon evident in the ♂ hyphae to produce many tiny, sinuous filaments. These are the antheridial hyphae, or ♂ sexual organ primordia. Following this sexual activation of the ♂ plant, large lateral branches appear on the ♀, and each of these rapidly enlarges to form a globose structure densely packed with protoplasm, the oogonial initial or ♀ sexual organ primordium. Once the oogonial initials are fully formed, the growth of antheridial hyphae in the vicinity is altered in such a way that subsequent growth is directed toward the mature oogonial initials. Once an antheridial hypha has reached the oogonial initial, its tip becomes somewhat enlarged, often branched, and appressed to the wall of the oogonial initial. The portion of the antheridial hypha in contact with the oogonial wall is then cut off from the supporting hypha by a cross wall and is now differentiated as the ♂ gametangium, or antheridium. Shortly thereafter, a similar differentiative process occurs in the oogonial initial by the formation of a basal septum, followed by a reorganization of the protoplasmic contents to form a number of large, uninucleate, ♀ gametes, or eggs. It now appears highly probable that meiosis occurs, both in the antheridium and in the oogonium, during this differentiative process. Finally, tiny tubes grow into the lumen of the oogonium from the antheridium, and a single ♂ nucleus is discharged into each ♀ gamete.

The initiation of the sexual process and the coordination of the several stages in the progression in *Achlya* have been shown to be mediated by a series of sexual hormones (Raper, 1939, 1951). The hormonal coordinating system, as interpreted from the evidence available some 15 years ago (Raper, 1955), considered specific secretions to operate at four successive stages during the progression.

1. Sexual activation of the ♂ and the quantitative control of the production of antheridial hyphae. This process is initiated by a hormone

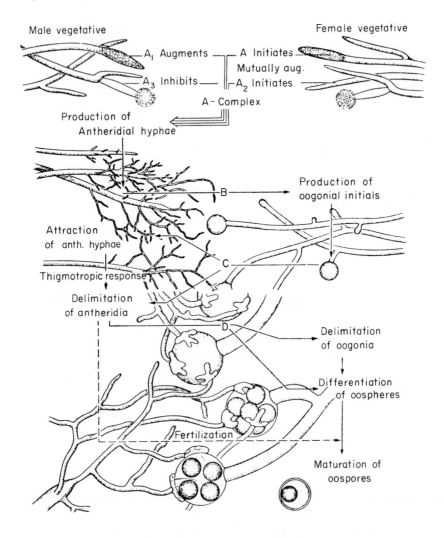

Male vegetative

A_1 Augments ——— A Initiates
Mutually aug.
A_3 Inhibits ——— A_2 Initiates

A - Complex

Female vegetative

Production of
Antheridial hyphae

B ——→ Production of
oogonial initials

Attraction
of anth. hyphae

Thigmotropic response

C

Delimitation
of antheridia

D ——→ Delimitation
of oogonia

Differentiation
of oospheres

Fertilization

Maturation of
oospores

Fig. 3.   A semidiagrammatic representation of the sexual progression in a heterothallic species of *Achlya* relating the sequence of morphological developments to the origins and specific activities of the several sexual hormones. (From Raper, 1955.)

from the vegetative ♀ plant, hormone A, but the response of the ♂ to this substance was thought to be modified by three additional secretions, two from the ♂ and one from the ♀.

2. Induction of oogonial initials. Secretion(s) from the activated male, i.e., only *after* the production of antheridial hyphae, elicits a specific response in the female, the *de novo* induction of heavy, lateral branches and their subsequent differentiation into oogonial initials.

3. Chemotropic attraction of antheridial hyphae and the differentiation of antheridia. The oogonial initials, in turn, are the foci of the
secretion(s) that chemotropically attracts the antheridial hyphae and
later, after the antheridial hyphae are applied to the oogonial wall, induces the differentiation of antheridia.

4. Differentiation of oogonia. Secretion(s) from the differentiated antheridia induces the final differentiation of the oogonium.

A number of details of this interpretation have recently been called
into question on the basis of new findings (Machlis, 1966). For example,
hormone A, absorbed on minute plastic spheres, "mock oogonia," beauti-
fully serves to attract antheridial hyphae, and hormone A alone has also
been shown to induce the differentiation of antheridial hyphal tips to
form antheridia (Barksdale, 1963, 1967). These effects, however, have
been seen only at concentrations of hormone A of the order of 10–100
times the normal physiological concentration of the hormone. In addition, the demonstration that hormone A *can* effect these responses does
not necessarily mean that in nature it does. The possibility that a distinct
hormone C, very similar to hormone A or even modified from it, is the
agent normally responsible for the chemotropic growth and differentiation of the ♂ element has in no way been discounted.

Of one thing, however, there is no question: hormone A is real. In fact,
it is the second plant sexual hormone to be isolated and tentatively identified (McMorris and Barksdale, 1967; Arsenault *et al.*, 1968). It is a 29-
carbon sterol, a close relative of stigmasterol, and very probably has the
following structure:*

Although it shares certain features in common with the hormonal system in *Achlya*, that in *Mucor* differs from it in certain basic respects. In
the first place, the type of sexuality involved is quite different, the two
compatible, interacting strains being morphologically indistinguishable.
They are thus not differentiated morphologically as ♂ and ♀, but physiologically — a situation termed incompatibility — and the two interacting
strains are simply distinguished as (+) and (−). The morphological mani-
festation of sexuality in *Mucor* are reasonably simple. When (+) and (−)

*Note added in proof. The formulas for sirenin of *Allomyces* and for antheridiol (hormone
A) of *Achlya* have been confirmed by the synthesis of both compounds (Edwards *et al.*, 1969;
Corey *et al.*, 1969).

strains are grown together, specialized lateral branches that are des-
tined to form the sexual organs originate from both (+) and (−) hyphae.
These sexual organ primordia are known as zygophores, and ordinarily,
as in a mating between compatible strains, (+) and (−) zygophores are
mutually attracted and grow in such a way as to meet tip-to-tip. Once
compatible zygophores have come together at their tips, they enlarge,
and a multinucleate gametangium is cut off at the tip of each. The two
gametangia soon fuse to form a multinucleate fusion product, the zygo-
spore.

The activity of a diffusible agent in the induction of zygophores in
Mucor was established in 1924, the first demonstration of a sexual hormone in any plant (Burgeff, 1924). Zygophores were produced on mycelia of both (+) and (−) strains of *Mucor mucedo* when the two plants were
physically separated by a permeable collodion membrane. More recent
work (Banbury, 1954, 1955; Plempel, 1957, 1960a,b, 1963a,b; Plempel
and Braunitzer, 1958; Plempel and Dawid, 1961) has specified the number, origins, and specific activities of several hormones that initiate and
coordinate the sexual progression in *Mucor* and has reported the isolation and characterization of the zygophore-inducing hormones.

The pattern of sexuality exhibited by *Mucor* accounts for a basic but
unusual feature of the hormonal mechanism. It was noted above that
there is no morphological differentiation between (+) and (−) strains, either in the vegetative phase or at any stage in the sexual progression.
The same sequence of developmental stages accordingly occurs simultaneously in both mates, and a corresponding complementary duplicity
also is found with the initiating and coordinating sexual hormones.

Each of the two strains secretes a hormone that actuates the other, in
the absence of any morphological response, to secrete a second specific
hormone. This second hormone induces the *de novo* production of zygophores, the secretion of each strain eliciting a response only in hyphae of
the other strain. The first morphological evidence of sexual activity thus
signals the completion of two distinct stages mediated by no less than
four specific secretions. The four substances involved to this point are
secreted into the substrate and pass from one plant to the other by
diffusion.

Once they are formed, each of the (+) and (−) zygophores secretes into
the air a specific substance that serves, by chemotropic action, to attract
the zygophores of the opposite mating type and to repel those of the same
mating type.

The zygophore-inducing hormones have been isolated from mixed cultures of (+) and (−) strains. The individual hormones, though demonstrably different by biological tests, cannot be chromatographically separated and thus are considered to be very similar in chemical structure.
The isolated hormone is a clear, yellow-red, viscous oil with the probable
formula $C_{20}H_{25}O_5$ (Plempel, 1963b).

A third hormonal mechanism that coordinates activities throughout the entire sexual progression occurs in the ascomycetous cup fungus, *Ascobolus stercorarius* (Bistis, 1956, 1957, 1965; Bistis and Raper, 1963; Bistis and Olive, 1968). In this species, as in *Mucor*, there are two morphologically indistinguishable classes, i.e., mating types A and a, and sexual interaction occurs only when a plant of one class interacts with one of the other. The basic pattern of sexuality here differs from that in *Mucor*, however, in that plants of both types are hermaphroditic, i.e., each produces functional ♂ and ♀ sexual organs. Plants of the two mating types are self-sterile but cross-fertile, however, and interact only when the male element of one class is brought to the female element of the other. In each mating between two compatible plants, two distinct, reciprocal crosses are accordingly possible. The sexual progression here, as in *Achlya* and *Mucor*, consists of a series of successive stages; also, as in *Mucor*, the same developmental stages occur simultaneously in the two mates.

The vegetative plant of *Ascobolus* consists of a dense, rapidly growing mycelium, the aerial hyphae of which become fragmented to produce vast numbers of specialized spores, known as oidia. In isolation, the oidia germinate within a few hours and develop into new vegetative plants. In the presence of a previously established mycelium of *Ascobolus* of either mating type, the oidia do not germinate. In the presence of a mycelium of the opposite mating class, however, the oidia are sexually activated to function as ♂ gametangia or antheridia, and small segments of the vegetative mycelium similarly become differentiated as antheridia under the same circumstances. The vegetative mycelium of the opposite mating type in the vicinity of these activated antheridia produce *de novo* ♀ sexual organs, or ascogonia, tightly coiled, multicellular filaments. The apical cell of each ascogonium becomes differentiated as the trichogyne, and this grows directly to a differentiated antheridium and fuses with it (Fig. 4). One or more male nuclei pass through the trichogyne into the ascogonium, where "fertilization" occurs. Following fertilization, there is a rapid proliferation of vegetative hyphae in the vicinity of the fertilized ascogonium, and this forms a protective sheath within which the subsequent development of the fertilized ascogonium occurs.

A number of specific secretions are known to be effective in this sequence, but the number of secretions and the specific responses that they elicit have not been determined in exact detail. Because of the parallel development in the two mates, only one sequence of hormones will be followed in detail (Fig. 5). Secretion(s) of class a vegetative mycelium, for example, inhibits the vegetative development of oidia of both classes. Secretion(s) of vegetative mycelium a − whether the same as the oidial-inhibiting secretion is unknown − can sexually activate oidia or induce differentiation of antheridia only in elements of the opposite or A vegetative mycelium. Secretion(s) of antheridia of class A causes three discern-

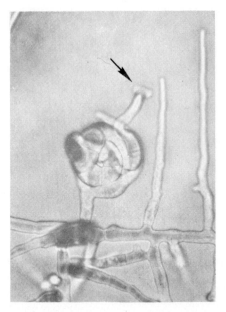

Fig. 4. The sexual apparatus of *Ascobolus stercorarius*. The tightly coiled ♀ gametangium, the ascogonium, terminates in an elongate cell, the trichogyne. Directional growth of the trichogyne is chemotropically controlled by secretion(s) of the ♂ gametangium, or antheridium, here a sexually activated oidium (arrow). (Courtesy of G. N. Bistis.)

ible developments in the *a* mycelium: (a) the induction of ascogonial initials, (b) the differentiation of these initials to form functional ascogonia, and (c) direction of the growth of the trichogyne to the secreting male element. The first two of these activities are rigidly class specific, i.e., the secretions of the antheridia have no effect whatever upon mycelia of the same mating type. Somewhat surprisingly, however, the third effect is not class specific, and the secretions of activated antheridia of either class attract the trichogynes of both classes indiscriminately. Secretions of the fertilized ascogonium induce rapid proliferation of nearby vegetative hyphae and these form a protective sheath. It should be noted here, as with *Mucor*, that essentially the same progression occurs in each of two morphologically identical plants, and here, as there, the hormones must exist in complementary duplicity. Of the hormonal mechanisms known in fungi, that in *Ascobolus* is in many ways the most complex, and much work is obviously needed to clarify the many details that at present remain obscure.

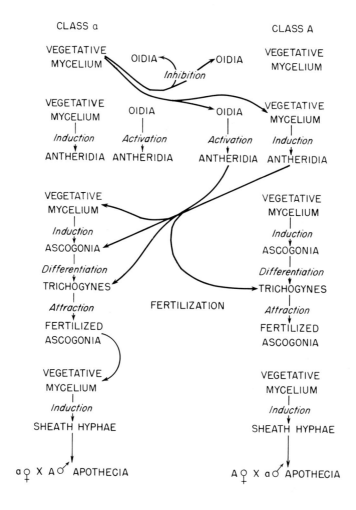

Fig. 5. The role of hormones in the sexual progression of *Ascobolus sterocarius*. The origins of the several hormones and responding elements are indicated by arrows. Only the hormones of the $a♀ × A♂$ sequence are shown, but a parallel set of hormones in the $A♀ × a♂$ sequence is required to account for the reciprocity of the sexual progression. Although the interaction between the two strains is reciprocal, the $a♀ × A♂$ sequence predominates in a mixed culture, and relatively few $A♀ × a♂$ apothecia are produced. (From Raper, 1969.)

III. INTERSPECIFIC CHEMICAL ECOLOGY — TRAPPING SUBSTANCES

Any impression that lower plants have expended all their ingenuity in the evolution of chemical systems for the regulation of sexual activities should be dispelled by brief descriptions of two interspecific systems in which fungi "deliberately" trap and consume animals (Duddington, 1968).

The first of these is a very inconspicuous water mold by the name of *Zoophagus* (Sommerstoff, 1911; Prowse, 1954). This is a very simple plant, actually a relative of *Achlya*, and in its vegetative state consists only of simple filaments with small lateral pegs. These lateral pegs, however, are tipped with a substance of unknown composition that rotifers apparently consider a confection. These tiny animals, evidently in an attempt to eat the "confection," become permanently affixed to the lateral pegs, which then produce short filaments that grow into the immobilized rotifers. The result is the effective conversion of rotifer meat into fungal protoplasm.

Finally, there is a group of fungi, the predacious Moniliales, that make their living by trapping nematodes (Drechsler, 1941; Duddington, 1962). This may seem a very unlikely activity for a sluggish fungus, when the extreme nervousness, activity, and probably extreme slipperiness of nematodes are appreciated. A typical member of this group consists of sparingly branched filaments bearing highly specialized lateral branches. Each of these lateral branches is a closed circle consisting of three cells. Nematodes, in addition to being nervous and active, are apparently very curious. When one enters one of these curious circular branches, it triggers an immediate response whereby each of the three cells suddenly inflates (Muller, 1958), an event signaling the end of the nematode as a free agent. Again, as with *Zoophagus*, absorptive fungal hyphae then grow into the prey.

The most significant aspect of this interspecific ecological relationship is the fact that certain species of nematode-trapping fungi produce no snares when grown in pure culture. When such an ill-equipped mycelium is presented with nematodes, however, the elaborate and highly specialized snares are formed in abundance within 24 hr. The correlation between the presence of nematodes and the formation of snares led to the concept that some metabolic product of the worm has a specific morphogenetic role in the development of the fungus. This was confirmed by the demonstration that the fluid from a culture of nematodes did, in fact, induce the fungus to form lateral branches that became differentiated as snares (Comandon and de Fonbrune, 1938; Lawton, 1957). The active principle(s) has been named *nemin* (Pramer and Stoll, 1959), and considerable progress has been made toward the characterization of the substance from the ascarid worms of the pig (Pramer, 1964). The active agent is not extractable in organic solvents, but it can be recovered in the proteinaceous fraction. That the active principle is a simpler compound normally associated with protein is indicated by (a) no loss of activity through the action of proteolytic enzymes and (b) the loss of activity through dialysis of the hydrolyzed protein.

IV. CONCLUSION

The above sample of intraspecific and interspecific interactions among lower plants that are mediated by specific chemical secretions is by no means exhaustive. Primary emphasis has been given a single type of system, the chemical regulation of sexual processes, to stress better the variety and complexity of the devices that have been evolved by primitive organisms to achieve the proper execution of a single, vital function. Equally varied arrays of systems are no doubt involved in each of many other aspects of the ecology–biology of microorganisms. Two additional vital functions of lower plants with which everyone is more or less familiar will serve to illustrate the principle of multiple means, i.e., divergent chemical systems, to common goals. The single, simple principle of inhibiting the growth or multiplication of competitors while leaving one's own vital processes unaffected underlies all cases of antibiosis, but consider the multitude of chemically distinct antibiotics and the wide range of metabolic mischief by which their effectiveness is achieved. Much the same range (although a totally different pattern of diversity) is evident in the expression of parasitism, an ecological relationship of particular importance to the achlorophyllous lower plants, the fungi. Chemical factors, abysmally ill-defined at present, must inevitably play essential roles in determining sensitivity versus resistance of the host and in every stage of the process of infection and pathogenicity. Parasite–host interactions ordinarily operate only through short distances, and extraorganismal effects such as active predation constitute only one extreme, though the most dramatic, aspect of parasitism.

These phenomena—the specific responses of organisms to metabolic products of other organisms either of the same species or of widely different species—are, in most cases, known only in general outline. Aside from the group of compounds known as antibiotics, only a very few of the active chemical agents have been chemically defined, and the actual targets of their specific activities are for the most part quite unknown. However, sufficient progress toward better definition of the field has been made in recent years to indicate the opportunities for the clarification of many basic biological phenomena that are promised by the study of the chemical ecology of the lower plants.

References

Arsenault, G. P., Biemann, K., Barksdale, A. W., and McMorris, T. C. (1968). The structure of antheridiol, a sex hormone in *Achlya bisexualis*. *J. Am. Chem. Soc.* **90**, 3635–3636.
Banbury, G. H. (1954). Processes controlling zygophore formation and zygotropism in *Mucor mucedo*. *Nature* **173**, 499.

Banbury, G. H. (1955). Physiological studies in the Mucorales. III. The zygotropism of zygophores of *Mucor mucedo. J. Exptl. Botany* 6, 235–244.

Bandoni, R. J. (1963). Conjugation in *Tremella mesenterica. Can. J. Botany* 41, 467–474.

Bandoni, R. J. (1965). Secondary control of conjugation in *Tremella mesenterica. Can. J. Botany* 43, 627–630.

Barksdale, A. W. (1963). The role of hormone A during sexual conjugation in *Achlya ambisexualis. Mycologia* 55, 627–632.

Barksdale, A. W. (1967). The sexual hormones of the fungus *Achlya. Ann. N.Y. Acad. Sci.* 144, 313–319.

Bistis, G. N. (1956). Sexuality in *Ascobolus stercorarius*. I. Morphology of the ascogonium; plasmogamy; evidence for a sexual hormonal mechanism. *Am. J. Botany* 43, 389–394.

Bistis, G. N. (1957). Sexuality in *Ascobolus stercorarius*. II. Preliminary experiments on various aspects of the sexual process. *Am. J. Botany* 44, 436–443.

Bistis, G. N. (1965). The function of the mating type locus in filamentous Ascomycetes. *In* "Incompatibility in Fungi" (K. Esser and J. R. Raper, eds.), pp. 23–31. Springer, Berlin.

Bistis, G. N., and Olive, L. S. (1968). Induction of antheridia in *Ascobolus stercorarius. Am. J. Botany* 55, 629–634.

Bistis, G. N., and Raper, J. R. (1963). Heterothallism and sexuality in *Ascobolus stercorarius. Am. J. Botany* 50, 880–891.

Brock, T. D. (1958a). Mating reaction in the yeast *Hansenula wingei. J. Bacteriol.* 75, 697–701.

Brock, T. D. (1958b). Protein as a specific cell surface component in the mating reaction of *Hansenula wingei. J. Bacteriol.* 76, 334–335.

Brock, T. D. (1965). Purification and characterization of an intracellular sex-specific mannan protein from yeast. *Proc. Natl. Acad. Sci. U.S.* 54, 1104–1112.

Burgeff, H. (1924). Untersuchungen über Sexualität und Parasitismus bei Mucorineen. I. *Botan. Abhandl. (K. Goebel)* 4, 5–135.

Coleman, A. W. (1962). Sexuality. *In* "Physiology and Biochemistry of Algae" (R. A. Lewin, ed.), pp. 711–727. Academic Press, New York.

Comandon, J., and de Fonbrune, P. (1938). Recherches experimentales sur les champignons predateurs de nematodes du sol. *Compt. Rend. Soc. Biol.* 129, 619.

Corey, E. J., Achiwa, K., and Katzenellenbogan, J. A. (1969). Total synthesis of *dl*-sirenin. *J. Am. Chem. Soc.* 91, 4318–4320.

Darden, W. H., Jr. (1966). Sexual differentiation in *Volvox aureus. J. Protozool.* 13, 239–255.

de Bary, A., and Strasburger, E. (1877). *Acetabularia mediterranea. Botan. Z.* 75, 713.

Drechsler, L. (1941). Predacious fungi. *Biol. Rev. Cambridge Phil. Soc.* 16, 265–290.

Duddington, C. L. (1962). Predacious fungi and the control of eelworms. *In* "Viewpoints in Biology" (J. D. Carthy and C. L. Duddington, eds.), Vol. I, pp. 151–200. Buttersworth, London and Washington, D.C.

Duddington, C. L. (1968). Predacious fungi. *In* "The Fungi" (G. C. Ainsworth and A. S. Sussman, eds.), Vol. III, pp. 239–251. Academic Press, New York.

Edwards, J. A., Miller, J. S., Sundeen, J., and Fried, J. H. (1969). The synthesis of the fungal sex hormone antheridiol. J. Am. Chem. Soc. 91, 1248–1249.

Förster, H., and Wiese, L. (1954a). Untersuchungen zur Kopulationsfähigkeit von *Chlamydomonas eugametos. Z. Naturforsch.* 9, 470–471.

Förster, H., and Wiese, L. (1954b). Gamonwirkung bei *Chlamydomonas eugametos. Z. Naturforsch.* 9, 548–550.

Förster, H., Wiese, L., and Braunitzer, G. (1956). Über das agglutinierend wirkende gynogamone von *Chlamydomonas eugametos. Z. Naturforsch.* 116, 315–317.

Hartmann, M. (1932). Neue Ergebnisse zum Befruchtungs- und Sexualitätsproblem. *Naturwissenschaften* 20, 567–573.

Lawton, J. R. (1957). The formation of constricting rings in nematode-catching Hyphomycetes grown in pure culture. *J. Exptl. Botany* 8, 50–54.

Levi, J. D. (1956). Mating reaction in yeast. *Nature* 177, 753–754.

Lewin, R. A. (1954). Sex in unicellular algae. *In* "Sex in Microorganisms" (D. H. Wenrich, I. F. Lewis, and J. R. Raper, eds.), pp. 100–133. Am. Assoc. Advance. Sci., Washington, D.C.

Machlis, L. (1958a). Evidence for a sexual hormone in *Allomyces*. *Physiol. Plantarum* 11, 181-192.

Machlis, L. (1958b). A study of sirenin, the chemotactic sexual hormone from the watermold *Allomyces*. *Physiol. Plantarum* 11, 845-854.

Machlis, L. (1958c). A procedure for the purification of sirenin. *Nature* 181, 1790-1791.

Machlis, L. (1966). Sex hormones in fungi. In "The Fungi" (G. C. Ainsworth and A. S. Sussman, eds.), Vol. II, pp. 415-433. Academic Press, New York.

Machlis, L., and Rawitscher-Kunkel, E. (1967). In "Fertilization" (C. B. Metz and A. Monroy, eds.), Vol. I. pp. 117-161. Academic Press, New York.

Machlis, L., Nutting, W. H., Williams, M. W., and Rapoport, H. (1966). Production, isolation, and characterization of serenin. *Biochemistry* 5, 2147-2153.

Machlis, L., Nutting, W. H., and Rapoport, H. (1968). The structure of sirenin. *J. Am. Chem. Soc.* 90, 1674-1676.

McMorris, T. C., and Barksdale, A. W. (1967). Isolation of a sex hormone from the water mold *Achlya bisexualis*. *Nature* 215, 320-321.

Moewus, F. (1933). Untersuchungen über Sexualität und Entwicklung von Chlorophyceen. *Arch. Protistenk* 80, 469-525.

Muller, H. G. (1958). The constricting ring mechanism of two predacious Hyphomycetes. *Brit. Mycol. Soc. Trans.* 41, 341-364.

Nickerson, W. J., and Thimann, K. V. (1941). The chemical control of conjugation in *Zygosaccharomyces*. *Am. J. Botany* 28, 617-621.

Nickerson, W. J., and Thimann, K. V. (1943). Chemical control of conjugation in *Zygosaccharomyces*. II. *Am. J. Botany* 30, 94-101.

Pfeffer, W. (1884). Locomotorische Richtungsbewegungen durch chemische Reize. *Untersuch Botan. Inst. Tubingen* 1, 363 pp.

Plempel, M. (1957). Die Sexualstoffe der Mucoraceae. *Arch. Mikrobiol.* 26, 154-174.

Plempel, M. (1960a). Die zygotropische Reaktion bei Mucorineen. I. *Planta* 55, 254-258.

Plempel, M. (1960b). Die Darstellung einer kristallinen Benzoesäure-Esters der Sexualstoffe von *Mucor mucedo*. *Naturwissenschaften* 47, 472-473.

Plempel, M. (1963a). Die Mucorineen-Gamone. *Naturwissenschaften* 50, 226.

Plempel, M. (1963b). Die chemischen Grundlagen der Sexualreaktion bei Zygomyceten. *Planta* 59, 492-508.

Plempel, M., and Braunitzer, G. (1958). Die Isolierung der Mucorineen Sexualstoffe I. *Z. Naturforsch.* 13b, 302-305.

Plempel, M., and Dawid, W. (1961). Die zygotropische Reaktion bei Mucorineen. II. *Planta* 56, 438-446.

Pramer, D. (1964). Nematode-trapping fungi. *Science* 144, 382-388.

Pramer, D., and Stoll, N. R. (1959). Nemin: a morphogenetic substance causing trap formation by predacious fungi. *Science* 129, 966-967.

Prowse, G. A. (1954). *Sommerstoffia spinosa* and *Zoophagus insidians* and *Rozellopsis inflata* the endoparasite of *Zoophagus*. *Brit. Mycol. Soc. Trans.* 37, 134-150.

Raper, J. R. (1939). Role of hormones in the sexual reaction of heterothallic Achlyas. *Science* 89, 321-322.

Raper, J. R. (1951). Sexual hormones in *Achlya*. *Am. Sci.* 39, 110-120.

Raper, J. R. (1955). Some problems in specificity in the sexuality of plants. In "Biological Specificity and Growth" (E. G. Butler, ed.), pp. 119-140. Princeton Univ. Press, Princeton, New Jersey.

Raper, J. R. (1957). Hormones and sexuality in lower plants. *Sym. Soc. Exptl. Biol.* 11, 143-165.

Raper, J. R. (1969). Growth and reproduction of fungi. In "Plant Physiology: A Treatise" (F. C. Steward, ed.), Vol. VI. Academic Press, New York. In preparation.

Rawitscher-Kunkel, E., and Machlis, L. (1962). Hormonal integration of sexual reproduction in *Oedogonium*. *Am. J. Botany* 49, 177-183.

Sommerstoff, H. (1911). Ein tierefangender Pilz (*Zoophagus insidians*, nov. gen., nov. sp.). *Oesterr. Botan. Z.* 61, 361-373.

Starr, R. C. (1968). Sexual differentiation in *Volvox*. *Proc. Natl. Acad. Sci. U.S.* 59, 1082-1088.

Tsubo, Y. (1957). On the mating reactions of a *Chlamydomonas*, with special references to clumping and chemotaxis. *Botan. Mag. (Tokyo)* **70**, 299–340.

Tsubo, Y. (1961). Chemotaxis and sexual behavior in *Chlamydomonas. J. Protozool.* **8**, 114–121.

Wiese, L., and Jones, R. F. (1963). Studies on gamete copulation in heterothallic chlamydomonads. *J. Cellular Comp. Physiol.* **61**, 265–274.

Yanagishima, N., Shimoda, C., Kumano, H., Takahashi, T., and Takao, N. (1968). Hormonal regulation of mating reaction in yeast. *Proc. 12th Intern. Genet. Congr., Tokyo* p. 65.

3

The Biochemical Ecology of Higher Plants

R. H. WHITTAKER

I. INTRODUCTION

I might open by referring to an unusual plant product – a straw man. We all know the technique: an imaginary or altered viewpoint of an opponent is constructed which can be demolished to give point to an article. This is not considered good form among scientists, but let me do something like it anyway. Let me advance, as background for a discussion of chemical ecology of higher plants, this view: there are two kingdoms of the living world, the plant and the animal kingdoms, which are sharply contrasted in their chemical characteristics. Animals are active, aggressive, and tissue-eating, producing from their metabolism a variety of wastes, toxins, and poisons that are obnoxious or dangerous. Plants, in contrast, are passive and benign, creating food by the ethereal process of photosynthesis, a nutrition without wastes. Plants are chemically rather pure, producing (with some exceptions) no obnoxious substances or toxins. Plants are so pure that, we were told a few years ago, their most characteristic substance, chlorophyll, can be put into dog food to remove the impure, animal odor from dogs. Such is my straw man of contrast. I am attributing it to no one in particular, but it is not so far from some popular views of plants and animals.

II. ALLELOPATHY — TWO CASES

Let us consider actual cases of chemical activities of higher plants. One may, for example, have a walnut tree in a garden. It will in some cases reward one with both nuts and a blighted or distinctive patch — an area nearly bare of undergrowth plants, or of grasses replacing other herbs. The effect is not simply a matter of shade or root competition by the walnut tree, for other species of trees may grow nearby without effects of such intensity on the plants beneath them. Likewise groves or forests of walnut are reported in some areas to be nearly barren of other vascular plants, in contrast to the rich undergrowth of some other deciduous forests.

To determine the meaning of such undergrowth effects, careful chemical detective work may be necessary. One needs to establish first that the effect is in fact chemical and not one of competition between plants for light, water, or nutrients. The source of the chemical in the plant should be found, and the route it follows from release from one plant to effect on another traced. Identification of the chemical or chemicals that inhibit other plants is to be sought, and the manner of inhibition of germination or growth determined. One wants finally to show, if possible under natural conditions, that the quantitative relations of the chemical agents identified as they occur in the soil are adequate to produce the observed degree of inhibition of other plants, which can be very difficult to prove. Some cases of the chemical effects of plants are accepted as sufficiently established when two points are demonstrated: (1) that an effective inhibitory chemical is being produced and occurs at a potentially effective concentration in the soil, and (2) that the inhibition is not an effect of plant competition for light, water, and nutrients, nor one of animal activities.

Cases in which the effects are chemical are referred to as *allelopathy*, and the chemicals responsible for them as *allelopathic* substances — chemicals that are released from a higher plant (directly or by way of decay processes) and that inhibit the germination, growth, or occurrence of other plants. Allelopathic effects of walnut trees (*Juglans* spp.) have been widely observed in Eurasia and North America (Stickney and Hoy, 1881; Cook, 1921; Massey, 1925; Brooks, 1951; Bode, 1958; Grümmer, 1961). The principal allelopathic chemical is identified as juglone (5-hydroxy-1,4-naphthoquinone) (Davis, 1928; Bode, 1958). The juglone occurs in a nontoxic form, hydroxyjuglone, in leaves, fruits, and other tissues. Rain washes it from living leaves and carries it to the soil; and it is released also, along with tannins, from dead leaves and fruits to the soil. Released into the soil in its oxidized form, juglone, it inhibits the growth of many undergrowth species (Bode, 1958). The effect is selective rather than consistent: while some species (broomsedge, ericaceous shrubs, many broad leaved herbs) are largely excluded, others (black raspberry and Kentucky bluegrass) are tolerant or "favored" and may form the ground cover under walnut trees (Brooks, 1951).

A second case involves the soft chaparral community in southern California, which has been intensively studied by C. H. Muller and his associates (Muller *et al.*, 1964; C. H. Muller, 1965, 1966, 1967; Muller and del Moral, 1966; W. H. Muller, 1965). The soft chaparral is a low shrubland of aromatic plants in a semiarid mediterranean climate; it is the California equivalent of such communities as the garrigue and tomillares around the Mediterranean Sea. A mint (*Salvia leucophylla*) and a sagebrush (*Artemisia californica*) dominate the community where it was studied by Muller. Both are aromatic, and on a still day the air is redolent with the fragrance of terpenes from these plants. The soft chaparral invades grasslands in this area. On certain clay soils the grass and herbs are at times virtually absent within the shrub patches and in belts 1–2 m wide surrounding them, while some inhibition of growth of the grassland plants may be observed several meters away from the shrub patches. It can be shown by various observations and experiments that the inhibition is not a matter of shade, or soil drought, or nutrient levels, or root competition, or animal effects and that it is not, in this case, a water-soluble substance washed from the leaves. Interaction or synergism of allelopathy with some of these other factors may, however, occur. Drought intensifies the allelopathy; in humid years inhibition of other plants in the belts surrounding the shrub patches is less conspicuous, whereas in dry years the same belts are striking for their barrenness. Root competition for soil moisture can, however, be excluded as the cause of the inhibition of germination of annual plants in the fringing belts (Muller, 1966).

The allelopathic chemicals are some of the terpenes responsible for the community's fragrance, notably cineole and camphor. These substances may be traced from the leaves, by volatilization into the atmosphere surrounding the shrubs, to the soil where they are adsorbed onto soil particles (Muller and Muller, 1964; C. H. Muller, 1965, 1966; Muller and del Moral, 1966). In experimental conditions the terpenes have a marked inhibitory effect on germination and seedling growth and on some soil bacteria (W. H. Muller, 1965; Muller and Hauge, 1967). In the field, they accumulate in the soil during the dry summer season, and in the spring they inhibit germination of annual plant seeds, even though the amount of moisture in the soil at that time may be quite adequate for germination. The terpenes also, by inhibiting respiration and growth of the seedlings which do germinate, increase the vulnerability of these to other environmental stress during the dry summer (W. H. Muller *et al.*, 1968).

III. ROUTES OF RELEASE FROM PLANTS

There are many other observations of allelopathic effects — consult reviews by Molisch (1937), Bonner (1950), Knapp (1954), Grümmer (1955, 1961), Rademacher (1959), Börner (1960), Evanari (1961), Rice (1967), and Muller (1966, 1967, 1969). They are of interest for the variety of routes of release of allelopathic substances.

Observations of allelopathic effects by way of rain-wash from leaves of trees include *Myrtus* and *Eucalyptus* (Yardeni and Evanari, 1952) and *Ailanthus* (Mergen, 1959). The false-flax weed (*Camelina alyssum*) reduces growth of flax by phenolic acids (p-hydroxybenzoic, vanillic, and probably ferulic) and other substances washed from the leaves (Grümmer and Beyer, 1960; Grümmer, 1961). In the hard chaparral of California, a taller shrubland related to the maquis of the Mediterranean, *Adenostoma fasciculatum* suppresses herb growth by water-soluble allelopathics (C. H. Muller *et al.*, 1968). In the same community *Arctostaphylos glauca* and *A. glandulosa* release effective allelopathics including arbutin, a glycoside of hydroquinone, and p-hydroxycinnamic acid, a phenolic acid (Hanawalt and Muller, 1968). Leaves of the Australian shrub *Eremophila mitchellii* release potentially inhibitory substances (Webb *et al.*, 1961). *Encelia farinosa*, an American desert semishrub, releases a toxic material (3-acetyl-6-methoxybenzaldehyde) studied by Bonner (1950; Gray and Bonner, 1948a,b) and believed by him to have allelopathic effect by way of rain-wash from leaves. Another semishrub, *Thamnosoma montana*, releases toxic furanocoumarins (Bennett and Bonner, 1953). Many desert semishrubs contain toxic materials (Bennett and Bonner, 1953; Muller and Muller, 1956); there are no evident allelopathic effects in most cases, though observations suggest that allelopathic suppression of herbs by shrubs may occur on certain soils (Muller, 1953; Muller and Muller, 1956). The European *Artemisia absinthium* has been shown to have allelopathic effects on other species (Bode, 1940; Funke, 1943; Evanari, 1961); an alkaloid, absinthin, excreted by glandular hairs on the leaves and washed from them to the soil is believed to be responsible. Fog drip transfers toxic compounds (chlorogenic, p-coumarylquinic, and gentisic acids) from leaves to the soil in stands of *Eucalyptus globulus* with meager undergrowth in California (del Moral and Muller, 1969).

Allelopathy by way of volatilization from the leaves has not as yet been established for aromatic plants other than those studied by Muller. The soils of a Mediterranean community (Rosmarino-Ericion) analogous to the soft chaparral are toxic to other plants, and a parallel mechanism may be suspected from the observations of Deleuil (1950, 1951a,b). Berries of mountain ash (*Sorbus aucuparius*) contain and release parasorbic acid, a lactone which inhibits growth of other plants under laboratory conditions (Cornman, 1946; Barton and Solt, 1949; Garb, 1961). Many seeds and fruits contain inhibitory materials which may have adaptive value in delaying germination until they have been leached; they might also (though the effect has not been shown in the field) inhibit germination of nearby seeds from which competing seedlings might come (Garb, 1961).

Other cases of allelopathy involve release from the roots. Peach root bark contains amygdalin, which breaks down in the soil to glucose, hydrocyanic acid, and benzaldehyde, of which the last, and possibly also the hydrocyanic acid, have allelopathic effects (Proebsting and Gilmore, 1940; Börner, 1960). Apple roots contain phlorizin, which similarly breaks down to toxic products in the soil (Fastabend, 1955; Börner, 1960). Cereal crops—oats, wheat, and rye—release toxic materials from roots; a principal inhibitory substance released from oat roots is scopoletin, a coumarin (Rademacher, 1941; Martin and Rademacher, 1960). Alkaloids of potential allelopathic effect are released from the roots of *Nicotiana, Datura*, and other plants (Winter, 1961). Roots of guayule (*Parthenium argentatum*) release *trans*-cinnamic acid (Bonner and Galston, 1944; Bonner, 1946, 1950. The *trans*-cinnamic acid has a stronger inhibitory effect on guayule itself than on other plants on which it was tested. This substance apparently acts to govern growth of roots in such a way that they avoid one another in occupying soil space (see figures in Muller, 1946); it is consequently an intraspecific chemical influence or plant pheromone, rather than an allelopathic.

A wide variety of substances are released into the soil from roots, some favorable and others unfavorable for the growth of other plants or of soil bacteria and fungi (Loehwing, 1937; Knapp, 1954; Börner, 1960; Martin and Rademacher, 1960; Woods, 1960; Winter, 1961). In many cases, release of allelopathic materials is partly or primarily through death and decay of roots, rather than by exudation from live roots. Because of the rapid turnover of root hairs and fine roots by death and replacement and sloughing of dead root bark, the relative significance of release through living surfaces and through death and decay is difficult to establish. Couch grass or quack grass (*Agropyron repens*), a widespread weed, inhibits growth of cultivated plants primarily by release of phenolic acids from decaying roots, along with an essential oil, agropyrene (Welbank, 1960; Grümmer, 1961). Bromegrass (*Bromus inermis*) and mustard (*Brassica oleracea*) release toxic materials from roots (Benedict, 1941; Varma, 1938). In many cases the allelopathics are released by decay of both above- and below-ground plant parts. Such is probably the case in the inhibition of their own growth and other species by *Helianthus* spp. (Curtis and Cottam, 1950; Wilson and Rice, 1968), horseweed (*Leptilon canadense = Aster canadensis*) (Keever, 1950), and Johnson grass, *Sorghum halepense* (Abdul-Wahab and Rice, 1967). In the last of these, the allelopathics have been identified as dhurrin (a cyanogenic glycoside), *p*-coumaric acid, and chlorogenic acid. In *Helianthus annuus*, substances with inhibitory effects are released by several routes, including decay of litter (chlorogenic acid and isochlorogenic acid), leaching from leaves (probably scopoletin and an α-naphthol derivative), and exudation from roots (Wilson and Rice, 1968).

Inhibitory effects of substances released from dead residues of plants have often been observed (Pickering, 1917, 1919; Bautz, 1953; Patrick and Koch, 1958; McCalla and Haskins, 1964; Guenzi *et al.*, 1967). In varied kinds of communities around the world, from cultivated fields, grasslands, and woodlands to temperate and tropical forests, phenolic acids are found in the soil from decay of plant parts (Winter, 1961; Whitehead, 1964; Wang *et al.*, 1967). Prominent among these in many observations are *p*-hydroxybenzoic, vanillic, ferulic, and *p*-coumaric acids. These most common phenolic acids may be linked by transformations in the soil under bacterial action—coumaric into hydroxybenzoic into benzoquinone and other materials, and ferulic into vanillic (Wang *et al.*, 1967). These substances, so general as to be normal components of soil, can be shown under experimental conditions to be inhibitory to plant germination and growth. Bacterial transformations prevent their accumulation in most soils beyond a steady-state level subject to seasonal fluctuation. It is probable, however, that the phenolic acids and their derivatives under bacterial action may exert *relative* allelopathic effects—effects expressed in degrees of inhibition which vary with levels of the various substances, with kind of soil and other environmental conditions, and with sensitivity of different plant species.

The importance of bacteria for the release and transformation of the phenolic acids is evident. Beyond this Rice (1964, 1965) has shown that an apparent allelopathic effect of an annual grass of succession, *Aristida oligantha*, is primarily indirect, i.e., by way of effects on bacteria. The decay products of the grass inhibit nitrogen-fixing bacteria in the soil, thereby retarding by nitrogen deficiency the invasion of the community by species other than *A. oligantha*, which is tolerant of very low soil nitrogen levels. Among substances released by *A. oligantha* and other species, gallotannic acid, gallic acid, and chlorogenic acid were found to be inhibitory to both higher plants and bacteria (Rice, 1965; Floyd and Rice, 1967). The common heather (*Calluna vulgaris*) illustrates a different version of allelopathy once removed—an effect of a vascular plant on a vascular plant by way of its effect on bacteria or fungi. Heather plants and litter release substances which inhibit mycorrhizal fungi, and can, by inhibiting mycorrhizal associations, inhibit invasion and growth of trees in heather moor. Removal of heather undergrowth in experimental strips in a spruce and pine plantation resulted in appearance of mycorrhizal (*Boletus*) sporophores and increased vigor of growth of the trees (Harley, 1952). It is a further step from dependence of an allelopathic effect on bacterial action on decay products, to one dependent on a toxin released from the metabolism of a bacterium or fungus utilizing decay products. Some antibiotics (i.e., substances released by bacteria or fungi which are inhibitory to other bacteria or fungi) are toxic to higher plants (Wright, 1951; Brian, 1957). Species of *Penicillium* growing in straw and other decaying plant litter liberate patulin, a substance toxic to higher plants and probably possessing some allelopathic effect (Behmer and McCalla, 1963).

We may draw certain conclusions from these observations. Substances which can produce allelopathic effects are virtually universal in plants and plant communities, and actual effects have been observed in a wide range of kinds of communities and climates. The effects come from most varied routes—from above-ground and below-ground plant parts, by leaching and volatilization, excretion and exudation, and release by decay, directly or through the agency of microorganisms. These varied routes are in large part continuous with one another. Given the fact that plants contain substantial amounts of potentially toxic materials, these substances can move by any of these routes in any combination into the soil in chemical forms which may be allelopathic. Plants do contain substantial amounts of toxic materials; these generally involve the wide range of groups of compounds referred to as secondary plant substances. Among those mentioned as having probable allelopathic effects are phenolic acids, coumarins and quinones, terpenes and essential oils, and alkaloids and organic cyanides.

IV. EFFECTS ON PLANT COMMUNITIES

It is probable that these substances have widely significant effects on the dynamics and composition of plant communities (Muller, 1966, 1969). I shall attempt to assess these community-level effects, though I must in the attempt move from observation to speculative interpretation.

Allelopathics have, first, prominent effects in plant succession. Soil development, overtopping and shading, root competition, and microclimatic change are all familiar as bases of the replacement of plant species by other species in succession. In the soft chaparral of California, however, allelopathic terpenes are partly causal (along with grazing and competitive effects of the shrubs) in the replacement of grassland by chaparral (Muller, 1969). In the hard chaparral, water-soluble allelopathics, not competitive effects, are primarily responsible for alternation of shrubs with annual plants in the fire cycle. Annual plants may be nearly absent from the mature chaparral (generally 10–30 years old) because of allelopathic inhibition. Fire in chaparral has the simultaneous effects of removing the source of the allelopathics, denaturing by heat the allelopathics accumulated in the soil, and triggering by heat the germination of some fire-adapted seeds. The fire is followed in the rainy season by a conspicuous blooming of annual plants where there were few or none before. The annual plants continue to appear each spring for some years, with gradually decreasing coverage as the shrubs grow back and allelopathics become effective, until few herbs are left in the mature chaparral (C. H. Muller *et al.*, 1968).

As has been mentioned, grasses (*Aristida oligantha* and *Sorghum halepense*) of old-field succession in Oklahoma produce phenolic substances, inhibitory to nitrogen bacteria and to seedlings of other plants, which consequently tend to retard invasion of grass stages by other species. *Helianthus annuus*, one of the dominants of the first stage of succession, releases substances allelopathic to other successional species; but *Aristida oligantha*, relatively tolerant of these substances, is able to replace *Helianthus* as dominant of a second stage (Wilson and Rice, 1968). Of twenty species from the old-field successions, sixteen species had some, and ten considerable, inhibitory activity against nitrogen-fixing and nitrifying bacteria. They therefore have a potential, if undemonstrated, indirect allelopathic effect (Rice, 1964, 1965; Abdul-Wahab and Rice, 1967; Rice and Parenti, 1967). Keever (1950) found that in old-field succession in North Carolina, horseweed (*Aster canadensis*) rapidly loses dominance after the first year because of toxicity effects of its own decay products, combined with competitive effects of other species on the horseweed. Guyot (1957) observed mosaiclike dominance patches of different successional populations in southern France, with apparent allelopathic effects of the dominants on occurrence of other species and with evidence of allelopathic self-toxicity on the part of *Hieracium pilosella*. Water-soluble substances inhibiting germination and growth of *Hieracium* and other species were extracted from the roots of *Hieracium* and from soil in which this species had grown (Becker *et al.*, 1950; Becker and Guyot, 1951). One judges that allelopathy may influence timing and sequence of succession in a number of ways: (a) by speeding replacement of one species by a successor, by allelopathic self-toxicity of the first species; (b) by allelopathic suppression of the first species by the second; (c) by slowing of species replacement by direct allelopathic effects of the dominant on potential invaders; (d) by indirect effects through decay products or inhibition of soil organisms; and (e) by influence on the sequence of species, through allelopathic effects of one species which determine what other species can invade its community and replace it.

Allelopathic autotoxicity such as that mentioned for *Aster* (Keever, 1950) and *Hieracium* (Guyot, 1957) is by no means uncommon among successional species. The centers of older shrub patches in the soft chaparral often show reduced coverage and plant vigor compared with the younger margins of the patches. Heavy accumulation of terpenes in the soil is the probable cause of this self-inhibition (Muller, 1966, 1969). Accumulation of toxic materials produces thinning of bromegrass (*Bromus inermis*) stands after some years of growth (Benedict, 1941). Related effects are more striking in the fairy-rings formed by sunflower species in some areas (Cooper and Stoesz, 1931; Curtis and Cottam, 1950). Experiments with these fairy-rings (Curtis and Cottam, 1950) rule out nutrient depletion and competition for water as causes. The sunflowers spread outward by rhizomes, and because of the toxicity of products of their own decay within the ring, show maximum vigor of growth where they invade new soil around the periphery of the fairy-ring. Self-toxicity seems an odd evolutionary strategy. One judges that the allelopathic substances have net survival value in relation to the full range of adaptive problems encountered by the plant. The substances presumably confer some adaptive advantage which outweighs (for successional species whose populations are in any case short-lived and nomadic) the apparent selective disadvantage of autotoxicity. Webb et al. (1967) describe autotoxic effects of a subtropical rain forest tree which germinates and grows beneath other species in the mixed stands in which it occurs.

Allelopathic effects of strong dominance may well appear in stable, as well as unstable communities. Where one finds strong single-species dominance, one often finds a meager undergrowth of few species. The community in question can be hard or soft chaparral (Muller, 1966, 1969), walnut forest, certain Eucalyptus plantations and certain oak woods (Yardeni and Evanari, 1952; del Moral and Muller, 1969; Muller, 1969), or various coniferous forests, such as stands of eastern hemlock (Tsuga canadensis), with few undergrowth species, in contrast to rich deciduous forests of similar environments (Whittaker, 1965). It is difficult in these cases to distinguish among effects of rigorous environment, competition, soil factors other than allelopathy, and direct or indirect allelopathic effects. There is evidence from chaparral and walnut stands, however, and inference from other communities to suggest that allelopathy is a major reason for the low species diversity. It is of interest that stands of a given Eucalyptus species may contain almost no other plants in the United States, but have a well-developed undergrowth in Australia. One infers that there has been evolutionary time for many Australian species to adapt to eucalypt allelopathy, whereas American species encountering effects of the introduced eucalypts are mostly excluded. Strong dominance by one plant species implies strong domination of soil chemistry by the products, including allelopathic ones, of that plant species' metabolism and decomposition. Species not adapted to the resulting distinctive allelopathic chemistry are not present in the community.

A more elusive problem is the subtle pattern in distribution of undergrowth in forests and other communities which are of mixed dominance. One observes in a forest patches of one species here and another species there, a few meters apart in environments not visibly different. Ecologists believe that light differences, root competition, wood decay remnants, differences in fungal biota, microrelief, dispersal accidents, and clonal history may all, in varied combinations, affect these intracommunity patterns. One should allow also for chemical relations among plants, broadening concern from allelopathics to leachates, exudates, and decay products in general. As Tukey (1966), Tukey and Mecklenburg (1964), Carlisle et al. (1966, 1967), and others have shown, a whole range of substances including nutrients, amino acids, various carbohydrates, and organic acids, among them of course phenolics, are leached from above-ground plant surfaces to the soil in quite significant amounts. The influence of these on soil chemistry exists in addition to the complex influences of roots on soil chemistry. Some correlations of distribution of autotrophic undergrowth species with canopy species have been observed. Correlations of herb species with desert shrub species have been discussed by Went (1942), Gray and Bonner (1948a), and Muller and Muller (1956). Tamm (1950) found that growth of moss on the forest floor was related to distance from canopy trees. Reiners (1967) showed correlation of distribution of different shrub species in the Long Island forest with oaks, or with pines.

Such correlations are partial and often far from obvious in the field without statistical treatment of distributional data. We can, however, draw lines from a number of points—these observations, allelopathic effects in communities of strong dominance, and the very wide occurrence of potentially allelopathic materials—and find the lines converging on an inference. On the forest floor there is a pattern of soil chemistry differing from place to place as it is influenced by leachates, exudates, decay products, and microbial metabolism. Soil bacteria, fungi, and undergrowth plants respond in their distributions to this chemical pattern. The associations of the species are generally not obvious because the responses are relative—partial correlations—and are intertwined with and obscured by responses to microrelief and other factors. Probably, however, the chemical effects are, in varying kinds of combination with other patterning effects, a significant part of the complex and subtle design of the undergrowth carpet of forests and other communities.

If so, then (carrying speculation a step further) chemical interrelations of species may be related also to the high species diversities of some communities. Undergrowth species may be thought responsive not simply to the chemical effects of one canopy species, but to the full range of soil chemistry as affected by different canopy species and combinations of canopy species where their effects overlap. I know of no reports of response to combinations of dominants in forests, but Deleuil (1954) states that a composite, *Hyoseris scabra*, grows in the presence of an onion, *Allium chamaemoly*, only if there is present a third species, *Bellis annua*, which mitigates for itself and the *Hyoseris* the allelopathic effects of the onion. The more complex the pattern of canopy species, the more complex the pattern of niches, as regards place-to-place differences in soil chemistry and soil microorganisms within the community offered to seedlings of dominants and to undergrowth plants.

Chemical adaptation of species to one another—in all directions among plants, animals, bacteria, and fungi—should be self-intensifying through evolutionary time. Given time and a stable environment, additional species enter the community and reach accommodation with species already present by means of niche differences that include chemical adaptations. Increasing numbers of species present imply increasing variety of niche possibilities of chemical and other interactions, in which still other species may find places in the community. Ehrlich and Raven

(1965) suggest that interaction and chemical coadaptation of plants and herbivores have been a major basis of the evolution of high species diversities in terrestrial communities. One may broaden the concept to include the chemical relations of plants with bacteria and fungi, and of plants with other plants affecting their chemical environment directly or by way of soil bacteria and fungi. The basis of the fabled richness of species in the tropical rain forest must include these chemical adaptations by plants to one another and to other organisms. Chemical effects of plants may thus have much to do with species diversity of plant communities at both ends – evolution through time of chemical and other coadaptations to produce the high species diversities of some communities (primarily those of more stable and favorable environments), and allelopathic restriction of diversities in cases of strong dominance as a special case at the other extreme (primarily in some unstable communities and less favorable environments).

V. SECONDARY PLANT SUBSTANCES

The chemicals responsible for these interactions among plants generally belong among the secondary plant substances (Muller, 1966). These have certain characteristics (though a very wide range of kinds of organic compounds is involved). They are not, so far as is known, essential to the basic protoplasmic metabolism of the plant. There is in most cases no evident reason, if the plant is considered by itself without reference to other organisms, why the plant should produce them at all. They are of irregular or sporadic occurrence, appearing in some plants or plant families and not in others – this fact reinforcing the view that they are not essential to plant metabolism. The occurrence of the same or related secondary compounds in related plant species makes these compounds important concerns of chemical taxonomy of plants. On the other hand, appearance of compounds, or related compounds, in plants of distant phylogenetic relation suggests that some secondary plant substances have often been independently evolved.

Extensive discussion of secondary plant substances and their metabolic pathways may be found in texts (Robinson, 1963; Davies *et al.*, 1964; Doby, 1965; Bonner and Varner, 1965) and monographs (Harborne, 1964, 1967; Pridham, 1967; Manske and Holmes, 1950–1968; Fieser and Fieser, 1959; Hoch, 1961; Kjaer, 1958). We may summarize some information bearing on their ecologic roles.

A. The Phenolic Grouping

Substances of this large grouping are aromatic in the sense of inclusion of one or more benzene rings in their structure. Some are derived from the aromatic amino acids (phenylalanine, tyrosine, tryptophan) which in turn are synthesized by the pathway through shikimic and pre-phenic acids; others are produced by condensation of acetate units.

Among the single-ring, six-carbon phenolic substances are hydroqui-none, pyrogallol, and phloroglucinol; arbutin is a glycoside of the first of these. Gallic acid and gentisic acid are seven-carbon phenols; vanil-lin, p-hydroxybenzaldehyde, and syringaldehyde are eight-carbon phe-nols. The last three are produced in the oxidation of lignin, and the nine-carbon phenols may also be released from lignin in the decomposition of plant tissues. Among the nine-carbon phenols are the cinnamic acids (trans-cinnamic, p-coumaric, caffeic, ferulic, and sinapic acids) and the coumarins (coumarin and scopoletin, lactones of trans-cinnamic and ferulic acid, respectively). Juglone is a ten-carbon phenolic quinone; chlorogenic acid is a depside combining a single benzene ring (of caffeic acid) with quinic acid. Most of the simpler phenolic substances so far mentioned have been implicated in allelopathic effects. Some phenolic compounds occur in the essential oils of plants, and phenolic compounds are responsible for some of the scents of plants. Phenolic quinones with repellent function occur in plants (Schildknecht et al., 1967) and arthro-pods (see Eisner, Chapter 8). There is evidence that phenolic compounds contribute to the protection of plants against infection by fungi (and, probably bacteria and viruses); certain phenolic compounds protective against fungi are produced by plants in response to infection (Cruickshank and Perrin, 1964). The lignins, from which some of the phenolic acids are derived, are high polymers of single benzene ring phenolic alcohols (coniferyl, synapyl, and p-hydroxycinnamyl alcohols), and some of the alcohols themselves occur in plants as glycosides (coniferin, syringin).

Fusion of a furan or pyran ring with the benzene ring of a coumarin produces furanocoumarins and pyranocoumarins; some of the former have been used as fish poisons and some extracted from desert shrubs have been found inhibitory to other plants (Bennett and Bonner, 1953). The lignans unite two benzene rings through side chains; some are of interest as drugs. Anthraquinones unite two benzene rings by two inter-mediate carbons which form with them a six-carbon ring and are each linked with an oxygen atom; anthraquinones occur as glycosides in sev-eral families (Rubiaceae, Rhamnacceae, Polygonaceae), and are of in-terest as purgatives and dyes. More widely distributed and significant are the flavonoids and related compounds, with two benzene rings united by a three-carbon chain (or a pyran ring, or other modification). Anthocyanins, flavones, and flavonols are most familiar among flavo-noid pigments of flowers and other plant parts; but the flavonoids in-clude a wide range of leucoanthocyanidins and other colorless sub-stances of uncertain evolutionary meaning. There is evidence that some flavonoids protect plants against parasitic fungi. Among flavonoid rela-tives, may be mentioned stilbene derivatives (with two-carbon, double-bond linkage between the benzene rings), which are of high toxicity to fungi, insects, fish, and mice, and rotenone (a more complex flavonoid ketone related to furanocoumarins) of noted toxicity to fish and insects. Many of the flavonoids, both pigmented and colorless substances, occur in plants as glycosides in solution in vacuoles.

Tannins are polymeric phenolic compounds. Hydrolysis of some of the tannins yields the simple, seven-carbon gallic acid, others give ellagic acid or other phenolic acids. Other tannins, those termed "condensed tannins," are of less clearly known structure involving catechins, phenolic compounds of two benzene rings closely related to leucoanthocyanidins. Tannins, which have strong protein-adsorbing properties, generally occur in vacuoles, or are otherwise separated from the protoplasm of the plants in which they occur. Tannins are responsible for the astringent or bitter taste of many plant tissues, which make them unpalatable to man and presumably to other animals, and also contribute to the protection of wood from decay by fungi.

B. The Terpenoid Grouping

A large variety of organic compounds can be interpreted using isoprene, a five-carbon, branched-chain hydrocarbon, as a building unit. The synthetic pathway for the group is believed to lead, however, not through isoprene (which has not been isolated from plants), but from acetate through mevalonic acid to the terpenes and their many derivatives.

Five-carbon hemiterpenoids occur in some plants as simple alcohols, aldehydes, and acids. Union of two of the five-carbon units, generally linked head-to-tail, forms a ten-carbon compound or monoterpene. In many such compounds (as well as higher terpenoids), two bonds between carbons link carbons from both units into a closed, six-carbon ring, but chains and other open arrangements also occur. Camphor and cineole, the pinenes abundant in the turpentine from pines, limonene, phellandrene, menthol, cymol, geraniol, and citronellol are among the many monoterpene hydrocarbons and alcohols, aldehydes, and ketones of these. Nepetalactone and related monoterpenes have defensive functions in both plants and insects (Eisner, 1964). A variety of sesquiterpenes, based on three five-carbon units, and diterpenes, based on four units, occur in plants. The lower terpenes (hemi-, mono-, sesqui-, and diterpenes) are the major components of essential oils of many plants and important among the plant scents to which animals respond. Lower terpenoids also form the resins of conifers and tropical plants, in some of which they occur in particular abundance (Langenheim, 1969). Essential oils and resins are segregated from protoplasm in globules within cells, or in dead cells or specialized ducts or glands. They are believed to contribute to the protection of plants from animals and fungi; their significance to plants would otherwise be obscure (except as scents of flowers and by-products related to the photosynthetic pigments).

The phytol moiety of chlorophyll is a diterpene, and the gibberellins are apparently diterpene derivatives. Several groups of terpenoid quinones occur in cell organelles and function in electron-transfer reactions. Triterpenes occur in plants less widely as glycosides, such as the bitter substances of the cucurbits. The carotenoid pigments are tetraterpenes (though some have fewer than 40 carbon atoms). Higher terpene polymers occur in various dicot plants of scattered taxonomic relations as rubber and gutta percha. Rubber occurs as microscopic particles suspended in milky sap, or latex, in cells, and vascular tissues.

Steroid compounds are based on a structure related to the triterpenes, with three six-carbon and one five-carbon rings, but fewer than 30 carbon atoms. Sterols occur widely in plants; however, their significance to the plants is generally unknown. Their significance to animals consuming the plants ranges from strongly positive (as precursors of vitamin D) to strongly negative. The cardiac glycosides or cardenolides are effective poisons occurring in some plants, those of the milkweed and dogbane families particularly; their aglycones (i.e., the compound which is linked to a sugar to form a glycoside) are sterols. A celebrated case of plant and animal chemistry involves the concentration of cardiac glycosides from food plants into the tissues of the monarch butterfly and its relatives (Brower and Brower, 1964; Reichstein et al., 1968), whereby they give the butterflies protection against predators. While a number of 24-carbon steroids (bufadienolides) occur in plants, substances of this group also occur in animals as toad poisons. Sapogenins are other plant steroids of *Yucca*, *Agave*, and other genera; like the cardiac poisons, they occur as glycosides (saponins). Sapogenins are toxic to animals through hemolysis and other effects on membrane permeability. Toxic sterols which include a nitrogen atom (and hence are also alkaloids) occur as glycosides in *Solanum*, *Veratrum*, and other genera. Among steroid and terpenoid discouragements to animals should also be mentioned the occurrence in plants of substances which disturb the endocrine control of insect life cycles: ecdysones (molting and metamorphosis hormones) and analogs in many plants (Kaplanis et al., 1967; Hoffmeister et al., 1967) and substances mimicking effects of juvenile hormones (inhibiting metamorphosis) in some (Sláma and Williams, 1966).

C. Alkaloids and Nitriles

The alkaloids are a heterogeneous grouping of nitrogenous bases. Common ring systems include benzenoid and terpenoid skeletons, pyridine, pyrrolidine (nicotine couples one of each of these), piperidine, tropane, and tropolone rings. Other diverse structures also occur, some of which are unrelated to the phenolic and terpenoid groupings. Those with benzene rings may be synthesized by a metabolic pathway from the aromatic amino acid tryptophan to the plant hormone indoleacetic acid, the vitamin nicotinic acid, and certain alkaloids (nicotine and the indole alkaloids). Alkaloids, like tannins, are responsible for the bitter unpalatability of many plant tissues. They also include many toxic materials and a number of highly effective poisons. Colchicine is of interest as a toxin affecting cell division, which has been put to use to achieve doubling of chromosome number in plants. Berberine gives protection against root-infecting fungi in *Mahonia* (Greathouse and Watkins, 1938). Alkaloids known for their effects on man include nicotine, caffeine, quinine, aconite, atropine, ephedrine, strychnine, serpentine, reserpine, morphine, codeine, mescaline, and lysergic acid. Man's use of alkaloids for flavor, mild stimulation, medicinal effect, or pleasurable self-destruction should not obscure a common theme: they are probably, though not necessarily in all cases, repellents and toxins, evolutionary expressions of quiet antagonism of the plant to its enemies. Some of the alkaloids are synthesized in the roots from which they may be both released into the soil and translocated to stems and leaves (Mothes, 1955, 1960). Many of the alkaloids occur in plant cells as glycosides.

Amines are simpler nitrogen-containing compounds which may be grouped with alkaloids or separated (for lack of a heterocyclic ring including the nitrogen). They occur frequently in plants and some have physiological effects on animals such as a strong effect on blood pressure. Isobutylamides combine isobutylamine with fatty acids by amide linkages; they occur in few plants, but include the insecticide pellitorine. A group of far more ecological significance are the nitriles or organic cyanides. They are glycosides of which those which are most common and known for ecological effect might be grouped both with alkaloids for occurrence of nitrogen and with phenolics for the benzene ring in the aglycone. The two which are most common and known for allelopathic effect are amygdalin (which breaks down to benzaldehyde, hydrocyanic acid, and the sugar gentiobiose) and dhurrin (*p*-hydroxybenzaldehyde, hydrocyanic acid, and glucose). Nitriles containing aglycone fractions other than the benzaldehydes are known. Some plants contain nitriles at concentrations toxic to animals, and related compounds releasing hydrocyanic acid are known among the defensive secretions of animals (Eisner and Meinwald, 1966; and Eisner, Chapter 8).

diotetryne nitrile

D. Other Groups

There remains a biochemical miscellany of other groups of substances of considerable interest, but less widespread occurrence.

Mustard oils are nitrogen- and sulfur-containing compounds occurring in plants of the crucifer family and some others. Having characteristically sharp odors and flavors and in some cases causing skin irritations, they probably give the plant protection against animals. They are also germination inhibitors and toxic to fungi. Urushiol and related com-

mention effect of burning

pounds responsible for the irritating effect of poison ivy are phenols with a single aromatic ring and long aliphatic side chains. Long-chain acetylenic alcohols and ketones occur in some plants and are responsible for the toxicity of the roots of some of the Umbelliferae. Other acetylenic compounds, some including sulfur-bearing thienyl rings, are widespread in the composites; certain of these compounds have been shown to be powerful insecticides and nematocides. Certain fatty acids have been shown to have antibacterial effects (Spoehr *et al.*, 1949), and some of the lipids, phospholipids, and glycolipids distinctive to particular groups of plants probably belong among the secondary substances.

Hemlock wild carrot

A number of amino acids which are not part of the normal alphabet of proteins have been found in plants (Fowden, 1958). Among these azetidine-2-carboxylic acid, a proline analog which occurs in the lily family, produces marked inhibition of plant growth and abnormality of animal development by its contribution to the synthesis of misshapen protein molecules (Fowden, 1963). Some simple lactones (parasorbic acid, ranunculin) have rings including one oxygen with four or five carbons; among them, some are fungicidal and some are skin irritants. A six-carbon, straight-chain aldehyde (2-hexenal) occurs as a repellent in the ginkgo tree (Major, 1967).

steric hindrance

Rhubarb leaves

Oxalic acid accumulates in some plant tissues as calcium oxalate crystals, or raphides, in some cases to concentrations affecting palatability. Silica particles or opal phytoliths are abundant in tissues of grasses and some other monocots and in the soils beneath them (Smithson, 1958; Baker, 1959; Jones *et al.*, 1963; Parry and Smithson, 1964). The abrasive effect of this plant sand on animal mouthparts offers relative protection against animal consumption, although there have evolved mammals with high-crowned teeth and insects able to consume the abrasive tissues.

VI. EVOLUTION OF WASTES AND REPELLENTS

The secondary substances have always offered students of the biochemistry of plants, supposedly simpler organisms than animals, an embarrassment of riches. There have consequently been efforts to find metabolic functions for them, or to interpret them as by-products of pathways to substances important in plant metabolism. The quantities of these compounds elaborated and secluded from protoplasmic function in many plants tends to defeat the view that they are generally needed for unknown metabolic functions, and to suggest that they are primarily by-products of metabolic systems producing other substances. It might thus be thought that photosynthetic pigments and steroids are the true aims of terpene metabolism, flavonoid pigments and lignins of phenolic metabolism, and plant hormones of alkaloid metabolism.

One is struck, in a survey of the secondary substances, by the extent to which plants are in fact protected from the effects of most of these compounds (Muller, 1966). They are rendered less toxic as glycosides (many compounds of various groups), are inactivated as polymers (rubber, tannins), are separated from the enzyme which breaks them into toxic products (amygdalin), become toxic on oxidation outside the cell (juglone), are secreted into special intracellular or intercellular compartments or surface glands (many terpenoids and other substances), or are kept inactive as crystals (calcium oxalate). Many phenolic, terpenoid, and other secondary substances, some of known fungicidal properties, are secreted into heartwood (Rennerfelt and Nacht, 1955).

One is further struck by the evidence of prevalent toxicity or repellent effects. The summary above has emphasized known toxicity to other organisms over cases where no toxicity is known. Nevertheless, the indications of unpalatableness, inhibitory effect, or action as poisons are impressive. Interpretation of secondary substances as wastes alone would suggest a fair amount of disorganization in the evolution of plant metabolism. It implies that plant metabolisms produce quantities of by-products so toxic to the plant itself that they must be segregated, but the chemical character of which differs in a more or less accidental way from one plant group to another. One would prefer a more convincing evolutionary interpretation.

The view has developed through observations of Dethier (1954), Fraenkel (1959), Ehrlich and Raven (1965) and others, that the secondary plant substances have their primary meaning as defenses against the plant's enemies. The evolution of these substances is probably not, in fact, comprehensible except in an ecological context, including organisms other than the plants producing them. The reasoning applies also to plant structures related to the secondary substances, notably duct systems and many of the glands—often quite striking little organs—which adorn the leaves and other surfaces of plants. Merz (1959) has shown that in *Senecio viscosus* the glands secrete substances protecting the plant against insect consumption. The glandular hairs of *Primula obconica* secrete a toxic quinone (Schildknecht et al., 1967). *Artemisia absinthium* has leaf glands which secrete alkaloids that are in all probability repellent to animals as well as allelopathic (Bode, 1940). The different roles attributed to the secondary substances are not mutually exclusive; one may grant, on the one hand, the significance of their relations to pigments, structural materials, and regulatory substances (including some functions yet to be discovered), and assert, on the other hand, their central place in the subject matter of chemical ecology.

Functions of secondary substances in relation to other organisms may be distinguished: they may be repellents (rendering the plant relatively unpalatable or toxic to animals), attractants (behavioral stimulants, including the pigments and scents by which animals are drawn to flowers), phytoncides (protecting the plant against pathogenic, parasitic, or decay-producing bacteria and fungi), or allelopathics (inhibiting other plants). It is not necessarily the case that any particular compound triples as repellent, phytoncide, and allelopathic. It is likely, however, that the secondary substances of a given plant will include compounds effective as both repellents and phytoncides (whether or not some compounds are both), some of which may also, if circumstances permit their accumulation in environment, become allelopathic. There is a possible advantage, for successional species particularly, in allelopathic effects which inhibit other competing plant species. It has not been shown, however, that this advantage is sufficient to produce selection for the production of secondary substances for allelopathic function. In general, it seems likely that allelopathic effects are secondary effects of secondary plant substances whose primary evolutionary meaning is the defense they offer against animals and microorganisms.

To be effective for defense, many of these substances must be relatively toxic and potentially toxic to the plant itself, hence their seclusion from protoplasmic function, their exertion of allelopathic effects on other plants, and the possibility of autotoxicity when they accumulate to high concentrations in the plant's own environment. Internally the plant must maintain tolerable, steady-state levels of potentially toxic materials in its tissues. Toxicity may be reduced by combining the substances into glycosides or polymers, or by other devices. With or without such relative detoxication, however, the plant must in many cases maintain a flow of potentially toxic materials from synthesis, through living cells, and out of these by secretion into dead compartments or tissues, or by release into environment through leaching, volatilization, or exudation.

To make disposal possible, the plant must be to some degree leaky, open to some loss of materials through its surfaces. Plants are generally open to such loss through their surfaces both below and above ground, and some plants require rainwash leaching for favorable growth (Tukey and Mecklenburg, 1964). The loss of foods and vitamins from plants appears nonadaptive. One judges that it occurs as a consequence of the adaptive value of open surfaces—for root intake and for discharge of surplus toxic substances. Muller's (1966) work suggests that there may be a partial environmental correlation in the manner of release into environment. Went (Chapter 4) considers that allelopathic substances are more characteristic of plants of dry climates, but I suspect, for reasons to be developed below, that they are quite general. All the major groups of secondary substances occur in plants of all major environments, even though families and other taxonomic groups of plants differ in evolutionary "choice" of the secondary substance groupings which are emphasized in their metabolism and defensive chemistry. It appears, however, that the volatile terpenes are more prominent in plants of semiarid and arid environments, the water-soluble phenolics in those of more humid environments. The observation suggests that there may be (relative, not essential) selective advantage to production of secondary substances in forms that can be excreted by volatilization in a dry climate, by leaching in a wet climate. The substances must in any case be released into the environment by one of those means, or by the death and decomposition of the plant, or both.

They are thus discharged into the environment like wastes, even if they are not simply wastes. One might outline their evolutionary logic thus: plants have need for various specialty compounds – as pigments, regulatory substances, skeletal materials, etc. Many of these are compounds of the major secondary substance groupings. Protoplasm is the most complex and highly perfected system we know, but it cannot be quite perfect. An enormous number of transformations, rate controls, and enzymes are involved. It is impossible that protoplasmic function should provide enough of every metabolite needed and not too much of some, should exclude metabolic byways yielding some unneeded materials and recycle every product, and should do this in the face of changing environmental conditions. There is not enough selective advantage to have brought evolution of enzymes and controls for the use or recycling of every metabolite produced.

Among the plant's wastes and by-products, some are likely to be toxic or repellent to other organisms. Selection will tend to increase the concentrations of these. Through evolutionary time mutations will occur producing additional by-products or chemical novelties, some of which will be repellent to the plant's enemies. These will be selected also, along with devices of inactivation for the more toxic of them. Toxic materials which are not inactivated in the plant, or which even with inactivation may reach toxic levels, will in many cases reach levels requiring excretion from the living plant tissues. Evolution may thus either add defensive function to excretory necessity or add excretory necessity to defensive function, for different compounds. There is no line of division between wastes and defensive substances in plant chemistry, although in higher plants (at least) selection has strongly edited the plant's metabolism of secondary substances toward defensive effectiveness.

As the plant evolves its chemical defenses under selective pressure from its enemies, the enemies evolve accommodation to the defenses. There is thus a tendency toward evolutionary intensification of the chemical distinctiveness of the secondary substances of a group of plants. There is the wavelike, or parallel evolution of groups of plants with groups of animals adapted to them as Ehrlich and Raven (1965) have shown for genera and subfamilies of butterflies in relation to families of plants. Adaptation by animals may include use of the plant's distinctive chemistry for behavioral cues by which the plant is located. Germination of spores of fungi and of seeds of vascular parasites and hemiparasites depends in some cases on exudates from the roots of a plant species with which they are associated (Evanari, 1961). Some of the same secondary substances may then figure in the interaction both as allomones (chemical agents adaptive to the organism producing them, in this case as repellents) and as kairomones (chemical agents adaptive to an organism other than their source, in this case as chemical cues by which enemies of the plant respond to it). (Brown *et al.*, 1969). Adaptation by the animals may also include, as shown for the milkweed–monarch interaction and suspected for others, use of the plant's defensive chemicals by the animal consumers for their defense against predators.

Evolution leads toward a balanced accommodation of the plant and its enemies, made relatively stable by genetic feedback (Pimentel, 1968). The stable coadaptation has made recognition of some aspects of the interactions difficult. The defensive importance of secondary substances has not generally been obvious to botanists, when some of the plants' tissues were in fact being eaten or infected. The protection is of course relative and almost never complete (though certain plants are very well protected indeed – Kaplanis *et al.*, 1967; Major, 1967). The importance of effects of animal consumption on plant populations and of effects of plant resistance on levels of animal populations have not been evident to ecologists. The fraction of the mature plant consumed is often (apart from man's unstable domestic species) small; but it is not necessarily small in the seedling stage. The fraction of the mature plant consumed may be small not because animal consumption is of small significance for plant populations and their evolution, but because of the accommodation of plant and animal (which may include relative or marginal unpalatableness of the plant, differing in degree of effect for different plant and animal individuals and with age and physiological condition). The accommodation may tend to hide its own significance from casual observation. Regulation of animal populations may normally involve predators and parasites, however, whether or not relative unpalatableness of the plant contributes to that regulation. Evolutionary accommodation leads toward formation of complexes of species adapted to interaction. Such a complex may include a plant with its secondary chemistry, a number of animals, fungi, and bacteria adapted to it (and in some cases using its chemistry to their own adaptive advantages), and parasites, predators, and sometimes mimics adapted, in turn, to interaction with the species adapted to the plant.

VII. CONCLUSION

Some comments on allelopathic effects suggest that these are peculiar details of natural history, found as special cases here and there. The reasoning of this paper is that allelopathy is an expression in effects on other plants of a major biological phenomenon – the secondary chemistry of defense which is universal among higher plants. Potentially allelopathic chemicals are normal parts of the environment of land plants.

The degree of actual, as distinguished from potential, significance to be ascribed to allelopathic effects is difficult to judge. I shall suggest, however, some points of judgment which seem to me important.

First, the known examples of allelopathic effects are only the most conspicuous cases, which have been observed because they stand out from a matrix of more general and less obvious influences. The ecologist seeking to evaluate allelopathy as a phenomenon is in the position of a geographer attempting to understand the topography of the ocean bottom from data on islands alone. Most allelopathic effects must lie below the surface of the community relationships we are able to observe. These effects will be mostly unrecognized both because they are involved in the complexities of other population effects in communities and because the chemicals responsible for them are invisible to us. Their significance is no more to be discounted because they are not evident than is that of other population processes which are not normally recognized. Second, most effects will be effects of degree — relative effects on population level and on extent of a population's distribution in a given direction — rather than obvious exclusions and complete incompatibilities of species. Third, the long-term effects will include successful adaptation — evolution by species of tolerance of the levels of allelopathic materials in soil which they encounter. The fact of evolved tolerance of allelopathic factors no more discounts their significance than successful adaptation to temperature, seasonal moisture limitation, or animal consumption discounts the ecological significance of these factors.

Such reasoning, the evidences and inferences on community effects discussed earlier, and the universal occurrence of secondary substances in higher plants lead this author to a conclusion: allelopathy is not a peculiarity of a few plants but a widespread and normal, although mostly inconspicuous, phenomenon of natural communities. One must regret the lack of certain research which would assist this provisional judgment. Data on relative effects of allelopathic substances on levels of plant populations, in cases of less striking allelopathy, are lacking. There are no studies of population responses to gradients or patterns of concentrations of allelopathic substances in communities. The possible significance of combined allelopathic effects of two or more species in excluding or favoring the populations of other species is unknown. There is little knowledge of the place of secondary substances in the economy of the plant, i.e., the fraction of its photosynthetic energy which is spent on these chemicals and the adaptive balance of this expenditure against loss of tissue and individuals to enemies of the plant. There is no study as yet of the implications of the various secondary substances in a given species to the different enemies and associates of that species, or of the role of chemical defense in the population dynamics of the plant and its enemies. It is easy to add to my inferences on the submarine topography of allelopathy a plea for further exploration.

It is not difficult to find parallels for these chemical interactions and antagonisms throughout the living world. They are reported not only between plants and between plants and animals, but within and, probably in all possible directions, between the major groups of organisms — plants, animals, fungi, protists, and monerans (bacteria and blue-green algae). They are general phenomena of life, and involve man as much as other organisms. Man utilizes for his own purposes many secondary plant substances and antibiotics — rubber and resins, caffeine and nicotine, many spices, scents, flavors, pigments, natural pesticides, drugs, and poisons. Man has developed his own armament of chemicals antagonistic to other species. Some, such as antibiotics and rotenone, are appropriated from other organisms, others, such as the chlorinated hydrocarbon and organophosphate poisons, were devised and produced by man himself. Man acts now as an unstable dominant, engaged in an accelerating loading of his own environment with combustion products, pesticides, chemical wastes, and other materials in an uncontrolled experiment in future toxicity of man to himself.

Allelopathic substances are not only significant in the function of plant communities, they are part of the extensive traffic in chemical influences relating organisms of all the major groups to one another in natural communities. We are accustomed to thinking of community metabolism in terms of two important fractions of that metabolism — energy flow, food chains, and production pyramids, on the one hand, and inorganic nutrient circulation, on the other. We are concerned here with a third major fraction: the role of organic materials that are significant for effects on an organism different from their source not primarily as food, but for other chemical effects. Central among such substances are wastes and repellents, but these intergrade with substances stimulatory to at least some other species, and substances serving as behavioral signals for another species. Many of these substances affect another species after release to the environment, but some serve as repellents without such release. I am suggesting the term *allelochemics* for these substances in general — chemicals significant to organisms of a species different from their source, for reasons other than food as such. Ecologists may thus think of community metabolism in terms of three major groups of substances — inorganic nutrients, foods, and allelochemics — by which the species in a community are linked with one another and their environment.

There is, finally, my straw man on the chemical purity of plants contrasted with animals. I do not think the living world is best viewed in terms of two contrasting kingdoms, plants and animals, but as four kingdoms by the Copeland (1956) system, or five by my own (Whittaker, 1969), treating the fungi as one. Biochemical antagonism is common to all the kingdoms. It is not true that plants are chemically pure; they are in this chemical combat up to their growing tips, with fascinating arrays of secondary chemicals present in them as a consequence. They are, like

other organisms, exploiting the possibilities for metabolic unattractive-
ness to potential enemies. Because of the importance of the higher
plants, their secondary substances are widespread in the environment—
not only in soils, but, though less obviously, in the atmosphere and in
fresh waters (see also Went, Chapter 4). If we do not notice the perva-
siveness of these secondary plant substances, wastes, repellents, and
toxins, in our environment and food, it is only because we have had millions of years to evolve tolerance of their occurrence, and time to learn to
avoid occurrences in food we cannot tolerate. It is to be regretted that, in
contrast, neither man nor many other organisms may have time to
evolve adaptation to man's present activities as an unstable species in
loading the environment with wastes, repellents, and toxins.

VIII. SUMMARY

Higher plants synthesize substantial quantities of substances repellent or inhibitory to other organisms. These substances move from the
plant into the soil by varied routes—by leaching or volatilization from
leaves, by exudation from roots, or by death and decay of both above-
ground and below-ground plant parts. In the soil they may exert allelopathic effects—inhibiting the germination and growth of other plants.
Allelopathic effects have significant influence on the rate and species
sequence of plant succession and on the species composition of stable
communities. Chemical interactions affect the species diversity of natural communities in both directions: strong dominance and intense allelopathic effects contribute to the low species diversity of some communities, whereas variety of chemical accommodations are part of the
basis (as aspects of niche differentiation) of the high species diversity
of others.

The chemicals responsible for allelopathic effects belong among the
secondary plant substances; they are phenolic, terpenoid or alkaloid
compounds, or of smaller groups including organic cyanides. Since these
substances are mostly somewhat toxic, most of them are inactivated in
or segregated from the protoplasm of the plant. High concentrations in
plants have resulted from selection for the defense these substances
offer against the animals, bacteria, and fungi that are the plant's natural
enemies. Evolution tends to produce a balanced accommodation of the
plant population and these enemies, including the chemical adaptation
of the species to one another. As the defensive substances reach the environment, they may also affect other plants (and in some cases the plant
population which produces them). Allelopathy is thus one expression of
the more general phenomenon of chemical interaction, and is probably
more widely significant in the function of natural communities than
actual observations yet establish.

References

Abdul-Wahab, A. S., and Rice, E. L. (1967). Plant inhibition by Johnson grass and its possible significance in old-field succession. *Bull. Torrey Botan. Club* **94**, 486–497.

Baker, G. (1959). Opal phytoliths in some Victorian soils and "red rain" residues. *Australian J. Botany* **7**, 64–87.

Barton, L. V., and Solt, M. L. (1949). Growth inhibitors in seeds. *Contrib. Boyce Thompson Inst.* **15**, 259–278.

Bautz, E. (1953). Einwirkung verschiedener Bodentypen und Boden-Extrakte auf die Keimung von Picea excelsa. *Z. Botan.* **41**, 41–84.

Becker, Y., and Guyot, L. (1951). Sur une particularité fonctionelle des exsudats racinaires de certains végétaux. *Compt. Rend.* **232**, 1585–1587.

Becker, Y., Guyot, L., Massenot, M., and Montegut, J. (1950). Sur la présence d'excrétats radiculaires toxiques dans le sol de la pelouse herbeuse à *Brachypodium pinnatum* du Nord de la France. *Compt. Rend.* **231**, 165–167.

Behmer, D. E., and McCalla, T. M., (1963). The inhibition of seedling growth by crop residues in soil innoculated with *Penicillium urticae* Bainer. *Plant Soil* **18**, 199–206.

Benedict, H. M. (1941). The inhibitory effect of dead roots on the growth of bromegrass. *J. Am. Soc. Agron.* **33**, 1108–1109.

Bennett, E. L., and Bonner, J. (1953). Isolation of plant growth inhibitors from *Thamnosoma montana. Am. J. Botany* **40**, 29–33.

Bode, H. R. (1940). Über die Blattausscheidung des Wermuts und ihre Wirkung auf andere Pflanzen. *Planta* **30**, 567–589.

Bode, H. R. (1958). Beiträge zur Kenntnis allelopathischer Erscheinungen bei einigen Juglandaceen. *Planta* **51**, 440–480.

Börner, H. (1960). Liberation of organic substances from higher plants and their role in the soil sickness problem. *Botan. Rev.* **26**, 393–424.

Bonner, J. (1946). Further investigation of toxic substances which arise from guayule plants: relation of toxic substances to the growth of guayule in soil. *Botan. Gaz.* **107**, 343–351.

Bonner J. (1950). The role of toxic substances in the interactions of higher plants. *Botan. Rev.* **16**, 51–65.

Bonner, J., and Galston, A. W. (1944). Toxic substances from the culture media of guayule which may inhibit growth. *Botan. Gaz.* **106**, 185–198.

Bonner, J., and Varner, J. E., eds. (1965). "Plant Biochemistry," 1054 pp. Academic Press, New York.

Brian, P. W. (1957). Effects of antibiotics on plants. *Ann. Rev. Plant Physiol.* **8**, 413–426.

Brooks, M. G. (1951). Effect of black walnut trees and their products on other vegetation. *West Va. Univ. Agr. Expt. Sta. Bull.* **347**, 1–31.

Brower, L. P., and Brower, J. V. Z. (1964). Birds, butterflies, and plant poisons: a study in ecological chemistry. *Zoologica* **49**, 137–159.

Brown, W. L., Eisner, T., and Whittaker, R. H. (1969). Transspecific chemical messengers: allomones and kairomones. *BioScience* (in press).

Carlisle, A., Brown, A. H. F., and White, E. J. (1966). The organic matter and nutrient elements in the precipitation beneath a sessile oak canopy. *J. Ecol.* **54**, 87–98.

Carlisle, A., Brown, A. H. F., and White, E. J. (1967). The nutrient content of tree stem flow and ground flora litter and leachates in a sessile oak (*Quercus petraea*) woodland. *J. Ecol.* **55**, 615–627.

Cook, M. T. (1921). Wilting caused by walnut trees. *Phytopathology* **11**, 346.

Cooper, W. S., and Stoesz, A. D. (1931). The subterranean organs of *Helianthus scaberrimus. Bull. Torrey Botan. Club* **58**, 67–72.

Copeland, H. J. (1956). "The Classification of Lower Organisms." 302 pp. Pacific Books, Palo Alto, California.

Cornman, I. (1946). Alteration of mitosis by coumarin and parasorbic acid. *Am. J. Botany* **33**, 217. (Abstr.)

Cruickshank, I. A. M., and Perrin, D. R. (1964). Pathological function of phenolic compounds in plants. *In* "Biochemistry of Phenolic Compounds" (J. B. Harborne, ed.), pp. 511–544. Academic Press, New York.

Curtis, J. T., and Cottam, G. (1950). Antibiotic and autotoxic effects in prairie sunflower. *Bull. Torrey Botan. Club* **77**, 187–191.

Davies, D. D., Giovanelli, J., and Ap Rees, T. (1964). "Plant Biochemistry." 454 pp. Davis, Philadelphia, Pennsylvania.

Davis, E. F. (1928). The toxic principle of *Juglans nigra* as identified with synthetic juglone, and its toxic effects on tomato and alfalfa plants. *Am. J. Botany* **15**, 620. (Abstr.)

Deleuil, G. (1950). Mise en évidence de substances toxiques pour les thérophytes dans les associations du Rosmarino-Ericion. *Compt. Rend.* **230**, 1362–1364.

Deleuil, G. (1951a). Origine des substances toxiques du sol des associations sans thérophytes du Rosmarino-Ericion. *Compt. Rend.* **232**, 2038–2039.

Deleuil, G. (1951b). Explication de la présence de certains thérophytes rencontrés parfois dans les associations du Rosmarino-Ericion. *Compt. Rend.* **232**, 2476–2477.

Deleuil, G. (1954). Action réciproque et interspécifique des substances toxiques radiculaires. *Compt. Rend.* **238**, 2185–2186.

del Moral, R., and Muller, C. H. (1969). Fog drip: a mechanism of toxin transport from *Eucalyptus globulus*. *Bull. Torrey Botan. Club* **96**, 467–475.

Dethier, V. G. (1954). Evolution of feeding preferences in phytophagous insects. *Evolution* **8**, 33–54.

Doby, G. (1965). "Plant Biochemistry." 768 pp. Wiley (Interscience), New York.

Ehrlich, P. R., and Raven, P. H. (1965). Butterflies and plants: a study in coevolution. *Evolution* **18**, 586–608.

Eisner, T. (1964). Catnip: its raison d'être. *Science* **146**, 1318–1320.

Eisner, T., and Meinwald, J. (1966). Defensive secretions of arthropods. *Science* **153**, 1341–1350.

Evanari, M. (1961). Chemical influences of other plants (allelopathy). *Handbuch Pflanzenphysiol.* **16**, 691–736.

Fastabend, H. (1955). Über die Ursachen der Bodenmüdigkeit in Obstbaumschulen. *Landwirtsch.-Angew. Wiss., Sonderh. Gartenbau* **4**, 1–95.

Fieser, L. F., and Fieser, M. (1959). "Steroids" 945 pp. Reinhold, New York.

Floyd, G. L., and Rice, E. L. (1967). Inhibition of higher plants by three bacterial growth inhibitors. *Bull. Torrey Botan. Club* **94**, 125–129.

Fowden, L. (1958). New amino acids of plants. *Biol. Rev. Cambridge Phil. Soc.* **33**, 393–441.

Fowden, L. (1963). Amino-acid analogues and the growth of seedlings. *J. Exptl. Botany* **14**, 387–398.

Fraenkel, G. S. (1959). The raison d'être of secondary plant substances. *Science* **129**, 1466–1470.

Funke, G. L. (1943). The influence of Artemisia Absinthium on neighboring plants. *Blumea* **5**, 281–293.

Garb, S. (1961). Differential growth-inhibitors produced by plants. *Botan. Rev.* **27**, 422–443.

Gray, R., and Bonner, J. (1948a). An inhibitor of plant growth from the leaves of *Encelia farinosa*. *Am. J. Botany* **35**, 52–57.

Gray, R., and Bonner, J. (1948b). Structure determination and synthesis of a plant inhibitor, 3-acetyl-6-methoxybenzaldehyde, found in the leaves of *Encelia farinosa*. *J. Am. Chem. Soc.* **70**, 1249–1253.

Greathouse, G. A., and Watkins, G. M. (1938). Berberine as a factor in the resistance of *Mahonia trifoliolata* and *M. swazeyi* to *Phymatotrichum* root rot. *Am. J. Botany* **25**, 743–748.

Grümmer, G. (1955). "Die gegenseitige Beeinflussung höhere Pflanzen – Allelopathy." 162 pp. Fischer, Jena.

Grümmer, G. (1961). The role of toxic substances in the interrelationships between higher plants. *Symp. Soc. Exptl. Biol.* **15**, 219–228.

Grümmer, G., and Beyer, H. (1960). The influence exerted by species of *Camelina* on flax by means of toxic substances. *Brit. Ecol. Soc. Symp.* **1**, 153–157.

Guenzi, W. D., McCalla, T. M., and Norstadt, F. A. (1967). Presence and persistence of phytotoxic substances in wheat, oat, corn, and sorghum residues. *Agron. J.* **59**, 163–165.

Guyot, A. L. (1957). Les microassociations végétales au sein du Brometum erecti. *Vegetatio* **7**, 321–354.

Hanawalt, R. B., and Muller, C. H. (1968). Inhibition of annual plants by *Arctostaphylos*. Conference on Plant-Plant Interactions, Santa Barbara, California.

Harborne, J. B., ed. (1964). "Biochemistry of Phenolic Compounds," 618 pp. Academic Press, New York.

Harborne, J. B. (1967). "Comparative Biochemistry of the Flavonoids," 383 pp. Academic Press, New York.

Harley, J. L. (1952). Associations between microorganisms and higher plants (mycorrhiza). *Ann. Rev. Microbiol.* **6**, 367–386.

Hoch, J. H. (1961). "A Survey of Cardiac Glycosides and Genins," 93 pp. Univ. of South Carolina Press, Charleston, South Carolina.

Hoffmeister, H., Heinrich, G., Staal, G. B., and van der Burg, W. J. (1967). Über das Vorkommen von Ecdysteron in Eiben. *Naturwissenschaften* **54**, 471.

Jones, L. H. P., Milne, A. A., and Wadham, S. M. (1963). Studies of silica in the oat plant. II. Distribution of the silica in the plant. *Plant Soil* **18**, 358–371.

Kaplanis, J. N., Thompson, M. J., Robbins, W. E., and Bryce, B. M. (1967). Insect hormones: alpha ecdysone and 20-hydroxyecdysone in bracken fern. *Science* **157**, 1436–1438.

Keever, C. (1950). Causes of succession on old fields of the Piedmont, North Carolina. *Ecol. Monographs* **20**, 229–250.

Kjaer, A. (1958). Secondary organic sulfur-compounds of plants. *Handbuch Pflanzenphysiol.* **9**, 64–88.

Knapp, R. (1954). "Experimentelle Soziologie der höheren Pflanzen," 202 pp. Ulmer, Stuttgart.

Langenheim, J. H. (1969). Amber: a botanical inquiry. *Science* **163**, 1157–1169.

Loehwing, W. F. (1937). Root interactions of plants. *Botan. Rev.* **3**, 195–239.

McCalla, T. M., and Haskins, F. A. (1964). Phytotoxic substances from soil microorganisms and crop residues. *Bacteriol. Rev.* **28**, 181–207.

Major, R. T. (1967). The ginkgo, the most ancient living tree. *Science* **157**, 1270–1273.

Manske, R. H. F., and Holmes, H. L. eds. (1950–1968). "The Alkaloids: Chemistry and Physiology," 11 Vols. Academic Press, New York.

Martin, P., and Rademacher, B. (1960). Studies on the mutual influences of weeds and crops. *Brit. Ecol. Soc. Symp.* **1**, 143–152.

Massey, A. B. (1925). Antagonism of the walnuts (Juglans nigra and J. cinerea L.) in certain plant associations. *Phytopathology* **15**, 773–784.

Mergen, F. (1959). A toxic principle in the leaves of ailanthus. *Botan. Gaz.* **121**, 32–36.

Merz, E. (1959). Pflanzen und Raupen. Über einige Prinzipien der Futterwahl bei Grossschmetterlingsraupen. *Biol. Zentr.* **78**, 152–188.

Molisch, H. (1937). "Der Einfluss einer Pflanze auf die andere, Allelopathie," 106 pp. Fischer, Jena.

Mothes, K. (1955). Physiology of alkaloids. *Ann. Rev. Plant Physiol.* **6**, 393–432.

Mothes, K. (1960). Alkaloids in the plant. *In* "The Alkaloids" (R. H. F. Manske and H. L. Holmes, eds.), Vol. 6, pp. 1–29. Academic Press, New York.

Muller, C. H. (1946). Root development and ecological relations of guayule. *U.S. Dept. Agr. Tech. Bull.* **923**, 1–114.

Muller, C. H. (1953). The association of desert annuals with shrubs. *Am. J. Botany* **40**, 53–60.

Muller, C. H. (1965). Inhibitory terpenes volatilized from *Salvia* shrubs. *Bull. Torrey Botan. Club* **92**, 38–45.

Muller, C. H. (1966). The role of chemical inhibition (allelopathy) in vegetational composition. *Bull. Torrey Botan. Club* **93**, 332–351.

Muller, C. H. (1967). Die Bedeutung der Allelopathie für die Zusammensetzung der Vegetation. *Z. Pflanzenkrankh. Pflanzenschutz* **74**, 333–346.

Muller, C. H. (1969). Allelopathy as a factor in ecological process. *Vegetatio* **17** (in press).

Muller, C. H., and del Moral, R. (1966). Soil toxicity induced by terpenes from *Salvia leucophylla. Bull. Torrey Botan. Club* **93**, 130–137.

Muller, C. H., Muller, W. H., and Haines, B. L. (1964). Volatile growth inhibitors produced by aromatic shrubs. *Science* **143**, 471–473.

Muller, C. H., Hanawalt, R. B., and McPherson, J. K. (1968). Allelopathic control of herb growth in the fire cycle of California chaparral. *Bull. Torrey Botan. Club* **95**, 225–231.

Muller, W. H. (1965). Volatile materials produced by *Salvia leucophylla*: effects on seedling growth and soil bacteria. *Botan. Gaz.* **126**, 195–200.

Muller, W. H., and Hauge, R. (1967). Volatile growth inhibitors produced by *Salvia leucophylla*: effect on seedling anatomy. *Bull. Torrey Botan. Club* **94**, 182–191.

Muller, W. H., and Muller, C. H. (1956). Association patterns involving desert plants that contain toxic products. *Am. J. Botany* **43**, 354–361.

Muller, W. H., and Muller, C. H. (1964). Volatile growth inhibitors produced by *Salvia* species. *Bull. Torrey Botan. Club* **91**, 327–330.

Muller, W. H., Lorber, P., and Haley, B. (1968). Volatile growth inhibitors produced by *Salvia leucophylla*: effect on seedling growth and respiration. *Bull. Torrey Botan. Club* **95**, 415–422.

Parry, D. W., and Smithson, F. (1964). Types of opaline silica depositions in the leaves of British grasses. *Ann. Botany (London) N.S.* **28**, 169–184.

Patrick, Z. A., and Koch, L. W. (1958). Inhibition of respiration, germination, and growth by substances arising during the decomposition of certain plant residues in the soil. *Can. J. Botany* **36**, 621–647.

Pickering, S. V. (1917). The effect of one plant on another. *Ann. Botany (London)* **31**, 181–187.

Pickering, S. V. (1919). The action of one crop on another. *J. Roy. Hort. Soc.* **43**, 372–380.

Pimentel, D. (1968). Population regulation and genetic feedback. *Science* **159**, 1432–1437.

Pridham, J, B., ed. (1967). "Terpenoids in Plants," 257 pp. Academic Press, New York.

Proebsting, E. L., and Gilmore, A. E. (1940). The relation of peach root toxicity to the re-establishing of peach orchards. *Proc. Am. Soc. Hort. Sci.* **38**, 21–26.

Rademacher, B. (1941). Über den antagonistischen Einfluss von Roggen und Weizen auf Keimung und Entwicklung mancher Unkräuter. *Pflanzenbau* **17**, 131–143.

Rademacher, B. (1959). Gegenseitige Beeinflussung höherer Pflanzen. *Handbuch Pflanzenphysiol.* **11**, 655–706.

Reichstein, T., von Euw, J., Parsons, J. A., and Rothschild, M. (1968). Heart poisons in the monarch butterfly. *Science* **161**, 861–866.

Reiners, W. A. (1967). Relationships between vegetational strata in the pine barrens of central Long Island, New York. *Bull. Torrey Botan. Club* **94**, 87–99.

Rennerfelt, E., and Nacht, G. (1955). The fungicidal activity of some constituents from heartwood of conifers. *Svensk Botan. Tidskr.* **49**, 419–432.

Rice, E. L. (1964). Inhibition of nitrogen-fixing and nitrifying bacteria by seed plants. *Ecology* **45**, 824–837.

Rice, E. L. (1965). Inhibition of nitrogen-fixing and nitrifying bacteria by seed plants. II. Characterization and identification of inhibitors. *Physiol. Plantarum* **18**, 255–268.

Rice, E. L. (1967). Chemical warfare between plants. *Bios (Mt. Vernon, Iowa)* **38**, 67–74.

Rice, E. L., and Parenti, R. L. (1967). Inhibition of nitrogen-fixing and nitrifying bacteria by seed plants. V. Inhibitors produced by *Bromus japonicus* Thunb. *Southwestern Naturalist* **12**, 97–103.

Robinson, T. (1963). "The Organic Constituents of Higher Plants: Their Chemistry and Interrelationships," 306 pp. Burgess, Minneapolis.

Schildknecht, H., Bayer, I., and Schmidt, H. (1967). Über Pflanzenabwehrstoffe IV. Struktur der Primelgestoffes. *Z. Naturforschg.* **22**, 36–41.

Sláma, K., and Williams, C. M. (1966). The juvenile hormone. V. The sensitivity of the bug, *Pyrrhocoris apterus*, to a hormonally active factor in American paper-pulp. *Biol. Bull.* **130**, 235–246.

Smithson, F. (1958). Grass opal in British soils. *J. Soil Sci.* **9**, 148–154.

Spoehr, H. A., Smith, J. H. C., Strain, H. H., Milner, H. W., and Hardin, G. J. (1949). Fatty acid antibacterials from plants. *Carnegie Inst. Wash. Publ.* **586**, 1–67.

Stickney, J. S., and Hoy, P. R. (1881). Timber culture (comments following article). *Trans. Wisconsin State Hort. Soc.* **11**, 156–168.

Tamm, C. O. (1950). Growth and plant nutrient concentration in *Hylocomium proliferum* (L.) Lindb. in relation to tree canopy. *Oikos* **2**, 60–64.

Tukey, H. B., Jr. (1966). Leaching of metabolites from above-ground plant parts and its implications. *Bull. Torrey Botan. Club* **93**, 385–401.

Tukey, H. B., Jr., and Mecklenburg, R. A. (1964). Leaching of metabolites from foliage and subsequent reabsorption and redistribution of the leachate in plants. *Am. J. Botany* **51**, 737–742.

Varma, S. C. (1938). On the nature of the competition between plants in the early phases of their development. *Ann. Botany (London)* N.S. **2**, 203–225.

Wang, T. S. C., Yang, T. -K., and Chuang, T. -T. (1967). Soil phenolic acids as plant growth inhibitors. *Soil Sci.* **103**, 239–246.

Webb, L. J., Tracey, J. G., and Haydock, K. P. (1961). The toxicity of *Eremophila mitchellii* Benth. leaves in relation to the establishment of adjacent herbs. *Australian J. Sci.* **24**, 244–245.

Webb, L. J., Tracey, J. G., and Haydock, K. P. (1967). A factor toxic to seedlings of the same species associated with living roots of the non-gregarious subtropical rain forest tree *Grevillea robusta. J. Appl. Ecol.* **4**, 13–25.

Welbank, P. J. (1960). Toxin production from *Agropyron repens. Brit. Ecol. Soc. Symp.* **1**, 158–164.

Went, F. W. (1942). The dependence of certain annual plants on shrubs in Southern California deserts. *Bull. Torrey Botan. Club* **69**, 100–114.

Whitehead, D. C. (1964). Identification of p-hydroxybenzoic, vanillic, p-coumaric and ferulic acids in soils. *Nature* **202**, 417–418.

Whittaker, R. H. (1965). Dominance and diversity in land plant communities. *Science* **147**, 250–260.

Whittaker, R. H. (1969). New concepts of kingdoms of organisms. *Science* **163**, 150–160.

Wilson, R. E., and Rice, E. L. (1968). Allelopathy as expressed by *Helianthus annuus* and its role in old-field succession. *Bull. Torrey Botan. Club* **95**, 432–448.

Winter, A. G. (1961). New physiological and biological aspects in the interrelationships between higher plants. *Symp. Soc. Exptl. Biol.* **15**, 229–244.

Woods, F. W. (1960). Biological antagonisms due to phytotoxic root exudates. *Botan. Rev.* **26**, 546–569.

Wright, J. M. (1951). Phytotoxic effects of some antibiotics. *Ann. Botany (London)* N.S. **15**, 493–499.

Yardeni, D., and Evanari, M. (1952). The germination inhibiting, growth inhibiting and phytocidal effect of certain leaves and leaf extracts. (Spanish summary.) *Phyton (Buenos Aires)* **2**, 11–16.

4

Plants and the Chemical Environment

F. W. WENT

I. INTRODUCTION

Since a plant is the product of interaction between its internal constitution and its environment, my subject could include a very considerable part of plant physiology. I will restrict it, however, to three aspects of the chemical environment which interest me most: the chemical environment as produced by surrounding plants, by surrounding air, and chemical changes produced by animals. The first and second of these subjects have an important bearing upon the composition of the vegetation as an ecosystem. The third should give us a better insight into developmental processes in general.

II. PLANT–PLANT INTERACTIONS

A. Observations in Nature

Plants do not grow haphazardly; they are arranged in very definite associations and communities. Whereas the more general groupings or associations are largely climatically and edaphically conditioned (forests, swamps, steppes, etc.), the smaller and more intimate groupings or communities differentiating the vegetation within climatically and edaphically uniform areas must be based on another type of control. The associations are mostly expressed as differences in life forms such as trees, shrubs, and herbs; the communities are based on differences in species. Such species groupings can be highly differentiated, as any naturalist knows who is sufficiently acquainted with a local flora. Before being close enough to a locality to recognize individual plants, he can already predict which species he will find in a particular area because of the specific groupings of plants. Let me give a few examples from my own experience.

Studying the epiphytic flora in a mountain forest in Java, I found that certain epiphytic orchids were usually or exclusively growing on certain host trees, and that there was no general progression in specificity among host trees. In other words, there was no progressive roughness of bark or decrease in light intensity with which the occurrence of these orchids was associated. On the contrary, when branches of adjoining trees were more or less intertwined and seemed to be equally suited for attachment by many orchids, the orchids remained specifically associated with their "own" host tree. For instance, where branches of *Cestrum aurantiacum* and *Saurauia penduliflora* grew intermixed, *Oberonia oxystophyllum* grew exclusively on the *Saurauia* branches, while *Liparis bilobulata* was restricted to *Cestrum* (Went, 1940). This specificity was so pronounced that trees could be identified by the orchids living on them.

In other areas different relationships are found. In our southwestern deserts there are very pronounced correlations between certain shrubs and particular herbs. For instance, *Rafinesquia neomexicana*, an annual composite, grows poorly in the open desert, thrives inside clumps of *Krameria canescens* and *Franseria dumosa*, but avoids shrubs of *Larrea divaricata* and *Encelia farinosa*. On the other hand, *Lepidium fremontii*, *Delphinium parishii*, and *Phacelia tanacetifolia* are most commonly found under *Larrea divaricata* (Went, 1942). A number of annual plants, such as *Baeria chrysostoma*, *Linanthus aureus*, and *Plantago insularis*, avoid shrubs and grow in the open between shrubs (Juhren *et al.*, 1956). There are many instances of specific mushrooms growing near special trees, such as *Boletus laricinus* near *Larix*, *Boletus felleus* near *Tsuga*, *Boletus luteus* near *Pinus*, and *Clavaria abietina* near *Picea*. But the case which has intrigued me most is that described over 40 years ago by Kooper (1927). He studied the weed vegetation developing after the plowing of rice and sugar cane fields in Java and found that in successive years the same spot always produced the same combination of weeds. He could distinguish between several dozen different weed communities, each with its own specific composition, sometimes having ten to a dozen "constants," i.e., species occurring in each locality where the community was found. From my own experience I can attest to the reality of these various weed communities. In different parts of Java I found them with the same species composition as described by Kooper from East Java. As an added point of interest, these weed communities retain the same composition from germination to maturity, meaning that as many plants matured as germinated in the first place, and that the community composition was not a result of competition or differential survival but of differential germination. Since then the same thing was found for the vegetation of annuals in the desert (Juhren *et al.*, 1956). In Kooper's case there was no correlation between inorganic chemical composition and physical properties of the soil and weed community, but sugar cane yield was positively correlated with the weeds.

B. Chemical Interpretation of Inhibitions

All of these examples have in common a very great specificity of plant occurrence, indicating a high degree of differentiation in the environment. It is impossible to envisage this differentiation in terms of physical factors such as light, temperature, or humidity. If different degrees of shading were involved, a whole succession of communities should be found around a lone tree in a field, which is not the case. Therefore, we are more or less forced to assume that the fine differentiations in plant occurrence must have an organic chemical basis, and research during recent decennia at least partially substantiates this conclusion.

As a first example of chemical control I would like to discuss *Parthenium argentatum* (guayule), a Mexican composite containing rubber, which in its natural habitat is very evenly spaced in the desert with a considerable distance between adjacent shrubs. When grown as a field crop at close spacing, the plants along the edges of the field, where they are bordered around only one half of their circumference by other plants, are twice as big; on the four corners they are four times as big as the plants in the center of the field. All these observations suggested mutual inhibition of *Parthenium* plants, and laboratory experiments (Bonner and Galston, 1944) confirmed the existence of a substance, produced by *Parthenium* roots, which inhibits the growth of other plants, including young *Parthenium* plants themselves. Chemical isolation of the most active of the root secretions produced *trans*-cinnamic acid as the main inhibitor. Thus there is good evidence that by secreting *trans*-cinnamic acid, *Parthenium* shrubs inhibit the growth of other plants in their environment, at least as far as their roots reach.

The inhibitory effect of *Encelia farinosa* plants on germination and growth of others in their immediate surroundings is at least partly attributable to 3-acetyl-6-methoxybenzaldehyde, which Gray and Bonner (1948) isolated from *Encelia* leaves.

There are numerous other cases where certain plants excrete substances causing inhibition of growth of other plants in their neighborhood. For example, *Artemisia absinthium* leaves decrease the growth of

Nepeta and others (Bode, 1940), and the soil is often bare for a considerable distance around *Eucalyptus* and *Juglans* trees. In the latter case,
the inhibition of other plants is attributed to juglone.

Another case of inhibition, involving barley (*Hordeum sativum*), has
been investigated by Overland (1966). This is a so-called smother crop,
since it keeps down weeds in fields, especially if it has been sown thickly.
This showed that its roots secrete inhibitors, which were tentatively
identified as hordein and gramin, two alkaloids. The interesting part of
this analysis is that wild barley, *Hordeum spontaneum*, closely related to
H. sativum, is growing in the wild state in almost pure stands, indicating
that even there it keeps down other plants. It can be imagined that one of
the reasons barley became a cultivated crop plant was that it produced
nice pure stands without having to be weeded or treated with herbicides.

There are many other known cases where one plant species inhibits
other plants around it, and generally it is assumed that this is due to inhibitory substances excreted by roots or given off by other parts of the
plant. Extensive reviews of the phenomenon, generally called allelopathy following Molisch's (1937) example in the first general book on the
subject, have been written by Grümmer (1955) and Evenari (1961). But
is it not peculiar that most allelopathic effects known are inhibitions?
One might expect this on the basis of the Darwinian concept of the struggle for existence, in which plants adopted a form of biological warfare to
eliminate competitors, particularly because plants have no other means
of aggression. But many examples can be quoted, in addition to those
mentioned in the introduction, where the presence of one plant facili-
tates or conditions the presence of other species, such as in plant communities. There one should expect stimulatory substances, yet hardly
any stimulators have been discovered among plant excretions. The question therefore arises: Are the inhibitors described earlier really agents in
biological warfare in nature?

There are several general remarks to be made about the inhibitors just
described.

1. They are found primarily in shrubs and plants from dry climates
(*Parthenium, Encelia, Hordeum*). There are many other suspected
cases of inhibitors produced by plants of dry climates: *Larrea, Eucalyptus, Myrtus, Triodia*, many more than in moist climates.

2. Whereas they inhibit the growth of a number of plants, the strong-
est inhibition in each case is against their own seedlings. Therefore, if
the inhibitor were primarily there to cut down competition, it would be
against the species itself. This, of course, does not make sense. I therefore submit the following hypothesis.

3. In arid climates the amount of water limits plant development.
Occasionally a heavy rain will replenish the soil water supply which
normally would result in vigorous growth of all shrubs. If this were all
above-ground growth, the root system in the following drought period
might not be adequate to supply enough water to the tops. Therefore a

mechanism by which the plant restricts its own growth under occasional favorable conditions would be of distinct advantage. In the desert, adequate root growth should always precede top growth. The root inhibitor mechanism seems to insure this, for in the greenhouse, with plenty of water to leach out the inhibitor, growth of these desert shrubs is rapid.

C. Chemical Interpretation of Stimulations

In animals the chemical environment plays a most important role in controlling their health and survival through variations in the mineral, vitamin, and energy contents of their food. Whereas variations in the mineral content of the soil can also have powerful consequences on plant growth, this discussion is nevertheless restricted to the effects of the organic chemical environment.

It is widely accepted that photosynthetic plants do not need "energy food" since they produce this themselves. Furthermore, since most plants can be grown in nutrient sand culture without any organic addenda, they obviously also have no requirements for vitamins. There are some interesting exceptions to this rule, however. The best documented case is that of the _Camellia_. This is a shrub of the forest undergrowth that fails to grow in normal sand culture. Only after the addition of minute quantities of thiamine (vitamin B_1) will growth in sand culture take place. Thus thiamine can be considered to be a vitamin for _Camellia_. That this is actually the case follows from the observations that thiamine is required for root growth. Normally it is produced in the leaf and transported to the roots. _Camellia_ leaves, however, have lost the ability to synthesize thiamine and since the roots of this plant need it just as much as those of other plants, _Camellia_ can only grow when thiamine is supplied exogenously (Bonner and Greene, 1938). With the multitude of fungi growing in the forest litter, there is an adequate supply of thiamine, and thus _Camellia_ as well as other possible forest plants with a thiamine requirement are capable of growth. There is also the case of at least one orchid, _Cypripedium_, which grows much better when an organic medium is provided (Went, 1957). Finally, it is also known that most plants grow exceedingly poorly when started in dunes or on sand. This may be partly due to the leaching of nutrient elements from such sandy soils in moist climates.

D. Biological Interpretation of Stimulations

There is an additional factor involved in growth responses which is often overlooked, namely mycorrhiza. I would like to include this in the organic chemical environment of plants, even though we do not know how much organic matter is contributed through this relationship. But we do know that on poor sandy soils most trees grow poorly unless they have mycorrhiza, and in many places (Puerto Rico, Australia) imported pines started to grow well only after their mycorrhizal fungi were imported as well.

Mycorrhiza is the symbiosis of plant roots with the mycelium of fungi. This phenomenon was discovered in temperate forests where pines, oaks, and beeches produce curiously shortened and thickened roots. These often resemble corals in their close branching, and are covered with a dense sheath of fungal hyphae. The fungus penetrates some distance into the cortex of the root, and its hyphae end intra- or intercellularly in bundles or much-branched "arbuscules." Although discovered in temperate forests, it occurs from the tropical jungle to the subarctic forest, from sea level to timberline, and from very high rainfall areas to the desert.

The relationship between root and fungus must be symbiotic; one seldom, if ever, finds roots invaded and killed by mycorrhizal fungi. Parasitic root fungi such as *Pythium* and *Fusarium* kill the whole root, and do not leave the cells penetrated by their mycelium alive. But curiously enough we do not have much information on what the symbiotic partners contribute to each other.

In the literature on mycorrhiza the root is usually given the task of providing the fungus with carbohydrates. The fungus in its turn supplies the root with nitrogen and phosphorus, in addition to special growth substances such as auxins. The fungus is also considered to replace the root hairs that are lacking in pines and oaks, and thus provides water and inorganic nutrients. All this contributes to the chemical environment of the root. However, nowhere in the mycorrhiza literature is there any consideration given to the role of litter and humus in the soil.

Soil litter is mainly broken down by fungi (Went and Stark, 1968), as observed in the Central Amazonian rain forest. In a montane *Pinus murrayana* forest there are hardly any bacteria or Actinomycetes in the litter, and decomposition seems to be almost exclusively performed by a Dematiae, a sterile mycelium, a *Penicillium*, and a *Mucor*. Except for the Dematiae which occur mainly in the upper litter layer, the other three fungi were isolated from litter, mycorrhiza, and rhizomorphs, thus connecting litter with the roots.

None of the fungi growing in the litter produce any fruiting bodies. In the Amazonian forest the wood-decomposing fungi (Polyporaceae, *Xylaria*) form mushrooms only where there are no roots in the neighborhood, such as higher up in dying tree trunks or on fallen branches which make no contact with soil. A similar situation is known in temperate forests, where mushrooms of *Boletus subtomentosus* form only when trenches are dug in the forest floor to isolate the roots in a block of soil from surrounding trees (Romell, 1939). Also, desert fungi such as *Podaxon* and *Montagnea* do not produce fruiting bodies in the neighborhood of shrubs, but mostly where roots and soil have been disturbed, e.g. along roadsides.

From these observations the inference can be made that the nutrients released by mycelial activity while decomposing litter, and which normally would be available for fruiting body growth, are diverted towards the mycorrhiza. This seems certain in the case of saprophytes, such as pine-drops (*Pterospora*) or the snowplant (*Sarcodes*). They produce dense masses of mycorrhizal roots that are entirely fed by rhizomorphs of fungi and on which adventitious buds develop which will grow out into next year's flower stalks.

There are other indications that the fungi can provide food to whichever structures they are connected with—mushrooms or mycorrhiza. In agar culture there are always droplets of an osmotically highly active solution excreted by hyphae from the sterile mycelia. This must be sugar. The same droplets appear on the hyphae growing out of rhizomorphs in a moist chamber. If those hyphae were intracellular, they presumably would excrete this sugar into the mycorrhiza. We know through the work of Knudson (1922) that for orchids the fungal symbiont can be replaced by sugar. Therefore, the fungi which live symbiotically with the roots of plants in mycorrhiza form an important part of the chemical environment of many plants. The sugar they provide to the root system of plants may supplement the internal sugar supply through the phloem. In some cases, such as the saprophytes, it is the only sugar supply. And in the deep shade of dense forests, even seedlings with few leaves can grow if supplied with sugars by their mycorrhizal fungi.

E. Litter and Humus

A new aspect to the yearly loss of photosynthate by trees and other plants through leaf drop and dead branches is introduced when one considers the role of litter and humus. Much of this loss, but we do not know how much, can be picked up again by fungi and returned to the tree by way of its roots. This certainly is the case with the minerals in the Amazonian jungle, where the soils are so poor that only complete recycling of the minerals contained in the forest litter can insure the continued growth of a luxuriant forest. Once this cycle (in which mycorrhiza provides the closing link) is broken by cutting and burning, no forest can reestablish itself on those soils.

This also changes our attitude toward the old humus theory of plant nutrition, so thoroughly dismissed by Liebig. From the aforegoing observations and considerations it is quite possible that the decomposition of organic matter in the soil by fungi provides a source of both mineral and organic nutrition through direct transfer by way of mycelium. At the same time the mycelium provides mechanical strength to the soil; it may be an important factor in soil structure.

F. Parasitic Interrelations

Parasites create a special chemical environment for the plant. In most cases their effects are deleterious. Many plant parasites produce toxins, killing the cells in their neighborhood. This destroys semipermeability, and the soluble cell contents ooze out of the killed cells, making it unnecessary for the parasite to penetrate the cell. In this way *Botrytis cinerea*, producing soft brown rot in so many plants, causes collapse of all soft tissues with which it comes in contact. It is suspected that the toxin in this case is a pectolytic enzyme (Brown, 1955). Slightly more interesting are the cases in which parasites produce growth inhibitors in the host plant. When infested with red spiders, many plants have a strongly reduced growth rate. The degree of growth inhibition is completely out of proportion to the size of the spider infestation; apparently the spiders inject a powerful growth inhibitor into the phloem. This means that less sugar is used up by the plant and more is available to the spider.

Sometimes the plant provides the specific environment the parasite needs. This has been investigated in the case of phanerogamic parasites such as *Striga* and *Orobanche*. Their seeds germinate only in the immediate vicinity of roots of their host plants, and it was found that extracts of these roots contain a germination promoter for *Striga* or *Orobanche*. Although *d*-xyloketose is effective in *Striga* germination, it does not seem to be the actual substance produced by *Sorghum* roots, its host plant (Brown *et al.*, 1952).

III. THE AERIAL ENVIRONMENT

Other aspects of the chemical environment of plants which have come under scrutiny recently are aerial environmental factors such as smog and the natural volatile emanations of plants. Nothing more will be said about smog than that photochemical reaction products of hydrocarbons, such as gasoline vapors, are highly toxic for many plants and restrict the number of trees and plants which can grow around our metropoli. Commercial production of spinach and some other leafy vegetables has been discontinued around Los Angeles because of their great sensitivity to smog, a number of pines are dying on account of smog, and certain tobacco varieties are killed as seedlings by smog.

A different type of smog which is due to natural causes is found as the heat or summer haze in all vegetated areas. To what extent it is toxic or influences plant growth has not been established as yet, but my guess is that terpenes and other volatile organic compounds, which are photochemically transformed into a blue haze, may have a measurable effect on plant growth. This certainly is borne out by observations and measurements of Muller (1966) who claims that bare areas around aromatic shrubs like *Salvia mellifera* are caused by their volatile emanations. He has shown that the germination of many seeds is inhibited or prevented if leaves of *Salvia* are enclosed with them in a petri dish.

The amount of volatile organic materials, predominantly terpenes, formed by plants is enormous. My guess is that a billion tons of volatiles per year are released over the whole world (Went, 1964). Thus far we have not found a plant which does not produce measurable amounts of terpenoid substances, whether they are algae, liverworts, or higher plants. No beneficial general physiological or metabolic role has been discovered for these terpenoids, and their ubiquity makes it unacceptable to consider them merely as waste products or metabolic errors. I suggest that one of their functions is polymerization on the plant surface, producing cuticular material and providing a membrane around the plant which partially seals it off from the dry environment. Since volatilized terpenes can be activated photochemically, probably through the production of free radicals, polymerization of terpenes can also occur in the atmosphere. This leads to the formation of blue hazes and results in a change of the radiation environment of plants.

In this production of volatiles lies an enormous and only partially explored field for future research, and one can make any number of guesses where this research will lead.

IV. PLANT–ANIMAL INTERACTIONS: GALLS

The theoretical significance of all the previous cases of modified chemical environment for plant development fades into the background compared with what research on insect galls could provide. Unfortunately very little concrete evidence has been collected as yet, but I want to discuss the problem because of its potential value in the solution of problems of growth and development. Galls are instances of localized abnormal growth in plants, produced by external agents. In the case of crowngall, undifferentiated tissue masses are growing on stems under the influence of a bacterium *Agrobacterium tumaefaciens*, and in this case excessive auxin production is the initial gall-producing stimulus. But similar to plant callus cultures, the auxin only causes undifferentiated callus growth. Nematodes can produce tremendous swellings on the roots of many plants, but again no specific structures develop. Certain fungi cause so-called witches brooms when through suspension of apical bud dominance, all lateral buds on the infected branch start to grow. The lateral branches repeat this procedure until a dense tangled mat of branches is formed. This behavior might be explained by auxin destruction in the apical buds of the witches broom.

Among arthropods there are many which cause abnormal growth in plants: excessive swelling of epidermal cells; irregularities in the leaf surface producing sacklike outgrowths in which aphids or mites are protected, and many other unspecific responses. But the most remarkable galls are produced by gallwasps (Hymenoptera: Cynipidae), gallmidges (Diptera: Cecidomyidae), sawflies (Hymenoptera: Tenthredinidae), and psyllids (Hemiptera: Psyllidae). Almost as remarkable as the gall structures produced is the fact that these gall insects restrict their activities to a small number of plant families and genera. More than 90% of all the hundreds of known Cynipid galls occur on two totally unrelated genera: oaks and roses. The galls produced are of very different types and on many different organs: buds, stems, leaves, roots, or anthers. A cursory look through Felt's book, "Plant Galls and Gall Makers" (1940) will show an enormous variety of gall shapes, on a restricted number of host plants.

We consider here only the organoid galls in which seemingly new organs unlike any organs normally found on the host species are formed. Often very complex new tissues are present in these galls, and a remarkable degree of tissue and shape differentiation occurs. Yet all this is caused by the presence of a tiny egg or larva, usually having a mass of only a fraction of a percent of the size of gall produced.

What makes these galls so unusually interesting is that in their case an extraneous stimulus, produced by the insect egg or developing larva, has a similar effect as the DNA or RNA normally present inside the cell or nucleus. In the case of the gall, we should be able to manipulate the insect egg or larva, removing it, reintroducing it, making extracts, injecting fractions, etc. Thus with an egg-laying gall insect and a micromanipulator we should be able to manipulate the external environment of a leaf or bud in a way similar to the way the nucleus and cytoplasm manipulate the internal environment of cell and tissue.

When one considers the potential of this approach, one wonders why it has not yet been carried out. Is there a catch? Yes, there are several. In the first place this research requires a very close working relationship between a classic entomologist versed in the raising of insects and a botanist who can grow and manipulate the host plant. A second difficulty is that complex galls which are most interesting to study are produced only once a year, with the gall insect producing eggs for not more than 1 week. Unless one finds a way of having a continuing gall and gall insect production, the gall research team would be out of work 50 weeks per year, and who wants to work as a lone wolf investigator in this day and age of DNA?

I would not have pressed this new approach to gall research as much if I had no good reason to expect great results from a new start. This is based on some old and new research on the *Pontania* gall of willow leaves. More than 80 years ago the grandmaster of gall research, M. W. Beijerinck (1888) had found that normally a gall would not develop when after inserting its ovipositor no egg was laid by the gall insect. This, however, was not the case with the *Pontania* gall. The sawfly sticks its ovipositor into a very young willow leaf, and places its egg halfway between the upper and lower epidermis next to the midrib. In the course of the next few weeks while the willow leaf enlarges to its final size, the leaf tissue around the egg grows a hundred times more than it does without the stimulus from the sawfly. It then produces a pea-sized swelling of solid tissue, usually pink-colored, with an internal gall chamber lined with soft parenchyma which is consumed by the *Pontania* larva. Contrary to any other gall producer, a good-sized gall is formed even when the sawfly fails to lay an egg after inserting its ovipositor. This was attributed by Beijerinck to a secretion from abdominal glands of the sawfly.

This was the situation as Hovanitz (1959) took up the problem again. A three-pronged approach was made. First tissue cultures were established of willows. Then the willows were cultured in a phytotron, and they were made to grow continuously, essential for uninterrupted insect culture. Then the *Pontania* sawflies were cultured on these willows, again successfully. Thus sawflies were available at any time of the year and the stage was set for a more detailed study of the gall formation processes. One of the early discoveries was that in the area studied (southern California) at least six distinct willow varieties could be found. Each variety had its own *Pontania* sawfly. While these all belonged to the same species they were adapted to an individual willow variety so that an unmatched combination of sawfly and willow produced poor or abortive galls. That meant a remarkable degree of specialization of both insect and host plant, suggesting that differentiation was due to highly specific interactions between insect and plant. This conclusion seemed to be borne out by an analysis of the abdominal secretion of the sawfly, which had an ultraviolet absorption pattern suggesting a nucleic acid. At this stage the work was discontinued; I still believe that it could be brought to an important conclusion and would tell us much more about the problems of morphogenesis and development.

References

Beijerinck, M. W. (1888). *Botan. Z.* **46**, 1–11, 17–27.

Bode, H. R. (1940). Uber die Blattausscheidungen des Wermuts und ihre Wirkung auf andere Pflanzen. *Planta* **30**, 567–589.

Bonner, J., and Galston, A. W. (1944). Toxic substances from the culture media of the guayule which may inhibit growth. *Botan. Gaz.* **106**, 185–198.

Bonner, J., and Greene, J. (1938). Vitamin B_1 and the growth of green plants. *Botan. Gaz.* **100**, 226–237.

Brown, R., Johnson, A. W., Robinson, E., and Todd, A. R. (1952). The *Striga* germination factor 2. Chromatographic purification of crude concentrates. *Biochem. J.* **50**, 596–600.

Brown, W. (1955). On the physiology of parasitism in plants. *Ann. Appl. Biol.* **43**, 325–341.

Evenari, M. (1961). Chemical influences of other plants (allelopathy). *Encyclopedia Plant Physiol.* **XVI**, 691–736.

Felt, E. P. (1940). "Plant Galls and Gall Makers." Hafner, New York.

Gray, R., and Bonner, J. (1948). Structure determination and synthesis of a plant growth inhibitor, 3-acetyl-6-methoxybenzaldehyde, found in the leaves of *Encelia farinosa. J. Am. Chem. Soc.* **70**, 1249–1253.

Grümmer, G. (1955). "Die gegenseitige Beeinflussung höherer Pflanzen-Allelopathie." Fischer, Jena.

Hovanitz, W. (1959). Insects and plant galls. *Sci. Am.* **201**, 151–162.

Juhren, M., Went, F. W., and Phillips, E. (1956). Ecology of desert plants. IV. Combined field and laboratory work on germination of annuals in the Joshua tree national monument, California. *Ecology* **37**, 318–330.

Knudson, L. (1922). Nonsymbiotic germination of orchid seeds. *Botan. Gaz.* **73**, 1.

Kooper, W. J. C. (1927). Sociological and ecological studies on the tropical weed vegetation of Pasuruan (the island of Java). *Rec. Trav. Botan. Neerl.* **24**, 1-255.

Molisch, H. (1937). "Der Einfluss einer Pflanze auf die andere-Allelopathie." Fischer, Jena.

Muller, C. H. (1966). Soil toxicity induced by terpenes from *Salvia leucophylla. Bull. Torrey Botan. Club* **93**, 332-351.

Overland, L. (1966). The role of allelopathic substances in the "smother crop" barley. *Am. J. Botany* **53**, 423-432.

Romell, L. G. (1938). A trenching experiment in spruce forest and its bearing on problems of mycotrophy. *Svensk Botan. Tidskr.* **32**: 89-99.

Went, F. W. (1940). Soziologie der Epiphyten eines tropischen Urwaldes. *Ann. Jardin Botan. Buitenzorg* **50**, 1-98.

Went, F. W. (1942). The dependence of certain annual plants on shrubs in southern California deserts. *Bull. Torrey Botan. Club* **69**, 100-114.

Went, F. W. (1957). The experimental control of plant growth. *Chronica Botan.* **17**, XVII, 343.

Went, F. W. (1964). The nature of aitken condensation nuclei in the atmosphere. *Proc. Natl. Acad. Sci. U.S.* **51**, 1259-1267.

Went, F. W., and Stark, N. (1968). The biological and mechanical role of soil fungi. *Proc. Natl. Acad. Sci. U.S.* **60**, 497-504.

5

Chemical Interactions between Plants and Insects

V. G. DETHIER

I. INTRODUCTION

It is self-evident that all life has essentially a common basis; the planet which serves as a substrate for life is nonuniform. Given these two facts plus the additional fact that life is characterized by change, the existence of diversity among organisms is an inescapable conclusion. These facts do not suffice, however, to explain the extraordinary richness of diversification. The inadequacy is manifest every time one observes the human physiognomy and contemplates the infinite variety of which he himself is a single example. Even if one were to discover some evolutionary meaning in facial variation, other puzzling examples abound as, for example, the infinite variation in fingerprints.

One possible explanation of the extent of diversity is that there is too much "noise" in life's mechanism of replication, that by the very nature of its operation it is incapable of exact duplication and is indeed even error prone. I am referring here to the randomness of mutations, to pleiotropy, and to so-called preadaptations. Another possible explanation of the apparent superfluity of variations is that it meets a need for organisms to come to terms not only with the nonuniformity of the inorganic environment, but also with all of the other organisms that are trying to make the same harmonious adjustment.

It is a canon of contemporary biology that organic replication is random and inexact and that inequalities of the nonliving and living envi-

ronment in any particular time and space favor one variation over another. Application of this canon nevertheless leaves unanswered many
of the questions engendered by diversity. One puzzling example in particular is the diversity of feeding habits of herbivores. Two characteristics of herbivorous behavior, namely, diversity and preference, have
been public knowledge for a long time, as the following quotation from a
nursery rhyme attests:

> Little boy blue come blow your horn.
> The sheep's in the meadow.
> The cow's in the corn.

or to borrow an example from carnivores:

> Jack Spratt could eat no fat.
> His wife could eat no lean.
> So betwixt the two of them
> They licked the platter clean.

The case of herbivorous diversity is of more than parochial interest
because the questions that it raises, were they answered, might provide
insight into the broader issues related to the nature of the common phenomenon of variation. Basically there are two patterns that diversity can
assume. If by diversity is meant difference, then species A can be totally
and exclusively different from species B or A can share *some* features in
common with B. Phrased in terms of feeding habits, A in the first case
might be a caterpillar that feeds only on carrot and B a caterpillar that
feeds only on tomato. In the second case, A would feed on carrot and lettuce and B on tomato and lettuce. Variations on this theme are possible.
For example, A might feed on plants 1 to 60 and B only on 60, or A on 1 to
60 and B on 30 to 90. All possibilities are exemplified in nature. One
range of the spectrum has been formulated as the problem of monophagy versus polyphagy.

II. SELECTIVE PRESSURES

The different patterns of feeding constitute an especially complex
form of diversity because they involve in each instance not one species
evolving against a nonuniform inorganic environment, but one species
evolving in the sphere of influence of other mutable species. It is not
always appreciated by students of insect behavior that plants and insects are contemporaneously evolving organisms (cf. Ehrlich and Rav-
en, 1965). Nor is it usually recognized that plants are not evolving primarily as "food" or "nonfood."
Too often in considering the feeding behavior of phytophagous insects,
the evolution of the plant is envisioned primarily in terms of the pressures exerted upon it by insect predators. It is not at all clear that predation, insect predation in particular, is the major pressure; the environ-

ment obviously is, and competing plant species are. There are very few examples (some crop plants excepted) of decimation by insects (Elton, 1938; Andrewartha and Birch, 1954). Microorganisms seem to be far more effective in this respect. The virtual extinction of the American chestnut and the American elm are cases in point, although it must be admitted that an insect vector is involved with the latter. Instances in which the insect is more directly involved include the defoliation of forest trees by gypsy moths, the infestation of spruces by spruce bud worm, the destruction of thousands of acres of prickly pear (*Opuntia*) in Australia by a tunneling caterpillar *Cactoblastis cactorum* (Folsom and Wardle, 1934), and the obliteration of areas of grasslands by migratory locusts. A striking case on a minor scale is the drastic reduction of seed set of lupine by larvae of the lycaenid *Glaucopsyche lygdamus* Doubleday (Breedlove and Ehrlich, 1968). A similar relation exists between *Oenothera* and larvae of the moth *Rhodoflora florida* (Dethier, unpublished observations). The insects in these cases can act as powerful selective agents in the plant populations.

It is apparent that plants are evolving in an atmosphere of multiple pressures, of which insect predation is but one. The avenues of escape from these pressures include morphological modifications, emigration to new habitats, and elaboration of defensive chemicals. Morphological modifications are of minor importance as a defense against predation. Movements into new habitats are ineffective against highly mobile, climatologically adaptable predators. A plant's best strategy against predators is to become inedible—not necessarily non-nutritious. This end is best accomplished by the elaboration of chemicals which are, for one reason or another, unacceptable to a predator; however, the selective preservation of specific chemicals cannot be evaluated solely in terms of insects. The activities of vertebrate herbivores are relevant. It is also significant that some plant chemicals are strikingly effective in suppressing the growth of competing plant species (Muller et al., 1964; Muller and Muller, 1964; Wells, 1964).

Insects are not evolving solely with respect to plants any more than are the plants with respect to insects; however, the plant is probably more critical to the insect than vice versa (pollination excepted). Evolution in the direction of botanical chemical diversification presents herbivores with a nonuniform food cover. All plants are not equally suitable as food. Again, it must be emphasized that suitability is not to be interpreted as solely nutritional in nature. From an insect's point of view all chemical innovations are relevant regardless of the "reason" for their selection. Insects are constantly being selected against by "resistant" plants. Concurrently the insect is evolving "resistance" to plants. I have deliberately employed the same word, "resistance," in both cases to emphasize the mutual evolutionary ebb and flow between the two organisms that are trying to outwit each other.

Of the two the herbivore would appear to have more tricks at its disposal. Whereas the principal evolutionary gambit of the plant is the elaboration of arrays of chemical compounds, common or exotic, the insect can adapt in a number of different ways. It can adapt biochemically at different stages in the sequence of digestion and assimilation; it can learn new feeding habits; it can move away in search of more suitable food. Morphological adaptations play as small a role in the evolution of feeding behavior of herbivores as they play in the development of resistance in plants. The oft-studied evolution of mouthparts into piercing, sucking, chewing, etc., signifies less what an insect ingests than how it ingests it, and the extreme structural modification of leaf miners has little bearing upon which species of plant will be consumed. Metabolic modifications, especially the development of specific enzymes and detoxification mechanisms, have greater relevance. Probably the most common and important modifications are the neural ones, sensory or central, which permit behavioral diversity.

Behavioral changes can be both rapid and reversible and thus constitute a formidably effective response to whatever evolutionary innovations plants essay. In view of this potential it would be astounding if behavior were not the preeminent moving force underlying diversity of feeding habits. Thus, in the confrontation between plants and phytophagous insects the antagonists move in different dimensions of time.

The versatility with which an insect responds evolutionarily to the chemical changes in plants must depend in part upon what genetic and biochemical latitude remains in it at any given time, because it too has been evolving in response to other pressures of which plants are only one. How the insect responds within its limits (that is, how it will be selected against) to the plant becoming inedible (i.e., unpalatable, toxic, or non-nutritious) depends upon what other services the plant performs. Selective pressures may be different in each case. Plants serve not only as food but as microhabitat, shelter, and protection. There are some striking examples of the ways in which a plant may offer protection. If, for example, it is inedible to vertebrate herbivores, the insects on it are protected from being themselves ingested. An interesting example of this is reported by Alpin *et al.* (1968). The cinnabar moth (*Callimorpha jacobaeae* L.) feeds on species of groundsel (*Senecio vulgaris* L. and *S. jacobaeae* L.) which contain compounds rendering the plants unacceptable to vertebrate herbivores. In pastures heavily grazed by cattle or rabbits the herbage surrounding the plants is closely cropped thus removing cover essential for certain arthropod predators and greatly reducing their toll of the eggs and larval stage of the moth (personal communication from Dempster and Harris to Alpin *et al.*, 1968). Another example is that of the arctiid moth *Seiractia echo* which feeds on cycad plants that contain cycasin (methylazoxymethanol-β-D-glycoside) and hydrolytic β-glycosidases. Cycasin and its aglycone (methylazoxymethanol) are potent toxins, carcinogens, and mutagens for vertebrates. The larvae have special detoxifying mechanisms enabling them to protect themselves against the lethal effects of their diet (Teas, 1967).

The plant could provide double protection. In addition to the sort mentioned above, it could supply insects with protective devices against their own predators by conferring unpalatability (Rothschild, 1964; Brower, 1958; Brower and Brower, 1964). Very recently, conclusive evidence has been advanced to show that some insects incorporate into their bodies plant chemicals that render them unpalatable or toxic to vertebrate predators. Larvae of the monarch butterfly (*Danaus plexippus*) that were fed on cabbage were eminently palatable to bluejays while those reared on a natural food plant (*Asclepias curassavica*) which contained cardenolides (cardiac glycosides) caused the jays to vomit. The jays rapidly learned to refuse these larvae (Brower *et al.*, 1967). Cardenolides similar to those found in milkweeds (Asclepiadaceae) upon which it feeds are also found in the tissues and in the defensive glands of an African grasshopper [*Poekilocerus bufonius* (Klug)] which is immune to ingestion by predators (von Euw *et al.*, 1967). The cinnabar moth (*Callimorpha jacobaeae* L.) also incorporates and even concentrates in its tissues poisonous alkaloids present in its food plants (*Senecio jacobaeae* L. and *S. vulgaris*) and is unacceptable to many vertebrate predators (Alpin *et al.*, 1968); however, the part these compounds play is at the moment enigmatic.

III. PLANTS AS FOOD

Despite these services offered by the plant, it nevertheless remains primarily a source of food. It had earlier been suggested (Fraenkel, 1953, 1959a,b) that green leaves are excellent sources of all food materials that insects require. House (1966) has pointed out, however, that the nutritional superiority of a food depends on nutrient balance, that is, the proportions of nutritionally important substances in the food with respect to the proportions required by the insect. It is also clear that different species of insects have different requirements (House, 1961, 1962). Laboratory studies with the sarcophagid *Agria affinis* (Fallen) showed that growth rate was decreased when the levels of amino acids in the larval diet were increased and was increased when the levels of other nutrients were increased. Proportion was the critical parameter. Similarly, with larvae of the sphingid *Celerio euphorbiae*, there was lower consumption and lower weight gain on an imbalanced diet (House, 1965). The diet contained half as much of the necessary vitamins and possible sources of vitamins and trace nutrients, a third more casein and amounts of certain amino acids, and the same quantity of sucrose, salts, and other ingredients as present in the 100% control diet. When a properly proportioned diet was diluted to 85, 70, and 50% of its nutrient content, the larvae consumed progressively more but gained the same weight on each diet.

The decreased consumption of the first diet was explained as a way of avoiding the metabolic difficulties that might ensue if, in the process of eating copiously to compensate for the deficiency of one ingredient, the larvae accumulated too much of another. The second case is a common phenomenon throughout the animal kingdom – decreased nutrient value per unit volume is compensated for by increasing intake.

It is clear from the extensive work of House and others that nutrient imbalance affects intake, that the nutrient requirements of insects differ subtly (rather than grossly) from one species to the next, and that many insects eat the food most suited to their requirements. The last statement should be rephrased to state that many insects are found feeding on the food most suited to their requirements.

The manifestly close relation that exists between some insects and the nutrient adequacy of the particular plants upon which they feed poses two fundamental questions: (1) Can insects discriminate among diets of various nutritive value, and, if so, by what means? (2) Are insects able to make up dietary deficiencies by recognizing diets containing the necessary supplements? One of the best attempts at answering these questions does not deal with a phytophagous species. House's (1967) experiments with larvae of *A. affinis* (Fallén) were designed in such a way that larvae had free access to four nutritionally different diets arranged in a 16-unit "latin square." At the beginning of the experiment one larva was placed on each of the 16 diets. After 5 hours the distribution of the larvae was recorded, and again twice a day for a total of 7 days. A significantly greater number of larvae were found on the most suitable food. The experiment particularly worthy of note is one summarized in House's Table IV (Table I). Diet *G* (unsatisfactory) contained the same amount of glucose as the satisfactory diet (*F*) but a lower concentration of amino acids; diet *H* (unsatisfactory) contained the same amount of amino acids as *F* but more glucose.

TABLE I

Total Numbers of Larvae Found Arrayed[a] on Four Media in a Preference Experiment[b]

Dietary media, composition (%)		Preference experiment	Feeding experiment
Amino acids	Glucose	Disposition of larvae, total numbers	Mature larvae (%), seventh day
(E) 0.75	2.1	96	4.7
(F) 1.125	1.5	255	48.7
(G) 0.75	1.5	87	0.7
(H) 1.125	2.1	117	28.4

[a]Three tests × 5 cages × 7 examinations.
[b]From House (1967).

The events leading to "selection" (i.e., selection of the food by the insect) are interesting. Initially the larvae wandered actively from one diet to another. They fed on unsatisfactory diets as well as on the suitable one. In House's words, "It seems most likely that food selection resulted from testing by trial and error until a satisfactory nutrient composition ... was found that abated a 'craving' or hunger arising from metabolic demands for certain nutrients that the other media did not supply adequately."

It would appear that the larvae could not in fact distinguish one diet from another on initial contact. (In the same vein, where an insect derives secondary advantages from eating plants containing highly toxic compounds, it does not necessarily follow that the useful compounds themselves are the ones that effect the preference.) After ingestion of suitable food, locomotion decreased so that the number of larvae on that food increased, whereas after ingestion of unsuitable food the larvae continued to wander. As a consequence, larvae never aggregated on unsuitable diets. Postingestive factors undoubtedly acted on locomotor centers, thus influencing locomotion and distribution. The nature of the relationship is not known, although there are suggestions from studies with the blowfly *Phormia regina* Meigen (Green, 1964) and locusts (Ellis and Hoyle, 1954; Hoyle, 1954) that blood-borne factors are involved. Green postulated a hormonal mechanism in the blowfly. Hoyle and Ellis and Hoyle related the concentration of potassium in the blood to feeding and activity; however, Chapman (1958) was not able to confirm these findings. Whatever the mechanism, modification of locomotion influences the distribution of insects on various foods. Sometimes the change in activity has nothing to do with feeding (cf. Ellis, 1951). In any case, it is not choice in the real sense. Furthermore, because of it being a postingestive effect, it is a much less efficient and more hazardous method of selecting diets than one based on an immediate, initial assessment of the suitability of food by gustatory or olfactory sensing.

A corollary of the proposition that insects select proper diets is the conclusion that animals suffering some nutritional deficiency tend to alleviate the deficiency by selecting a suitable diet. This concept of response to specific hungers demands reexamination in light of the recent work of Rodgers and Rozin (1966) with thiamine-deficient rats. It has been known for more than 30 years that thiamine-deficient rats prefer food containing thiamine. It was assumed that the animals recognized the needed component. Militating against this explanation was the failure of an immediate response to thiamine to appear in deficient rats. Even more damning was the observation that pairing thiamine initially with a distinctive flavor and then switching it to a different flavor left the rats still preferring the initial diet (Harris *et al.*, 1933; Scott and Verney, 1947). The usual paradigm for studying specific hungers consists of offering the deficient animal a choice between the old familiar nutritionally inadequate diet and a novel, new, enriched diet. As Rodgers and Rozin have shown, deficient rats prefer new, novel diets. In nature this propensity would enhance the likelihood of making up the deficit. The preference for novelty, whether or not it supplies the missing nutrient, is probably innate. In short, it is possible that the diet containing the missing substance is preferred, not because of its specific properties, but because of its novelty. Sustained ingestion may be due to a learned preference.

Whether or not a similar explanation can apply to insects is not known. Two capabilities are required of the insect: the appreciation of novelty, and learning. The only careful study of novelty was concerned with the response of houseflies toward novel visual stimuli (Mourier, 1965). It demonstrated that flies landed on and explored most readily squares of tile that were new and novel to them and that the novelty wore off after approximately 20 minutes. The appreciation of novelty is, therefore, not beyond the potential of insect behavior.

With respect to learning we are concerned only with a modification of feeding behavior as a consequence of experience. This capacity in phytophagous insects has been demonstrated by Jermy et al. (1968). First, newly hatched unfed larvae of Manduca sexta (Johan) and Heliothis zea (Boddie) were fed on an artificial diet devoid of plant material until the fourth or fifth instar. Larvae of each species were then divided into groups. One group of M. sexta was fed in tomato (Lycopersicon esculentum Mill.), one on tobacco (Nicotiana tabacum L.), and one on Jerusalem cherry (Solanum pseudocapsicum L.); one remained on diet. One group of H. zea remained on diet, one fed on cauliflower (Brassica oleracea var. botrytis DC.), one on geranium (Pelargonium hortorum Bailey), and one on dandelion (Taraxacum officinale Weber). When allowed to choose from all three plants in a preference test, the larvae clearly favored the plant upon which they had fed. The induced preference was specific for the inducing plant species and was not merely a change in the general threshold of food acceptability. Preference could be induced by feeding for as briefly as 24 hr.

It was not obliterated by two larval moults and subsequent feeding on an artificial diet. It is likely, therefore, that the information on which the induced preference was based was stored in the nervous system. The central nervous system (as opposed to the peripheral) seems to be a likely place; however, electrophysiological analyses made by Schoonhoven (1967a) have demonstrated that the response of gustatory receptors to various plants is changed (is different from normal) if larvae have been fed on artificial diet. It is possible, therefore, that induced changes in preference could come about as a consequence of changes in the chemoreceptors.

In addition to aggregation, other criteria of choice often used are optimal growth, survival, and successful reproduction. In the sense of the species these are indeed signs of a preferred plant. They indicate that a harmonious and optimal relationship has developed between the insect and the plant (at least from the insect's point of view). They do not, however, reflect the choice of the individuals; nor is it possible on the basis of these kinds of observations to assert that the choice was nutritionally based. The distinction being made here is important because in the context of the species one is dealing with natural selection, while in the context of the individual one is concerned with the mechanisms by which choice is effected. In seeking for the identity of what is evolving in the insect, the latter is the prime consideration.

There are very few convincing experiments demonstrating that the choice of a plant by an individual is in fact directed by the nutrient characteristics of that plant. There are numerous examples of the stimulating effectiveness of nutrient compounds, but there are not many examples of these being actual preferences for these compounds. The most universally accepted compounds are certain carbohydrates, especially sucrose and glucose. The very fact of "the sweet tooth" being ubiquitous means that its role in mediating specific preferences is suspect.

From time to time other essential nutrients are revealed to be gustatory stimuli (see reviews by Thorsteinson, 1960; Beck, 1965). Among them are amino acids, sterols, phospholipids, ascorbic acid, and B vitamins. Just as often, however, essential nutrients are without stimulating effect. After an intensive study of the influence of nutrient chemicals on the feeding behavior of the Colorado potato beetle [*Leptinotarsa decemlineata* (Say)], Hsiao and Fraenkel (1968a,b,c) concluded (at least with respect to this insect) that while many nutrient compounds enhance its biting responses, only a few elicit significant feeding responses. The wide distribution of these compounds throughout the plant kingdom argues against their being specific for the potato beetle.

In contrast, the number of examples of secondary plant substances that affect food selection by acting as attractants, repellents, arrestants, stimulants, or deterrents continues to grow. Only a few need be mentioned here. A more complete listing may be found in numerous reviews (e.g. Thorsteinson, 1960; Dethier, 1966). The commercial silkworm *Bombyx mori* (L.) is attracted by citral, linalyl acetate, linalol, and terpinyl acetate; it is induced to bite by β-sistosterol, isoquercitrin, morin, inositol, sucrose; it is induced to swallow by cellulose, silicate, and phosphate (Hamamura, 1965). According to Nayar and Fraenkel (1962), a closely related compound rather than β-sistosterol itself is one of the stimulants for biting. The catalpa sphinx [*Ceratomia catalpae* (Bdv.)] feeds only on four species of *Catalpa*, all of which contain catalposides (mixtures of 15 glycosides). Catalposides plus glucose induced feeding (Nayar and Fraenkel, 1963). Fifty percent of newly hatched larvae of *Pieris brassicae* (L.) will feed on a diet devoid of mustard glycosides, but the glycosides enhance intake by 20%. Larvae fed on cabbage will not transfer to diet (David and Gardiner, 1966a,b). Larvae of the moth *Plutella maculipennis* (Curtis) respond to sinigrin if sucrose is present

(Thorsteinson, 1953). The cabbage aphid [*Brevicoryne brassicae* (L.)] will feed and reproduce successfully on the abnormal host (*Vicia faba*) if the leaves are made to absorb sinigrin via the stems (Wensler, 1962). The Colorado potato beetle is still an enigma. Jermy (1961a) concluded that food selection by adults is based on the absence of repellents. Stürckow and Löw (1961) reported that alkaloids from *Solanum* do indeed deter feeding. Ritter (1967) argued that food selection by adults was based on a mixture of common taste substances (sugars, amino acids, etc.).

On the other hand, Hsiao and Fraenkel (1968c) observed that adults will feed and mature successfully on *Asclepias* when no alternative is present but will refuse it when potato is available. From this they concluded that the several acceptable species of *Asclepias* lack deterrents and that potato must have a token stimulant. Both nutritive substances and a token are required to release feeding. The bark beetle *Scolytus multistriatus* (Marsham) is stimulated to feed by a pentacyclic triterpene isolated from the bark of *Ulmus americana* (Baker and Norris, 1967).

Most of the studies of gustatory stimulants have dealt with the effects of single compounds. In 1953 Thorsteinson discovered that mixtures of sucrose plus sinigrin or other mustard glycosides were more stimulating to *Plutella* than either constituent alone. Since then numerous examples of synergism have been discovered (see reviews of Thorsteinson, 1960; Dethier, 1966). The majority represent true synergism in the sense that the effect of the mixture exceeds that expected of simple addition. Gothilf and Beck (1967) reported an instance in which stimulation resulted from mixing two substances which singly were without any effect whatsoever. The combination, potassium ions and the neutral lipid fraction of wheat germ oil, acted as a feeding stimulant for larvae of the cabbage looper *Trichoplusia ni* (Hübner). The phenomenon is not restricted to any particular chemical or physiological class of compounds. Thus, synergism occurs among nutrients (Dethier *et al.*, 1956; Omand and Dethier, 1969), among various kinds of token stimuli (Silverstein *et al.*, 1968), and between the two (*loc. cit.*). In view of these observations, experiments designed to bioassay compounds that have no nutritional value per se (e.g., secondary plant substances) should, as an indispensable prerequisite, have sugar present (Thorsteinson, 1960; Nayar and Fraenkel, 1962).

Secondary plant substances do not act exclusively as stimulants. Their importance as repellents and deterrents has long been known, but originally they were thought, in this capacity, to be concerned only with limiting the range of plants consumed by polyphagous insects. Gupta and Thorsteinson (1960), Thorsteinson (1960), and Jermy (1958, 1961a,b,c) proposed, however, that their role was not so restrictive. According to Jermy (1966) an overwhelming majority of nonhost plants of monophagous and oligophagous insects contain feeding inhibitors. The proposition is supported by many observations of which the following are examples: oligophagous insects will on occasion eat plants that do not contain the normal feeding stimulants; many newly hatched larvae of *P. brassicae* will eat a bland diet devoid of all plant material; oligophagous and some monophagous insects become more polyphagous as starvation progresses. Thus, plant chemicals must be conceived of as acting in either of two capacities, as deterrents to feeding or as stimulants to feeding.

IV. THE CHEMORECEPTIVE SYSTEMS OF INSECTS

All things considered, the case for the initial encounter between an insect and a plant being the important one for choice is convincing. Whether the stimuli are nutrients, toxins, or secondary plant substances is irrelevant insofar as the act of ingestion is concerned. It is misleading to juxtapose the terms because "secondary plant substances" denotes a category of plant chemicals while "nutrients" and "toxins" refer to the end result of ingestion. Furthermore, it must be remembered that an insect does not ingest only the compound that stimulates it to eat; it ingests the entire plant tissue of which that stimulant is merely a part. Since it has been well demonstrated that the adequacy of a diet depends upon its completeness and balance, to read significance into the fact that any given stimulant also possesses nutritional value fails to lay due emphasis on the prime effect of that compound. It is worth reiterating that insofar as the mechanism of individual choice is concerned, the postingestive effects of the stimulants are less important than the initial and immediate effect.

In seeking, therefore, what it is in the insect that evolves, the prime candidates are the chemoreceptive systems which detect plant chemicals and encode this information and the decoding command centers in the central nervous system. At the moment we possess no knowledge of the central mechanisms involved in feeding preferences of insects. Sensory systems are reasonably well understood even though the absolute volume of information is small. The most thoroughly studied systems in phytophagous insects are those of the larvae of the commercial silkworm (*Bombyx mori* L.) (Ishikawa, 1963; Ishikawa and Hirao, 1963), the large cabbage butterfly (*Pieris brassicae* L) (Schoonhoven, 1967b), and the tobacco hornworm [*Manduca sexta* (Johan.)] (Schoonhoven and Dethier, 1966). The Colorado potato beetle has also been studied (Stürckow, 1959). It is probably not too misleading to accept the situation presented by lepidopterous larvae as indicative of the general state of affairs in phytophagous insects.

Lepidopterous larvae have two clearly delineated chemoreceptive systems: a gustatory and an olfactory. The gustatory system consists of two sensilla styloconica on the galea of each maxilla. Among the components of each sensillum are four chemoreceptors of considerable specificity. These are listed in Table II. In terms of specificity, the complement of each sensillum is different; there are species differences as well. The extent of receptor specificity and species differences has not been thoroughly explored. In *P. brassicae* there is in each sensillum a receptor sensitive to low (10^{-5}–10^{-3} *M*) concentrations of mustard oil glycosides. The order of effectiveness for the cell in the lateral sensillum is glucocapparin > sinigrin = glucotropaeolin > sinalbin = glucosinalbin; for the cell in the medial sensillum, sinalbin > glucosinalbin = glucotropaeolin. The threshold of the medial cell is higher than that of the lateral. Both cells have a high level of specificity for these glycosides but do give some response to sodium cyanate at higher (100×) concentrations.

The olfactory organs are the antennae and maxillary palpi. There are 16 olfactory cells in each antenna and 19 to 24 in each palpus. With this complement of approximately 40 cells on each side of the head, the caterpillar is able to sense a wide variety of odors. The receptors are not narrowly specific, nor are they uniquely tuned to specific odors. Each one responds to many odors by differentially increasing or decreasing its rate of spontaneous firing (Schoonhoven and Dethier, 1966). The response spectrum of each cell is different, but considerable overlap occurs. In other words, the input from all of the receptors combined forms, for each odor, a different pattern that can be discriminated. Discrimination is based, therefore, on central decoding of total sensory patterns rather than on fine sensory filtering (Dethier, 1967).

TABLE II

Comparison of Chemoreceptor Cell Types in the Medial and Lateral Sensillum Styloconicum of Some Lepidoptera Larvae[a]

Genera	Medial sensillum	Lateral sensillum
Manduca	Water and salt	Water
	Sucrose and glucose	Salt
	Acid	Sucrose and glucose
		Inositol[b]
Galleria	Water	Water
	Salt	Salt
	Sucrose	
Philosamia	Salt	Water
	Salt	Salt
	Glucose	Sucrose
		Glucose
Bombyx	Water	Salt
	Salt	Sucrose
	Salt	Glucose
	Repellents	Inositol

[a]From Schoonhoven and Dethier (1966).
[b]The existence of an inositol receptor here is uncertain.

It can be concluded that in Schneider's (1969) terminology, phytopha-gous insects possess two categories of chemoreceptors; specialists and generalists. Accordingly, the information received by the central ner-vous system arrives in the form of either unique private-line messages or patterns. The central nervous system may act upon this input by com-manding acceptance or rejection. Input that elicits acceptance in one species may elicit rejection in another. Furthermore, a chemical stimu-lus acting on a single receptor at one concentration may influence the central nervous system to order acceptance, while the same chemical stimulating the same receptor at a different concentration may cause the central nervous system to order rejection. This duality has been demonstrated behaviorally and electrophysiologically with sodium chlo-ride stimulating labellar receptors in the blowfly *Phormia regina* Mei-gen (Dethier, 1968). From these considerations it can be seen that much of the specificity that underlies feeding preferences reflects receptor specificity (directly or through patterning), and much reflects specificity in command interneurons.

The following considerations suggest how the interaction between insect and plant may operate. Let us imagine three species of plants the chemical compositions of which are represented by (1) $R_1R_2R_3R_4R_5$, (2) $R_6 AR_7R_8R_9$, and (3) $R_{10}R_{11}R_{12}R_{13}R_{14}$. Insect I, for example, could prefer plant (2) because it contains the stimulant A; it could reject (1) and (3) be-cause they lack A, because they contain deterrents R_1-R_5, $R_{10}-R_{14}$, or for both reasons. Insect II could prefer plant (3) because it is insensitive to compounds $R_{10}-R_{14}$ but sensitive to deterrents R_1-R_9. It might or might not be able to discriminate among R_1-R_9. In the first example, insect I would require a receptor specifically tuned to compound A. We know that receptors with these characteristics exist: the sinigrin receptor of *P. brassicae* (Schoonhoven, 1967b), the grass receptor of *Locusta* (Schneider, 1965), the glucose receptor of *B. mori* and *Philosamia cyn-thia* (Ishikawa, 1963; Ishikawa and Hirao, 1963; Schoonhoven and De-thier, 1966), the carrion receptor of *Calliphora* and *Necrophorus*, the queen substance receptor of *Apis* (Schneider, 1969), and various acid and salt receptors. In the second example, insect II would require as re-ceptors either a generalist indiscriminantly sensitive to R_1-R_9, or recep-tors with overlapping sensitivities to R_1-R_9 so that a patterned code could be produced, or many specific receptors each sensitive to a single R. Evi-dence from induction experiments (Jermy *et al.*, 1968) contradicts the postulate that all R's represent one vast indiscriminate modality. Insects can become specifically adapted to one particular R. There is no evidence that any insect possesses a different receptor for each kind of R. Evi-dence that discrimination among different kinds of R's is possible has been presented in the discussion of olfaction (see also Dethier, 1967). There is also ample evidence that one species of insect may be chemi-cally blind to an R that is highly stimulating to another. *Phormia*, for example, apparently has no receptors sensitive to the various glycosides of Cruciferae that are so stimulating to the cabbage caterpillar (Dethier, unpublished observations). Detailed comparisons of the sensitivities of closely related phytophagous insects with different feeding preferences would be illuminating.

V. THE *RAISON d' ÊTRE* OF FEEDING DIVERSITY

We now come to the crux of the problem of monophagy and polyphagy. What is the raison *d'etre* for such diversity of feeding habits? The insinuation lurking behind this oft-asked question is that there is some advantage to there being many different kinds of diets, that there is an advantage to being polyphagous or monophagous as the case may be. When we attempt, however, to construct a ledger of credits and debits, no obvious advantages of either state appear in the balance. Nor can there be compelling disadvantages; if there were, the species concerned could not afford the luxury of their diet and would be placed in a weak competitive position. Their very existence testifies to the suitability of their particular feeding behavior. The relationship does not signify that other equally suitable relationships are impossible.

The most obvious explanation of diversity is that it involves some nutritional basis; however, very early in the history of this work the view was taken that a nutritional explanation was unsatisfactory (Dethier, 1947). Fraenkel (1953) asserted that no evidence existed for the hypothesis that feeding specificities could be attributed to nutrient differences among plants. Although his contention that all plants are equally nutritious for all phytophagous insects cannot be sustained in view of the evidence (House, 1962), the case against the causal role of nutrition is strong. House had suggested that monophagous insects are restricted in their diets because it is metabolically advantageous to be so. At the same time it has been alleged that some insects eat more than one species of plant because a varied diet is nutritionally better than any one plant alone. This assertion still leaves unanswered the question of what determines the choice of plants that are eaten.

Arguments that attempt to explain monophagy (and oligophagy) in terms of fine nutritional balance between plant and insect must be contradicted in the case of polyphagy, or alternatively, polyphagous insects must be strikingly different in their requirements. It does not appear possible to explain diversity of feeding habits on a nutritional basis. The hypothesis implies that some species are metabolically polyphagous and others metabolically monophagous. Evidence has already been presented to show that, at the very least, there are exceptions to this generalization. Waldbauer (1962) has shown that extirpation of the maxillary taste receptors of caterpillars permits utilization of a wider range of plants. Wensler (1962) has shown that aphids will do well on nonhost plants if token stimuli are supplied. Hsiao and Fraenkel (1968c) have shown that the Colorado potato beetle will do well on species of *Asclepias*. The nutritional hypothesis further assumes that metabolic idiosyncracies existed and then the insects sought suitable hosts. This implies that the nutritional suitability of hosts could be recognized. The hypothesis says nothing about plants that are not eaten. Nor does it follow that one type of feeding habit is more or less suitable than another.

On the basis of the ideas presented in foregoing sections, namely, that the initial encounter between insect and plant is crucial, that both plant and insect are evolving against a background of multiple pressures, that plants evolve by synthesizing different chemicals and insects by developing different sensory capabilities and central decision-making capabilities, a hypothesis of congruency is proposed here to explain the origin of feeding diversity. It assumes that changes in the neural systems of insects and the chemical systems in plants occur randomly by mutation. If there is a change in receptor sensitivity and/or central interpretation such that somewhere in the plant kingdom there is a chemical(s) that will stimulate, the plant possessing that chemical will be eaten. Similarly, if there are neural changes such that chemicals formerly acting as repellents or deterrents are no longer detected, the plants involved will be eaten. If there are neural changes such that formerly nonstimulating chemicals can now be detected and/or sensory input is now interpreted as unacceptable, the plant involved will not be eaten. In short, mutational neural changes are proposed which involve the addition, substraction, and substitution of capabilities. To explain the hypothesis further, the following greatly oversimplified model is presented. Let the critical receptor sites in four species of insects be represented by the geometric forms in the top row, cf. Fig. 1. For the sake of emphasis, the model rep-

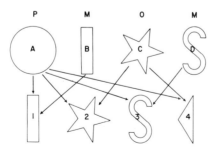

Fig. 1. A scheme to represent relations between phytophagous insects and their food plants. The top row of figures (A–D) represent receptor sites in the chemoreceptors of the insect. The bottom row of figures (1–4) represent chemicals in the plant. Arrows indicate which chemicals can interact with each receptor site. P, M, and O indicate polyphagy, monophagy, and oligophagy, respectively.

resents receptor sites; in practice one can substitute molecular sites representing metabolic pathways or detoxification mechanisms or any combination of the three limiting characters. The fact remains, however, that the first barrier to be overcome in the insect/plant relationship is a behavioral one. The insect must sense and discriminate before nutritional and toxic factors become operative. Let the bottom row represent various plant chemicals (secondary plant substances, nutrients, or both) arising by mutation in four species of plants. Each row was generated randomly by mutation. A comfortable fit of any figure in the bottom row into any figure in the top row indicates that the plant is fed upon by the insect. Thus, if a figure in the top row cannot accept any figure in the bottom row, no feeding occurs, the mutation is lethal. If a figure (A) in the top row accepts all figures in the bottom row, the result is polyphagy; if a figure (C) accepts some, the result is oligophagy; if it accepts only one (B or D), the result is monophagy. In fact, although our knowledge of receptor properties and plant chemicals is scanty, some actual data can be substituted to bring reality to the scheme (Table III). Similarly, one can postulate receptor sites for deterrents.

For the model to approximate reality the assumption must be made that the mutations involved represent quantum jumps. Thus it is assumed that no intermediates arise between A, B, C, and D (Fig. 1). This is not an unreasonable assumption when one considers for example, that mutations can change the chemoreceptor sensitivity of *E. coli*, which is normally sensitive (behaviorally) to glucose, galactose, ribose, aspartate, and serine, to give rise to organisms that are attracted to all compounds except galactose or other strains attracted to all compounds except serine (Hazelbauer *et al.*, 1969). The single-gene basis of tasters and nontasters among human beings is well established. Among phytophagous insects there is not enough available information about the genetics of feeding habits to test the assumption; however, there are a number of documented cases of sudden irreversible shifts in feeding habits [e.g., the shift of the apple maggot fly from *Crataegus* to *Vaccinium* (Woods, 1915)] to suggest that decisive changes can be brought about as assumed.

It is further assumed that reproductive isolation follows any fortuitous association that is initiated so that the mutant is not competing with its normal allele. The survival of the new mutant is related, therefore, to the availability of food rather than to competitive pressures from the normal allele. Ways and means of accomplishing isolation have been proposed (cf. Dethier, 1954).

According to this hypothesis changes between polyphagy and monophagy in either direction are equally likely, as are changes from one form of monophagy to another. It might be argued that a mutation from polyphagy to monophagy would present the mutant with fewer opportunities to feed than its normal allele has. This is apt to be so only if the food of the monophagous mutant is in short supply. In fact the food of monopha-

TABLE III
Receptor Sites and the Chemicals that Interact with Them

Site:	Sinigrin receptor	Specific site predicted	Specific site predicted	Specific site predicted
Insect:	Cabbage butterfly (P. brassicae)[a]	Catalpa sphinx (C. catalpae)[b]	Tobacco hornworm (M. sexta)[c]	Bark beetle (S. multistriatus)[d]
Chemical:	Sinigrin	Catalposides	Uncharacterized glycoside	A pentacyclic triterpene

[a]Schoonhoven (1967b).
[b]Nayar and Fraenkel (1963).
[c]Yamamoto and Fraenkel (1960).
[d]Baker and Norris (1967).

gous insects is almost invariably a very abundant and ubiquitous plant species.

If the ingested material is not at least marginally nutritious or if it is toxic, the mutation proves to be lethal. If it is not lethal, a new feeding habit has been established. It need not be better nutritionally than that of its normal allele because nutrition is only one function of the plant. Eating is a compromise among many factors. As Gordon (1961) has remarked, insects live on suboptimal nonutopian diets. Nature does not always provide the best. There are, as ethology has shown, such things as superoptimal stimuli. One can, Shakespeare to the contrary notwithstanding, "paint the lily" and "throw a perfume on the violet."

VI. CONCLUSION

The diverse associations between insects and their host plants arose as a consequence of the interaction of two independently mutating systems. This view can explain why there are some plants (e.g. ferns) which possess feeding deterrents even though they evolved long before phytophagous insects (Soo Hoo and Fraenkel, 1964). It can explain why there exist plants which could serve adequately as food but do not. Diversity and restrictions in the feeding habits of phytophagous insects represent different ways of accomplishing the same end with equal effectiveness against a constantly changing background of botanical chemical innovation.

References

Alpin, R. T., Benn, M. H., and Rothschild, M. (1968). Poisonous alkaloids in the body tissues of the cinnebar moth (Callimorpha jacobaeae L.). Nature 219, 747–748.
Andrewartha, H. G., and Birch, L. C. (1954). "The Distribution and Abundance of Animals," pp. 589–612. Univ. of Chicago Press, Chicago, Illinois.

Baker, J. E., and Norris, D. M. (1967). A feeding stimulant for *Scolytus multistriatus* (Coleoptera: Scolytidae) isolated from the bark of *Ulmus americana*. *Ann. Entomol. Soc. Am.* **60**, 1213–1215.

Beck, S. D. (1965). Resistance of plants to insects. *Ann. Rev. Entomol.* **10**, 207–232.

Breedlove, D. E., and Ehrlich, P. R. (1968). Plant–herbivore coevolution: lupines and lycaenids. *Science* **162**, 671–672.

Brower, L. P. (1958). Bird predation and foodplant specificity in closely related procryptic insects. *Am. Naturalist* **92**, 183–187.

Brower, L. P., and Brower, J. V. Z. (1964). Birds, butterflies, and plant poisons: a study in ecological chemistry. *Zoologica* **49**, 137–159.

Brower, L. P., Brower, J. V. Z., and Corvino, J. M. (1967). Plant poisons in a terrestrial food chain. *Proc. Natl. Acad. Sci. U.S.* **57**, 893–898.

Chapman, R. F. (1958). A field study of the potassium concentration in the blood of the red locust *Nomadacris septemfasciata* (Serv.) in relation to its activity. *Animal Behaviour* **6**, 60–67.

David, W. A. L., and Gardiner, B. O. C. (1966a). The effect of sinigrin on the feeding of *Pieris brassicae* larvae transferred from various diets. *Entomol. Exptl. Appl.* **9**, 95–98.

David, W. A. L., and Gardiner, B. O. C. (1966b). Mustard oil glucosides as feeding stimulants for *Pieris brassicae* larvae in a semi-synthetic diet. *Entomol. Exptl. Appl.* **9**, 247–255.

Dethier, V. G. (1947). "Chemical Insect Attractants and Repellents." McGraw-Hill (Blakiston), New York.

Dethier, V. G., (1954). Evolution of feeding preferences in phytophagous insects. *Evolution* **8**, 33–54.

Dethier, V. G. (1966). Feeding behaviour. *In* "Insect Behaviour" (P. T. Haskel, ed.), Symp. No. 3, pp. 46–58. Roy. Entomol. Soc., London.

Dethier, V. G. (1967). Feeding and drinking behavior of invertebrates. *In* "Handbook of Physiology" (C. F. Code, ed.), Sect. 6, Vol. I, pp. 79–86. Am. Physiol. Soc., Washington, D. C.

Dethier, V. G. (1968). Chemosensory input and taste discrimination in the blowfly. *Science* **161**, 389–391.

Dethier, V. G., Evans, D. R., and Rhoades, M. V. (1956). Some factors controlling the ingestion of carbohydrates by the blowfly. *Biol. Bull.* **111**, 204–222.

Ehrlich, P. R., and Raven, P. H. (1965). Butterflies and plants: a study in coevolution. *Evolution* **18**, 586–608.

Ellis, P. E. (1951). The marching behavior of hoppers of the African Migratory Locust in the laboratory. *Anti-Locust Res. Centre Bull., London* **7**, 1–48.

Ellis, P. E., and Hoyle, G. (1954). A physiological interpretation of the marching of hoppers of the African Migratory Locust (*Locusta migratoria migratorioides* R. & F.). *J. Exptl. Biol.* **31**, 271–279.

Elton, C. (1938). Animal numbers and adaptation. *In* "Evolution: Essays on Aspects of Evolutionary Biology" (G. R. Beer, ed.), pp. 127–137. Oxford (Clarendon), London and New York.

Folsom, J. W., and Wardle, R. A. (1934). "Entomology With Special Reference to its Ecological Aspects," p. 272. McGraw-Hill (Blakiston), New York.

Fraenkel, G. S. (1953). The nutritional value of green plants for insects. *Trans. Internat. Congr. Entomology 9th Congr. Amsterdam 1951.* Vol. 2, p. 290.

Fraenkel, G. S. (1959a). The *raison d'être* of secondary plant substances. *Science* **129**, 1466–1470.

Fraenkel, G. S. (1959b). The chemistry of host specificity of phytophagous insects. *Proc. 4th Intern. Congr. Biochem., Vienna, 1958* **12**, 1–14.

Gordon, H. T. (1961). Nutritional factors in insect resistance to chemicals. *Ann. Rev. Entomol.* **6**, 27–54.

Gothilf, S., and Beck, S. D. (1967). Larval feeding behaviour of the cabbage looper, *Trichoplusia ni. J. Insect Physiol.* **13**, 1039–1053.

Green, G. W. (1964). The control of spontaneous locomotor activity in *Phormia regina* Meigen. II. Experiments to determine the mechanism involved. *J. Insect Physiol.* **10**, 727–752.

Gupta, P. D., and Thorsteinson, A. J. (1960). Food plant relationships of the diamond-back moth *Plutella maculipennis* (Curt.). *Entomol. Exptl. Appl.* 3, 241-250.

Hamamura, Y. (1965). On the feeding mechanism and artificial food of silkworm, *Bombyx mori. Mem. Konan Univ. Sci. Ser.* 8, Art. 38, 17-22.

Harris, L. J., Clay, J., Hargreaves, F., and Ward, A. (1933). Appetite and choice of diet. The ability of the vitamin B deficient rat to discriminate between diets containing and lacking the vitamin. *Proc. Roy. Soc. (London)* B113, 161-190.

Hazelbauer, G. L., Mesibov, R. E., and Adler, J. (1969). *Proc. Natl. Acad. Sci. U.S.* (in press).

House, H. L. (1961). Insect nutrition. *Ann. Rev. Entomol.* 6, 13-26.

House, H. L. (1962). Insect nutrition. *Ann. Rev. Biochem.* 31, 653-672.

House, H. L. (1965). Effect of low levels of the nutrient content of a food and of nutrient imbalance on the feeding and the nutrition of a phytophagous larva, *Celerio euphorbiae* (Linnaeus) Lepidoptera: Sphingidae). *Can. Entomologist* 97, 62-68.

House, H. L. (1966). Effects of varying the ratio between the amino acids and the other nutrients in conjunction with a salt mixture on the fly *Agria affinis* (Fall.). *J. Insect. Physiol.* 12, 299-310.

House, H. L. (1967). The role of nutritional factors in food selection and preference as related to larval nutrition of an insect, *Pseudosarcophaga affinis* (Diptera, Sarcophagidae), on synthetic diets. *Can. Entomologist* 99, 1310-1321.

Hoyle, G. (1954). Changes in the blood potassium concentration of the African Migratory Locust (*Locusta migratoria migratorioides* R. & F.) during food deprivation, and the effect on neuromuscular activity. *J. Exptl. Biol.* 31, 260-270.

Hsiao, T. H., and Fraenkel, G. (1968a). The influence of nutrient chemicals on the feeding behavior of the Colorado potato beetle, *Leptinotarsa decemlineata* (Coleoptera: Chrysomelidae). *Ann. Entomol. Soc. Am.* 61, 44-54.

Hsiao, T. H., and Fraenkel, G. (1968b). The role of secondary plant substances in the food specificity of the Colorado potato beetle. *Ann. Entomol. Soc. Am.* 61, 485-493.

Hsiao, T. H., and Fraenkel, G. (1968c). Selection and specificity of the Colorado potato beetle for solanaceous and non-solanaceous plants. *Ann. Entomol. Soc. Am.* 61, 493-503.

Ishikawa, S. (1963). Responses of maxillary chemoreceptors in the larva of the silkworm, *Bombyx mori*, to stimulation by carbohydrates. *J. Cellular Comp. Physiol.* 61, 99-107.

Ishikawa, S., and Hirao, T. (1963). Electrophysiological studies of taste sensation in the larvae of the silkworm, *Bombyx mori. Sanshi Shikensho Hokoku* 18, 297-357.

Jermy, T. (1958). Untersuchungen über Auffinden und Wahl der Nahrung beim Kartoffelkäfer (*Leptinotarsa decemlineata* Say). *Entomol. Exptl. Appl.* 1, 179-208.

Jermy, T. (1961a). On the nature of oligophagy in *Leptinotarsa decemlineata* Say (Coleoptera: Chrysomelidae). *Acta Zool. Acad. Sci. Hung.* 7, 119-132.

Jermy, T. (1961b). Néhány szervetlen só rejektiv hatása a burgonyabogár (*Leptinotarsa decemlineata* Say) imágoira és larváua. (The rejective effect of some inorganic salts on Colorado beetle adults and larvae.) *Ann. Inst. Prot. Plant. Hung.* 8, 121-130.

Jermy, T. (1961c). Über die Nahrungsspezialisation phytophager Insekten. *Ber. Wiss. Pflanzenschutzkonf. Budapest, 1960* 2, 327-332.

Jermy, T. (1966). Feeding inhibitors and food preference in chewing phytophagous insects. *Entomol. Exptl. Appl.* 9, 1-12.

Jermy, T., Hanson, F. E., and Dethier, V. G. (1968). Induction of specific food preference in lepidopterous larvae. *Entomol. Exptl. Appl.* 11, 211-230.

Mourier, H. (1965). The behavior of house flies (*Musca domestica* L.) towards "new objects." *Videnskab. Medd. Dansk Naturh. Foren.* 128, 221-231.

Muller, C. H., and Muller, W. H. (1964). Antibiosis as a factor in vegetation patterns. *Science* 144, 889-890.

Muller, C. H., Muller, W. H., and Haines, B. L. (1964). Antibiosis as a factor in vegetation patterns. *Science* 143, 471.

Nayar, J. K., and Fraenkel, G. (1962). The chemical basis of hostplant selection in the silkworm, *Bombyx mori* (L.). *J. Insect Physiol.* 8, 505-525.

Nayar, J. K., and Fraenkel, G. (1963). The chemical basis of host selection in the Catalpa sphinx. *Ceratomia catalpae* (Boisduval) (Lepidoptera, Sphingidae). *Ann. Entomol. Soc. Am.* 56, 119-122.

Omand, E., and Dethier, V. G. (1969). An electrophysiological analysis of the action of carbohydrates of the sugar receptor of the blowfly. *Proc. Natl. Acad. Sci. U.S.* **62**, 136–143.

Ritter, F. J. (1967). Feeding stimulants for the Colorado beetle. *Mededel. Rijksfacult. Landbouwwetenschap. Gent* **32**, 291–305.

Rodgers, W., and Rozin, P. (1966). Novel food preferences in thiamine-deficient rats. *J. Comp. Physiol. Psychol.* **61**, 1–4.

Rothschild, M. (1964). An extension of Dr. Lincoln Brower's theory on bird predation and food specificity, together with some observations on bird memory in relation to apposematic colour patterns *Entomologist* **97**, 73–78.

Schneider, D. (1969). Insect olfaction: deciphering system for chemical messages. *Science* **163**, 1031–1037.

Schoonhoven, L. M. (1967a). Loss of hostplant specificity by *Manduca sexta* after rearing on an artificial diet. *Entomol. Exptl. Appl.* **10**, 270–272.

Schoonhoven, L. M. (1967b). Chemoreception of mustard oil glucosides in larvae of *Pieris brassicae. Koninkl. Ned. Akad. Wetenschap., Proc., Ser. C* **70**, 556–568.

Schoonhoven, L. M., and Dethier, V. G. (1966). Sensory aspects of hostplant discrimination by lepidopterous larvae. *Arch. Neerl. Zool.* **16**, 497–530.

Scott, E. M., and Verney, E. L. (1947). Self-selection of diet: VI. The nature of appetites for B vitamins. *J. Nutr.* **34**, 471–480.

Silverstein, R. M., Brownlee, R. G., Bellas, T. E., Wood, D. L., and Browne, L. E. (1968). Brevicomin: principal sex attractant in the frass of the female western pine beetle. *Science* **159**, 889–890.

Soo Hoo, C. F., and Fraenkel, G. (1964). The resistance of ferns to the feeding of *Prodenia eridania* larvae. *Ann. Entomol. Soc. Am.* **57**, 788–790.

Stürckow, B. (1959). Über den Geschmackssinn und den Tastsinn von *Leptinotarsa decemlineata* Say (Chrysomelidae). *Z. Vergleich. Physiol.* **42**, 255–302.

Stürckow, B., and Löw, I. (1961). Die Wirkung einiger *Solanum* – alkaloidglykoside auf den Kartoffelkafer, *Leptinotarsa decemlineata* Say. *Entomol. Exptl. Appl.* **4**, 133–142.

Teas, H. J. (1967). Cycasin synthesis in *Seirarctia echo* (Lepidoptera) larvae fed methylazoxymethanol. *Biochem. Biophys. Res. Commun.* **26**, 686–690.

Thorsteinson, A. J. (1953). The chemotactic responses that determine host specficity in an oligophagous insect (*Plutella maculipennis*). *Can. J. Zool.* **31**, 52–72.

Thorsteinson, A. J. (1960). Host selection in phytophagous insects. *Ann. Rev. Entomol.* **5**, 193–218.

von Euw, J., Fishelson, J. A., Parons, J. A., Reichstein, T., and Rothschild, M. (1967). Cardenolides (heart poisons) in a grasshopper feeding on milkweeds. *Nature* **214**, 35–39.

Waldbauer, G. P. (1962). The growth and reproduction of maxillectomized tobacco hornworms feeding on normally rejected non-solanaceous plants. *Entomol. Exptl. Appl.* **5**, 147–158.

Wells, P. V. (1964). Antibiosis as a factor in vegetation patterns. *Science* **144**, 889.

Wensler, R. J. D. (1962). Mode of host selection by an aphid. *Nature* **195**, 830–831.

Woods, W. C. (1915). Blueberry insects in Maine. *Bul. Maine Agr. Exptl. Sta.*, **244**, 249–288.

Yamamoto, R. T. and Fraenkel, G. S. (1960). Assay of the principal gustatory stimulant for the tobacco hornworm, Protoparce sexta, from solanaceous plants. *Ann. Entomol. Soc. Am.* **53**, 499–503.

6

Hormonal Interactions between Plants and Insects

CARROLL M. WILLIAMS

I. INTRODUCTION

Some 10 years ago *Life* magazine published a prophetic article entitled "The Ultimate Weapon in an Ancient War." In that article Albert Rosenfeld (1958) summed up the situation in the following words:

> Because insects are relatively resistant to radioactivity, there have been gloomy predictions that once men have wiped out their own kind with nuclear weapons, insects will inherit the earth.

The article cheerfully goes on to say:

> Now amid all this pessimism has come a bright
> hope. Scientists have discovered the basis for what
> seems to be the ultimate insecticide—a product as
> deadly as DDT but without DDT's shortcomings.

The strategy embodied in these glad tidings is to destroy the insects by turning their own hormones against them. In retrospect, there is precedence for so doing. For nearly 30 years we have been killing the weeds with 2,4D and related plant hormones against which the broad-leafed plants have not been able to evolve any resistance. More recently, we aspire to control the human population with hormonally active materials against which humans have little defense. And now the day appears to be near at hand when insect pests can be destroyed on a species-by-species basis without any damage to other forms of life.

The possibility of using insect hormones as insecticides arose fortuitously as a by-product of studies of insect physiology—from a research effort sufficiently pure and impractical to scandalize any Congressional Committee (Williams, 1956, 1967). One would have to search long and hard to find a clearer example of the strange and sometimes wonderful fruits of the untainted vintage of pure research.

Until a few years ago, the hormonal approach to the selective control of insects appeared to be a novel concept. Now, as we shall see, the strategy appears to be an ancient art invented by certain plants and practiced by them for tens of millions of years.

II. GENES AND METAMORPHOSIS

Our story has much to do with metamorphosis, a term which is simply defined as "change in form." According to that definition, metamorphosis is little short of ubiquitous. Who among us has not changed form during the past decade or two? In point of fact, the change in form which humans undergo at puberty is a *bona fide* metamorphosis involving as it does the activation of countless genes which direct the maturation of the primary and secondary sex characters.

The so-called hemimetabolous insects (such as cockroaches, locusts, and the true bugs) undergo a metamorphosis which has much in common with that seen in human beings at puberty. After days, weeks, or months of juvenile growth, the larval insect undergoes a swift maturation of its reproductive system accompanied in many cases by the formation of functional wings (Snodgrass, 1954; Wigglesworth, 1954, 1964; Williams, 1952a). Here again it is necessary to conjecture that many "larval genes" are switched off and many "adult genes" are switched on (Wigglesworth, 1959; Williams, 1958, 1961).

A clear trend in the evolution of the higher insects is a progressive specialization of the juvenile organism away from the ancestral pathway

Fig. 1. Three stages in the metamorphosis of the Polyphemus silkworm (*Antheraea polyphemus*). The fifth instar silkworm transforms into the pupa and the latter into the adult moth.

leading to the adult. The apogee of this trend is seen in the higher holo-metabolous insects in which larval specializations have proceeded to the point where the juvenile insect seems to have little in common with the adult that it will ultimately form.

Consider, for example, the development of the Polyphemus silkworm shown in Fig. 1. The genetic "construction manual" for building a Polyphemus moth is obviously divided into three distinct chapters. The first chapter provides the genetic information for constructing the silkworm itself. That accomplished, the second chapter tells how to rework the cells and tissues of the silkworm to form an essentially new organism, the pupa. The third and final chapter tells how to rework the cells and tissues of the pupa to form the full-fledged moth. This analogy serves to emphasize that the more advanced forms of metamorphosis involve the derepression and acting out of what is little short of successive batches of genetic information. We may think of the genome as being subdivided into three different "gene sets" corresponding to the successive chapters in the construction manual (Williams, 1963a).

III. ENDOCRINE CONTROL OF METAMORPHOSIS: ECDYSONE

The construction manual, to continue the analogy, is read by the individual cells which even in the smallest insects are numbered in the tens of millions. The metamorphosis of the insect as a whole is obviously the outcome of a mosaic of metamorphoses at the cellular level. And it is worth inquiring how these millions of cells are coordinated in the playback of their successive sets of genes.

The answer to this question is clear-cut. The coordination of the cellular community is the job of the insect's endocrine system. It turns out that a certain hormone, "ecdysone," is necessary for the synthetic acts prerequisite for growth and metamorphosis. It is synthesized by the prothoracic glands located in the insect's anterior end. When ecdysone is secreted and acts unopposed, it promotes developmental reactions accompanied by the derepression and implementation of the next set of genes in the programmed life history. The larval cells pupate; the pupal cells undergo adult differentiation.

What if the prothoracic glands stop secreting ecdysone? In that event, growth and metamorphosis come to an abrupt halt and the immature insect enters a state of developmental standstill. Nature has exploited this state of affairs to provide for the overwintering of larval or pupal insects in the dormant condition termed diapause. After a long winter's nap, ecdysone is again secreted and development resumes where it had left off.

Now, let us examine some of the things that ecdysone can do. When a pupal silkworm is subdivided into anterior and posterior ends, the front end containing the prothoracic glands can go ahead and metamorphose into the anterior end of an adult moth. By contrast, the posterior end continues to live for many months but cannot metamorphose because it has no source of ecdysone. If active prothoracic glands are implanted to supply the hormone, the abdomen responds by transforming into the abdomen of a moth (Williams, 1952b).

How far can one carry this sort of thing? Figure 2 illustrates what I call the "reduced insect." In this preparation, two segments have been isolated from the abdomen of a pupa of *Antheraea pernyi* and sandwiched between plastic slips. But before sealing the preparation with melted wax, I removed the nervous system, the gonads, the Malpighian tubules, the gut, and a large amount of the fat body. Yet, as shown in Fig. 3, the implantation of active prothoracic glands caused the reduced insect to transform into the corresponding abdominal segments of an adult moth.

A. Extraction and Characterization of Ecdysone

Seventeen years ago, the German scientists, Adolf Butenandt and Peter Karlson launched a massive effort to extract and characterize ecdysone. They finally succeeded in isolating 25 mg of pure crystalline α-ecdysone from a ton of silkworms. They also obtained a trace (0.33 mg) of a more polar material, β-ecdysone (Butenandt and Karlson, 1954; Karlson, 1956). In assays carried out at Harvard University, the pure materials were found to do all things anticipated of the hormonal secretion of the prothoracic glands (Williams, 1954).

Eleven years elapsed before the chemistry of α-ecdysone was finally worked out by x-ray diffraction (Huber and Hoppe, 1965; Karlson *et al.*, 1965). To everyone's surprise, it proved to be the exotic sterol illustrated in Fig. 4. Its resemblance to cholesterol (Fig. 5) is self-evident. Unlike most organisms, including human beings, insects cannot synthesize the sterol ring system. That being so, cholesterol (or certain closely related plant sterols) have long been known to be growth factors for insects. Professor Clayton (Chapter 10) has much to say about this curious state of affairs.

Once the structure of ecdysone was known, chemists on three continents undertook the difficult task of synthesizing the molecule. An American group (Harrison *et al.*, 1966; Siddall *et al.*, 1966) and a team of German and Swiss scientists (Kerb *et al.*, 1966) simultaneously announced successful syntheses. Subsequently, a third synthesis was described by Japanese chemists (Mori *et al.*, 1968). Meanwhile, β-ecdysone was found to have the same structure as α-ecdysone except that an additional hydroxyl group is present at C-20 (Hampshire and Horn, 1966). Unfortunately, β-ecdysone has been given a number of other trivial names: 20-hydroxyecdysone, ecdysterone, crustecdysone, and isoinokosterone. Its synthesis was recently described by Huppi and Siddall (1968) and by Kerb *et al.* (1968).

In addition to α- and β-ecdysones, two other ecdysones have been extracted and characterized: 20,26-dihydroxyecdysone, from the tobacco

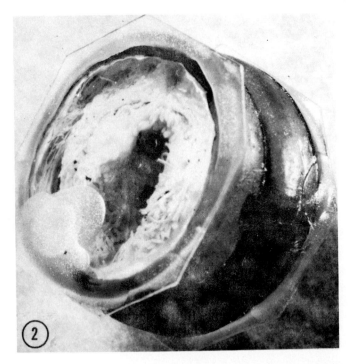

Fig. 2.   A "reduced insect" consisting of two segments of the abdomen of a diapausing pupa of *Antheraea pernyi* isolated and sandwiched between plastic slips. Before sealing the preparation with melted wax, the gut, gonads, central nervous system, Malpighian tubules, and a mass of fat body were excised.

hornworm *Manduca sexta* (Thompson *et al.*, 1967) and 2-deoxycrustecdysone, from the crayfish (Galbraith *et al.*, 1968).

B. The Phytoecdysones

The synthesis of ecdysone was so difficult and the yield so small that it seemed as though only vanishingly small amounts would ever be available for study. This discouraging prospect was changed overnight by a series of remarkable findings at Tohoku University in Japan. There, Professors Nakanishi and Takemoto discovered that certain plants contain amazing amounts of ecdysonelike materials, including authentic β-ecdysone. (For summaries see Kaplanis *et al.*, 1967; Ohtaki *et al.*, 1967; Staal, 1967a; Takemoto *et al.*, 1967a; Nakanishi, 1968.) These findings were soon confirmed and extended by several laboratories in Europe and the United States, and there can be little doubt that certain plants have gone in for the synthesis and accumulation of impressive amounts of these complicated, hormonally active sterols.

Consider, for example, that Butenandt and Karlson obtained 25 mg of α-ecdysone from a ton of silkworms. This amount of β-ecdysone can be recovered from 25 gm of the dried leaves or roots of the yew tree, *Taxus baccata* (Staal, 1967a,b), or from less than 2.5 gm of the dried rhizomes of the common fern, *Polypodium vulgare* (Jizba *et al.*, 1967).

C. Distribution of the Phytoecdysones

Table I presents a summary of Staal's (1967a) census of ecdysone activity in 73 species of gymnosperms, including representatives of no less than eight families. Thirteen of the 73 species provided active extracts; as indicated in Table I, all thirteen were obtained from only two families — the Podocarpaceae and the Taxaceae.

An even more massive study has been carried out in Japan by the Takeda Pharmaceutical Industries (Imai *et al.*, 1968). Extracts were prepared from 186 of the 188 families of higher plants in Japan, including 738 genera and 1056 species. As summarized in Table II, ecdysone activity was noted for 54 species distributed among 18 families. The rest, including 1002 species and 168 families, showed no detectable activity.

In another large-scale operation, Takemoto *et al.* (1967b) found activ-

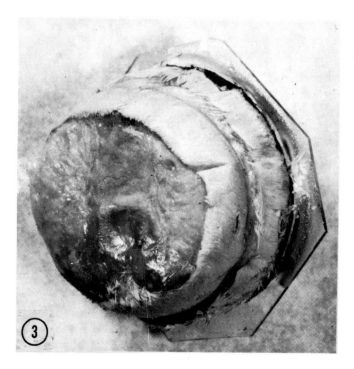

Fig. 3. One month after the implantation of a pair of prothoracic glands, the pupal segments have transformed into the scale-covered segments of the moth. (The upper plastic slip and the old pupal cuticle have been removed.)

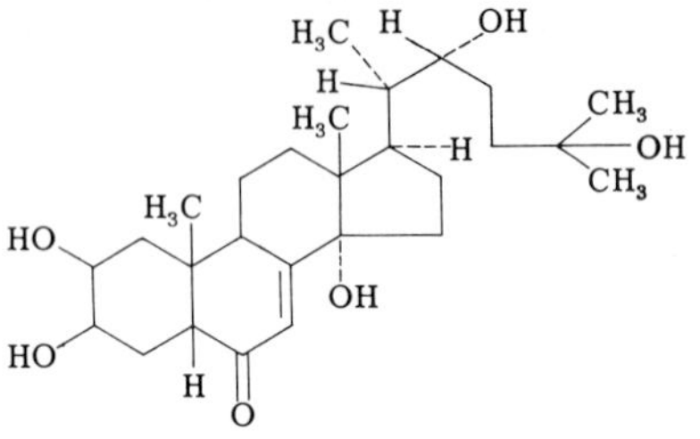

Figs. 4 and 5. Structural formulas of α-ecdysone and its parental compound, cholesterol.

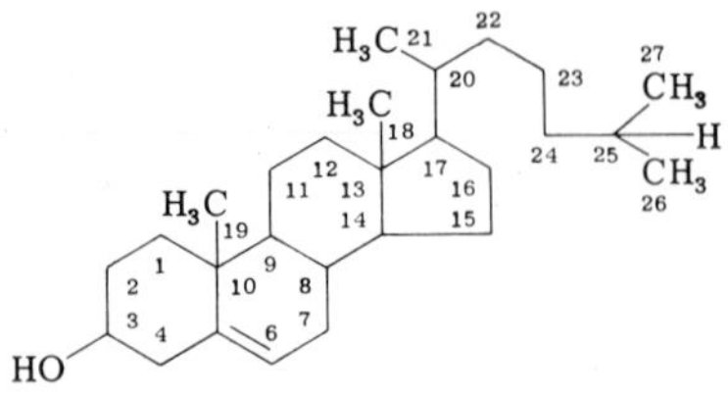

ity in extracts prepared from 22 of 43 species of ferns and 27 of 81 species of gymnosperms and angiosperms. No activity was detected in extracts of 14 species of mushrooms or 7 species of seaweeds.

The only clear pattern that emerges from these investigations is the high incidence of phytoecdysones in the ferns (Polypodiaceae) and in two families of gymnosperms (the Taxaceae and Podocarpaceae).

By latest count a total of 28 different phytoecdysones have been isolated and chemically characterized, and this number seems certain to increase. The most ubiquitous appears to be β-ecdysone and the least ubiquitous, α-ecdysone. In point of fact, the latter has been reported from only two species of ferns (Kaplanis *et al.*, 1967; Heinrich and Hoffmeister, 1967).

D. Endocrine Activity of the Phytoecdysones

Many of the phytoecdysones are superhormones in the sense of acting in lower concentrations than do α- or β-ecdysones. For example, cyasterone is twenty times as active as α-ecdysone when injected into silkworm pupae; in fly larvae it is twice as active (Ohtaki *et al.*, 1967). This greater potency is due to the resistance of cyasterone to inactivation. Thus, when injected into silkworm pupae, the time for 50% inactivation is 7 hr for α-ecdysone and 32 hr for cyasterone (Ohtaki and Williams, 1970).

There is general agreement that the ecdysones and phytoecdysones are inactive when topically applied to the unbroken skin in small volumes of volatile solvents such as methanol or acetone (Ohtaki *et al.*, 1967; Staal, 1967a). When the solvent evaporates, the hormone is deposited as a crystalline moiety on the outside of the cuticle. Yet, according to Sato *et al.* (1968), the ecdysones can gain entry if the test insect is totally immersed for 10 sec in a methanolic solution of hormone. This simple "dipping test" has been used by the Japanese investigators in tens of thousands of assays carried out on ligated abdomens of the rice-stem

TABLE I

Phytoecdysone Activity of Gymnosperms[a]

Family	Number of species	Number of species providing active extracts
Araucariceae	3	0
Cephalotaxaceae	2	0
Pinaceae	19	0
Cupressaceae	20	0
Taxodiaceae	9	0
Cycadaceae	1	0
Podocarpaceae	11	7
Taxaceae		
Taxus	5	5
Torreya	3	1
	73	13

[a] Data from Staal (1967a).

borer (*Chilo suppressalis*). In our experience the ecdysones can also penetrate the unbroken skin when topically applied in certain nonvolatile solvents such as undecylenic acid, α-tocopherol, or caprylic acid (Williams, 1968a). Administered in this way, the dose must be increased about tenfold.

TABLE II

Screening of Japanese Plants for Phytoecdysones[a]

Category of plant	Family	No. of active species
Ferns	Polypodiaceae	22
	Osmundaceae	1
	Lycopodiaceae	1
Gymnosperms	Taxaceae	3
	Podocarpaceae	2
	Cephalotaxaceae	1
	Cupressaceae	1
Angiosperms	Liliaceae	3
	Iridaceae	1
	Amaranthaceae	2
	Caryophyllaceae	3
	Ranunculaceae	1
	Malvaceae	2
	Stachyuraceae	2
	Cistaceae	1
	Labiatae	3
	Solanaceae	1
	Compositae	4
		54

[a] Data from Imai *et al.* (1968).

E. Pathological Effects of Excessive Ecdysone

When injected into diapausing Cynthia pupae in critically low doses, the phytoecdysones provoke the termination of dormancy and the formation of apparently normal moths. But for all materials except α-ecdysone, a doubling of the critical dose results in the formation of grossly abnormal and nonviable creatures such as illustrated in Figs. 6–9 (Kobayashi et al., 1967a,b; Williams, 1968b).

The pathological effects of excessive hormone are associated with a remarkable acceleration of development during the first few days after the injection. This fact is illustrated in Fig. 10, where the upper horizontal line records the sequence of events when normal development is triggered by the injection of a low dose of cyasterone (0.2 μg). The lower horizontal line illustrates the timing of these same events when extremely abnormal development is provoked by high doses of cyasterone (10 μg). Excessive hormone drives the early phase of metamorphosis so fast that, counting from the moment of injection, the sequence of events which normally requires 11 days is compressed into 4 or 5. Then, on about the third day of development, the precocious deposition of new cuticle locks all epidermal tissues in whatever stage they have managed to attain. As diagrammed in Fig. 10, the normal pace of development is resumed after about the third day. But the damage has already been done and the insect will never form a viable moth. The phenomenon is of great interest and constitutes the first clear-cut case of hyperhormonism in any invertebrate.

F. Effects of Ecdysone Ingestion

Though α- and β-ecdysones appear to have little or no effects when fed to insects, the same cannot be said of some of the phytoecdysones or certain synthetic ecdysone analogs. Robbins et al. (1968) have described experiments in which housefly larvae were reared on a sterile synthetic

---

Fig. 6.   Ventral view of the anterior end of an abnormal Cynthia moth formed by a pupa injected with 10 μg ponasterone A. In all individuals here represented, the old pupal cuticle has been removed.

Fig. 7.   Ventral view of a naked Cynthia abdomen formed by a pupal abdomen injected with 20 μg β-ecdysone.

Fig. 8.   Ventrolateral view of an abnormal Cynthia moth formed by a pupa injected with 10 μg ponasterone B.

Fig. 9.   Ventral view of a Cynthia "moth" formed by a pupa injected with 20 μg cyasterone. Extensive areas of pupal cuticle have reformed.

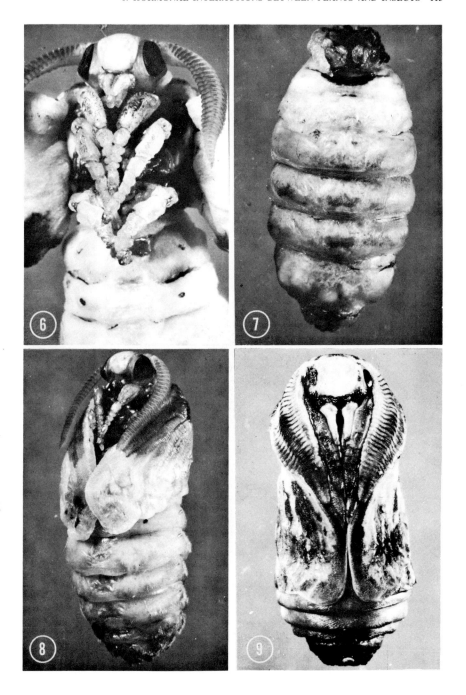

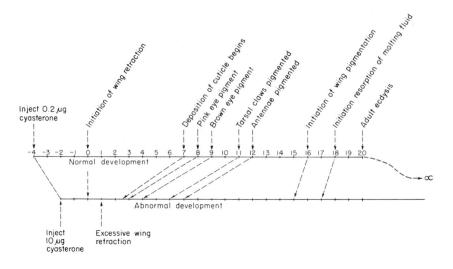

Fig. 10. The top line records the timetable of successive developmental events at 25°C when normal metamorphosis of a Cynthia pupa is provoked by the injection of a physiological dose of ecdysone (in this case 0.2 μg cyasterone). The lower horizontal line illustrates the timing of these same events when extremely abnormal development is provoked by the injection of an excessive dose of phytoecdysone (in this case 10 μg cyasterone).

diet containing β-ecdysone or ponasterone A. The former was inactive at the highest dose tested (150 parts per million), whereas ponasterone A at this same dose level caused severe derangement of growth and metamorphosis. Even more active was a synthetic ecdysone lacking hydroxyl groups in the side chain; this material was four times as lethal as ponasterone A.

Riddiford (1970) has reported similar experiments in which Cecropia silkworms were reared on a synthetic diet containing ponasterone A. As summarized in Table III, the development of nearly all individuals was blocked by doses as low as 1 part per billion parts of diet. Death usually occurred during or immediately after molting or at the time of metamorphosis. These results are truly sensational in that the effective dose was lower by several orders of magnitude than that required of any previously known insecticide. Evidently, the Cecropia silkworm is a million times more sensitive than the housefly to ponasterone A.

An abiding mystery is the nontoxic properties of β-ecdysone, the most widespread of the phytoecdysones, when orally administered. A possible explanation is provided by Staal (1968). Thus, in experiments carried out on the pyrrhocorid bug, *Dysdercus*, β-ecdysone proved to be a powerful feeding deterrent when administered in the drinking water in concentrations as low as 1 part per million. Until additional experiments of this

TABLE III

Development of Cecropia Silkworms on a Synthetic Diet Containing Ponasterone A[a]

Conc. in diet (ppm)	No. eggs	Percent attaining the indicated stage[b]						
		Larva₁	Larva₂	Larva₃	Larva₄	Larva₅	Pupa	Adult
0 (Controls)	28	100	96	93	93	89	82	82
0.2	26	100	69	65	35	31	12	0
0.01	50	100	74	66	56	48	32	16
0.001	50	100	80	79	58	46	12	8

[a]Data from Riddiford (1970).
[b]The indicated scoring system was complicated by numerous "intermediates" between successive larval instars.

type can be carried out on other insects, we can only speculate as to whether some and perhaps all of the phytoecdysones may function as potent deterrents and antifeeding agents for insects.

IV. GROWTH WITHOUT METAMORPHOSIS

Up to this point we have considered insect metamorphosis and, more particularly, the role of ecdysone. But the life story of insects is not one of ceaseless change and revolution. In point of fact, the strategy of insect development is to postpone metamorphosis until all necessary materials are at hand to construct a sexy adult.

As illustrated in Fig. 11, the Cecropia moth larva grows for about 6 weeks prior to metamorphosis; during that period it molts its cuticle on four separate occasions and increases its initial mass some 5000-fold.

It is easy to show that ecdysone is the necessary stimulus for larval growth and molting. Evidently there is at work a conservative force which suppresses metamorphosis—which blocks "growing up" without interfering with growth in an unchanging state.

V. JUVENILE HORMONE AND THE STATUS QUO

This conservative force proves to be "juvenile hormone"—an agent we have not had to mention up to this point. The hormone is synthesized by a pair of tiny cephalic glands, the corpora allata, which are also responsible for regulating its release into the blood. The action of juvenile hormone is to alter the cellular response to ecdysone—to suppress new synthetic acts without interfering with the use and reuse of genetic information already at the disposal of the cells and tissues. If this "brake" on progressive differentiation is removed by excising the corpora allata, the immature larva reacts to ecdysone by undergoing precocious metamorphosis to form a miniature pupa.

Fig. 11. The five larval stages of the Cecropia silkworm. The first larval stage (immediately after hatching from the egg) is the tiny black object perched on the metathoracic tergum of the mature fifth stage larva. Second, third, and fourth stage larvae are clinging to the twig. During the five larval instars, the overall increase in mass is up to 5000-fold.

Under normal circumstances metamorphosis is postponed until late in larval life when the corpora allata lose their ability to secrete juvenile hormone. Then, for the first time in the life history, ecdysone can act in the presence of little or no juvenile hormone. The net effect is to turn off the larval genes and turn on the pupal genes. The "reading frame" racks forward to the next gene set.

Particularly spectacular are the effects of juvenile hormone on the transformation of a pupa into an adult – a terminal phase of metamorphosis which proceeds in the presence of ecdysone and the absence of juvenile hormone. If juvenile hormone is supplied by the implantation of active corpora allata, the pupa is prevented from forming an adult. If an excess of hormone is caused to be present, the pupa molts into a second pupal stage (Fig. 12).

The phenomenon has its counterpart in the lower insects in which the mature larva normally transforms into a winged adult without traversing a pupal state. Here again, the implantation of active corpora allata at the outset of the final larval stage is fully effective in suppressing adult differentiation.

A. Cecropia Juvenile Hormone

For 20 years juvenile hormone remained a will-o-the-wisp resisting all efforts to extract or obtain it apart from the living insect. Then about 13 years ago a rich depot of the hormone was discovered in the abdomens of male Cecropia moths (Williams, 1956, 1963b). This was indeed a strange finding and to this day only the closely related male Cynthia moth has been found to accumulate the hormone in this way. To extract the hormone all one has to do is to excise the abdomens from male Cecropia or Cynthia moths, blend them in diethyl ether, then filter the solution, wash the filtrate with water, and evaporate the ether. One thereby obtains a golden oil, about 0.2 ml per abdomen, which shows impressive hormonal activity. The active principle in the golden oil proved to be a heat-stable, water-insoluble, uncharged substance (Williams, 1956; Williams and Law, 1965).

When injected into pupae of the moths, the crude extract duplicated all the effects previously realized when juvenile hormone was supplied by the implantation of active corpora allata. Indeed, it soon appeared that it was not necessary to inject the hormone: it sufficed merely to place the oily extract on the unbroken skin through which it promptly penetrated. The net result was the formation of nonviable creatures in which some cells had undergone metamorphosis and others had not.

This derangement of metamorphosis coupled with the extract's activity on topical application provided the first indication that juvenile hormone had potentialities as an insecticide (Williams, 1956).

Fig. 12. . Ventral view of a second pupal stage of the Polyphemus silkworm provoked by the implantation of four pairs of active corpora allata into a previously chilled pupa. On one side, the old pupal cuticle has been removed to reveal the new pupal cuticle that has formed.

B. Characterization of Cecropia Juvenile Hormone

Until 1965 all attempts to isolate the pure hormone from Cecropia oil were unsuccessful (Williams and Law, 1965). Then, after 3 years of intensive effort, this difficult task was finally accomplished by a team of scientists headed by Herbert Röller (Röller and Bjerke, 1965) at the Uni-

versity of Wisconsin. Soon thereafter its chemical structure was elucidated by mass spectrometry and other analytical studies performed on
less than 300 μg of pure material (Röller et al., 1967).

The empirical formula of Cecropia hormone is $C_{18}H_{30}O_3$, corresponding
to a molecular weight of 294. It proves to be the methyl ester of the epoxide of a previously unknown fatty acid derivative (Fig. 13). The apparent
simplicity of the molecule is deceptive. It has two double bonds and an
oxirane ring, and can therefore exist in eight different geometric isomers
plus two optical isomers, making a total of sixteen possible configurations.

C. Synthesis of Cecropia Hormone

The Wisconsin workers succeeded in synthesizing small amounts of
the racemic hormone from which they separated and assayed the individual isomers (Dahm et al., 1967, 1968). In this manner they were able
to show that the two double bonds in the authentic hormone are trans,
trans, whereas the oxirane ring is cis.

According to Röller (1968) and Röller and Dahm (1968) the trans configuration of both double bonds is crucial for biological activity. By contrast, the stereochemistry of the oxirane ring seems to be of secondary
importance. The authentic juvenile hormone, as well as the dl-hormone,
is active at submicrogram levels when assayed on diverse insects including representatives of the Coleoptera, Lepidoptera, Hemiptera, and
Orthoptera. Curiously enough, the synthetic ethyl ester proves to be eight
times as active as the native methyl ester.

Very recently, three independent and highly ingenious steroselective
syntheses of dl-hormone have been accomplished by chemists at Harvard (Corey et al., 1968), Stanford (Johnson et al., 1968), and the Syntex
Laboratory (Zurflüh et al., 1968). Table IV summarizes my assays of the
Harvard-synthesized hormone on four species of insects.

It is food for thought to observe that the Cecropia hormone is maxi-
mally active when assayed on pupae of the closely related saturniid,

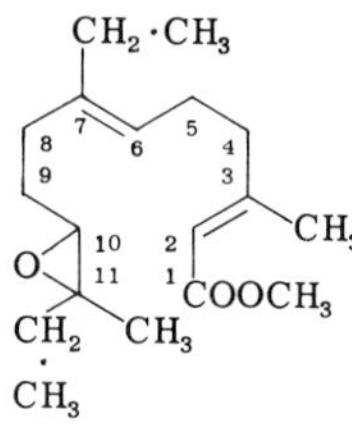

Fig. 13. The structural formula of Cecropia juvenile hormone. Both double bonds are trans,
whereas the oxirane ring is cis (Röller et al., 1967).

TABLE IV

Biological Assays of *dl* Cecropia Juvenile Hormone on Four Different Species

Species	Weight (gm)	Vehicle	Mode of administration	Hormone required for 3+ reaction(μg)	Critical dose (μg/gm live wt.)
Antheraea polyphemus (Cramer) (previously chilled pupae)	5	50 μl Olive oil	Injected	0.01–0.10	0.002–0.020
Tenebrio molitor L. (newly molted pupae)	0.130	1 μl Acetone	Topical	0.05–0.20	0.4–1.5
Pyrrhocoris apterus L. (newly molted 5th stage larvae)	0.020	1 μl Acetone	Topical	0.5	25
Oncopeltus fasciatus (Dallas) (newly molted 5th stage larvae)	0.022	1 μl Acetone	Topical	1.0	46

Antheraea polyphemus. Per unit of live weight it is 100-fold less active for pupae of the holometabolous *Tenebrio molitor* (Coleoptera), and 10,000-fold less active when assayed on the hemimetabolous bugs, *Pyrrhocoris* or *Oncopeltus* (Hemiptera).

D. Chemical Diversification of Juvenile Hormone

Despite the quantitative uncertainties that attend biological assays, evidence of this sort suggests that the Cecropia hormone may not be the juvenile hormone of all insects. Sláma and Williams (1966a) have argued that during the millions of years of insect evolution the detailed chemistry of juvenile hormone has also evolved and diversified to give a number of related molecules that serve as juvenile hormone for different kinds of insects.

Additional support for this point of view is the recent isolation from Cecropia oil of a second juvenile hormone which is identical in biological activity and structure to that described by Röller except that a methyl (rather than an ethyl) group is present at C-7. This variant of the hormone is responsible for about 13 to 20% of the endocrine activity of the oil (Meyer *et al.*, 1968).

Evolutionary changes in the chemistry of juvenile hormone would, of necessity, correspond with homologous changes in the receptor mechanisms in the cells and tissues. It seems altogether likely that this biochemical "retuning" of the target organs includes steric changes in the receptor sites to accommodate the altered conformation of the hormone.

E. Juvenile Hormone Analogs

Long before the chemistry of the Cecropia hormone was worked out, Schmialek (1961) detected traces of juvenile hormone activity in the sesquiterpene alcohol, farnesol, and its aldehyde, farnesal. This important discovery paved the way for the synthesis of a number of far more active farnesol derivatives such as farnesyl methyl ether (Schmialek, 1963) and farnesyl diethylamine (Karlson and Nachtigall, 1961; Karlson, 1963). Still more active was methyl-10, 11-epoxyfarnesoate prepared by Bowers and his co-workers at Beltsville (1965). Indeed, as noted in Fig. 14, this compound differs by only two carbons from the Cecropia hormone isolated by Röller and by only one carbon from the second Cecropia hormone isolated by Meyer. When subjected to biological assay, most of these synthetic materials showed high activity for certain species and extremely low activity for certain other species.

Meanwhile, at the Harvard Laboratories, John Law and the author prepared a synthetic mixture by a simple one-step process in which hydrogen chloride gas was bubbled through a chilled ethanolic solution of farnesenic acid (Law *et al.*, 1966). Without any purification, this mixture was far more active than crude Cecropia oil and fully effective when tested on all kinds of insects ranging from the wingless Thysanura to the most highly evolved Hymenoptera (Williams, 1966; Spielman and Williams, 1966; Vinson and Williams, 1967). Indeed, when assayed on Hemiptera, the crude mixture was 1000 times as active as pure Cecropia hormone. Its potency for so many kinds of insects can be attributed to the presence of a dozen or so active compounds including hydrochlorinated derivatives of ethyl farnesoate. This broad spectrum of activity coupled with the simplicity of the synthetic method suggests that the Law-Williams mixture has immediate promise as an insecticide.

Broad spectrum materials, such as the Law-Williams mixture and the Cecropia hormone, are selective in the sense of killing only insects.

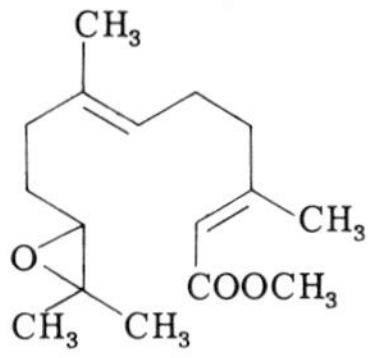

Fig. 14. Methyl-10,11-epoxyfarnesoate—a juvenile hormone analog synthesized by Bowers *et al.* (1965). It differs from authentic Cecropia hormone by only two methyl groups.

However, they fail to discriminate between the 0.1% of species that qualify as pests and the 99.9% that are either innocuous or downright helpful. Therefore, any large-scale or reckless use of the broad spectrum materials could constitute an ecological disaster of the first rank. However, there is nothing to prevent their immediate use in, say, the protection of stored products.

F. "Paper Factor"

The real need is for juvenile hormones that are tailor-made to attack only certain predetermined pests. This possibility first emerged from studies carried out in collaboration with the Czechoslovakian biologist, Karel Sláma.

In 1964 Sláma came to Harvard bringing with him his favorite experimental animal—the European bug, *Pyrrhocoris apterus* (Pyrrhocoridae). To our considerable mystification, the bugs failed to undergo normal development when we attempted to rear them in the Harvard laboratory. Instead of metamorphosing into sexually mature adults at the end of the fifth larval stage, they underwent a supernumerary larval molt to form giant sixth stage larvae, some of which underwent yet a further larval molt to form seventh stage larvae. The phenomenon is illustrated in Fig. 15. All individuals ultimately died without being able to complete metamorphosis.

Among tens of thousands of *Pyrrhocoris* which Sláma had cultured in Prague, the spontaneous formation of a sixth stage larva had never been encountered. Metamorphosis had always taken place at the end of the fifth larval stage—a result directly attributable to the inactivation of the corpora allata and the cessation of juvenile hormone secretion at the outset of the fifth larval stage. The formation of sixth stage larvae had been provoked only in experiments in which juvenile hormone was supplied by the implantation of active corpora allata. For these reasons it seemed certain that the Harvard cultures of *Pyrrhocoris* had access to some unknown source of juvenile hormone.

Among many possibilities that we examined, attention finally focused on the fragment of paper toweling (Scott Brand 150) which had been placed in each petri dish to provide a surface upon which the bugs could walk around. In Prague, Sláma had always used filter paper. We were astonished to find that when the toweling was replaced by a corresponding piece of Whatman's filter paper, the entire phenomenon vanished and all individuals developed normally.

The abovementioned finding seemed incomprehensible. In correspondence with the Scott Paper Company, we sought to inform ourselves as to the chemicals added to "Brand 150." All we learned was that the toweling was made from paper pulp.

Twenty other brands of toweling, napkins, and bathroom tissues were

Fig. 15. The effects of contact with "paper factor" are illustrated by these four specimens of the bug, *Pyrrhocoris apterus*. On the left is a fifth stage larva which normally transforms into the winged adult (*third from left*). When exposed to the juvenile hormone analog, the larva continues to grow without metamorphosis to form a sixth stage larva (*second from left*) which, in turn, may molt to a still-larger seventh stage larva (*extreme right*).

assembled and tested by allowing freshly molted, fifth stage *Pyrrhocoris* larvae to walk upon them. Twelve of the twenty showed great activity. This suggested that the active factor was widespread in paper products. Therefore, additional tests were performed on newspapers and journals. Surprisingly enough, all American newspapers and journals proved to be highly active. By contrast, papers of European or Japanese manufacture were usually inert.

In Fig. 16, John Law and the author are shown extracting the active material from a column packed with "Scott Brand 150" towels. This might be called a "poor man's" procedure. We isolated the hormonal material – several grams of it from a few hundred towels. By distillation we got back the methanol, and got back the towels, too, which were cleaner and better than ever. The extract, when dissolved in acetone and topically applied to any part of freshly molted fifth instar *Pyrrhocoris* larvae, was fully effective in blocking metamorphosis (Fig. 17).

We were amazed to find the "paper factor" to be active for only one family of insects, the Pyrrhocoridae. Even the Lygaeidae, the very next family among the Hemiptera, appeared to be totally unaffected. This was the first indication of the existence of juvenile hormonal materials with selective action on particular kinds of insects.

Fig. 16   C. M. Williams and Professor John H. Law extracting ''paper factor'' from a large glass column packed with paper towels.

Fig. 17.   Dead and dying *Pyrrhocoris* larvae that had been treated with "paper factor" at the outset of the fifth larval stage.

G. Botanical Origins of Paper Factor

To make a long story short, the paper factor was finally tracked down to its source in certain evergreen trees which are the source of American paper pulp. High activity was recorded for extracts of balsam fir (*Abies balsamea*), eastern hemlock (*Tsuga canadensis*), Pacific yew (*Taxus brevifolia*), and tamarack (*Larix laricina*). Extracts of shortleaf pine (*Pinus echinata*) and European larch (*Larix decidua*) showed only traces of activity, while extracts of red spruce (*Picea rubens*) were inactive.

These results documented the botanical origins of the active material. Evidently the high activity in American paper products is mainly derived from balsam fir (*Abies balsamea*)—a principal pulp tree indigenous to the northern United States and Canada. The tree synthesizes the active material, and the latter then accompanies the pulp all the way to the printed page.

The common grades of paper are manufactured from pulps which are made from wood, either by grinding (mechanical pulps) or by chemical processing of wood chips (chemical pulps). Carlisle and Ellis (1967) report that hormonal activity is found only in papers containing mechanical pulp and that chemical pulp is completely inactive. It may also be noted that Canada Balsam, an article of commerce obtained as an exudate from the balsam fir, is completely inactive.

Two years ago, Bowers *et al.* (1966) isolated the active material from the wood of the balsam fir and named it "juvabione." It proves to be the methyl ester of a certain unsaturated fatty acid derivative whose kinship with the other juvenile hormone analogs is self-evident (see Fig. 18). The Prague scientists (Cerny *et al.*, 1967) confirmed the presence of juvabione in fir wood and, in addition, found the related compound, dehydrojuvabione, which displayed the same selective action on insects of the family Pyrrhocoridae. Although juvabione and dehydrojuvabione are, to date, the only compounds with juvenile hormone activity isolated from

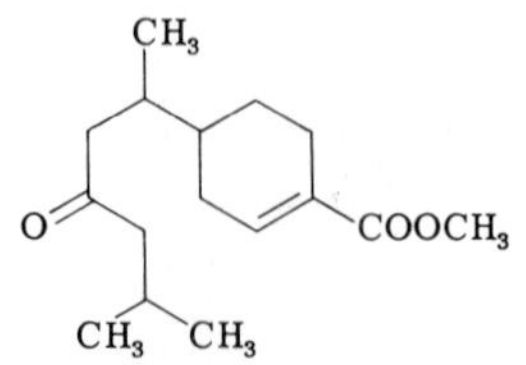

Fig. 18. The structural formula of "paper factor" (methyl todomatuate; "juvabione"), a highly selective juvenile hormone analog isolated from the wood of the balsam fir (Bowers *et al.*, 1966).

plants and identified, Bowers (1968) reports that among 52 species of plants chosen at random, 6 have provided extracts with juvenile hormone activity in the *Tenebrio* assay, and, of these, two extracts showed considerable activity.

VI. JUVENILE HORMONE AND EMBRYONIC DEVELOPMENT

Up to this point we have considered the lethal derangement of metamorphosis brought about by materials with juvenile hormone activity. The overall result is that immature larvae fail to form viable adults. Moreover, the maturation of the gonads is almost completely suppressed. Only in the absence of juvenile hormone can the cells and tissues undertake the new synthetic acts prerequisite for metamorphosis and sexual maturation.

The transformation of an insect egg into a first-stage larva is a metamorphosis every bit as impressive as anything occurring during postembryonic life. That being so, it is worth inquiring as to whether juvenile hormone can also interfere with the sequential gene action prerequisite for embryonic development. Three years ago, Sláma and Williams (1966b) showed that this is indeed the case. When we placed *Pyrrhocoris* eggs in contact with paper containing juvabione, embryonic development proceeded to a certain stage and then stopped with the still living embryo short of completion. The most affected eggs contained a small disc of embryonic tissue afloat on a mass of unincorporated yolk. In somewhat more advanced cases, the embryonic mass showed the rudiments of certain appendages but not of others. Sometimes parts of appendages had formed in the absence of the remainder. Needless to say, none of the affected eggs ever hatched.

Freshly laid eggs were found to be most sensitive to inhibition by juvabione. This finding suggested that the eggs might be yet more sensitive prior to oviposition. To test this possibility a crude extract of juvabione was topically applied to a series of adult female *Pyrrhocoris*. Without exception, the eggs deposited by the treated animals behaved as if they had been directly exposed to juvabione; that is, development proceeded to a certain stage and then stopped. This implies that after topical application to adult females, the hormonal analog is absorbed, translocated to the reproductive tract, and somehow gets inside the eggs. These simple experiments provided the first indication that the embryonic development of insects is subject to inhibition by the very same materials that can inhibit metamorphosis.

Subsequently, Riddiford and Williams (1967) studied the matter in further detail and found that the Law-Williams synthetic mixture was effective in blocking the embryonic development of Cecropia moth eggs as well as the eggs of three species of bugs. There is now general agreement that juvenile hormone analogs constitute incredibly effective ovicides and chemosterilants.

VII. VENEREAL DISSEMINATION OF JUVENILE HORMONE

Karel Sláma and his co-workers in Prague (Masner *et al.*, 1968) recently described an ingenious use of juvenile hormone as a chemosterilant. Thus, in the case of adult female *Pyrrhocoris* bugs, they found that permanent sterility is achieved by the topical application of 1 μg of the dihydrochloride of methyl farnesoate (see Fig. 19). The females seemed unaffected by this treatment; they mated repeatedly, and each week deposited a fresh clutch of eggs that failed to hatch.

The trick is to apply the hormonal analog not to females, but to males. According to the Prague investigators, males treated with 100 μg and caged for 4 hr with an equal number of females transfer up to 5 μg of hormone to each female in the process of mating. Males treated with 1000 μg transfer up to 7 μg. Manifestly, all exposed females are permanently sterilized.

The hormonal analog used in these experiments is by no means the most active analog for *Pyrrhocoris*; for example, the ethyl ester of this same molecule is ten times more potent. Therefore, the amount applied to each male remains fully effective when reduced to 10 μg.

The venereal dissemination of juvenile hormone insecticides combines the advantages of the classic "sterile male technique" (Knipling, 1955, 1959, 1962) with several additional advantages. For example, the new method remains fully effective in sterilizing females that have already mated with normal males or that subsequently do so. Moreover, like the sterile male technique, the venereal dissemination permits one to take careful aim at a single pestiferous species without posing any threat to other species. That being so, broad spectrum analogs such as the Law-Williams mixture can be used with impunity since their selectivity of action is automatically assured by the machinations of sex.

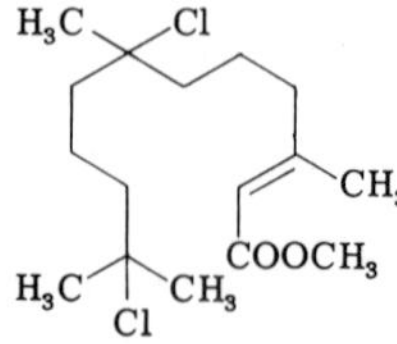

Fig. 19. The structural formula of 7,11-dihydrochloromethyl farnesoate, a juvenile hormone analog prepared by Romanuk *et al.* (1967). The ethyl ester of this compound is one of several highly active materials in the Law-Williams mixture (Law *et al.*, 1966).

VIII. CONCLUSION

The fast-breaking story of the phytoecdysones and the phyto-juvenile hormones has constituted one surprise after another. In the least it has served to transform – to metamorphose – the science of insect endocrinology.

Have the plants in question undertaken these exorbitant syntheses just for fun? Have the genes for these synthetic operations been carried along for millions of years as excess baggage?

I think not. Present indications are that certain plants and more particularly the ferns and evergreen trees have gone in for an incredibly sophisticated self-defense against insect predation – a method of insect control that we are just beginning to comprehend.

References

Bowers, W. S. (1968). Conference on plant–insect interactions. *BioScience* 18, 791–799.

Bowers, W. S., Thompson, M. J., and Uebel, E. C. (1965). Juvenile and gonadotropic hormone activity of 10,11-epoxy farnesenic acid methyl ester. *Life Sci.* 4, 2323–2331.

Bowers, W. S., Fales, H. M., Thompson, M. J., and Uebel, E. C. (1966). Identification of an active compound from balsam fir. *Science* 154, 1020–1021.

Butenandt, A., and Karlson, P. (1954). Über die Isolierung eines Metamorphose-Hormons der Insekten in kristallisierter Form. Z. *Naturforsch.* 9b, 389–391.

Carlisle, D. B., and Ellis, P. E. (1967). Abnormalities of growth and metamorphosis in some pyrrhocorid bugs: the paper factor. *Bull. Entomol. Res.* 57, 405–417.

Cerny, V., Dolejs, L., Lábler, L., and Sorm, F. (1967). Dehydrojuvabione – a new compound with juvenile hormone activity from balsam fir. *Tetrahedron letters* pp. 1053–1057.

Corey, E. J., Katzenellenbogen, J. A., Gilman, N. W., Roman, S. A., and Erickson, B. W. (1968). Stereospecific total synthesis of the *dl*-C_{18} Cecropia juvenile hormone. *J. Am. Chem. Soc.* 90, 5618–5620.

Dahm, K. H., Trost, B. M., and Röller, H. (1967). The juvenile hormone. V. Synthesis of the racemic juvenile hormone. *J. Am. Chem. Soc.* 89, 5292–5294.

Dahm, K. H., Röller, H., and Trost, B. M. (1968). The juvenile hormone. IV. Stereochemistry of juvenile hormone and biological activity of some of its isomers and related compounds. *Life Sci.* 7, 129–137.

Galbraith, M. N., Horn, D. H. S., and Middleton, E. J. (1968). Structure of deoxycrustecdysone, a second crustacean moulting hormone. *Chem. Commun.* pp. 83–85.

Hampshire, F., and Horn, D. H. S. (1966). Structure of crustecdysone, a crustacean moulting hormone. *Chem. Commun.* pp. 37–38.

Harrison, I. T., Siddall, J. B., and Fried, J. H. (1966). Steroids CCXCVII. Synthetic studies on insect hormones. Part III. An alternative synthesis of ecdysone and 22-isoecdysone. *Tetrahedron Letters* pp. 3457–3460.

Heinrich, G., and Hoffmeister, H. (1967). Ecdyson als Begleitsubstanz des Ecdysterons in *Polypodium vulgare* L. *Experientia* 23, 995.

Huber, R., and Hoppe, W. (1965). Zur Chemie des Ecdysons. VII. Die Kristall- und Molekülestrukturanalyse des Insektenverpuppungshormons Ecdyson mit der automatisierten Faltmolekülmethode. *Chem. Ber.* 98, 2403–2424.

Huppi, G., and Siddall, J. B. (1968). Steroids CCCXXXVI. Synthetic studies on insect hormones, Part VI. The synthesis of ponasterone A and its stereochemical identity with crustecdysone. *Tetrahedron Letters* pp. 1113–1114.

Imai, S., Toyosato, T., Fujioka, S., Sakai, M., and Sato, Y. (1968). Screening of plants for compounds with insect moulting activity. *Chem. Pharm. Bull.(Tokyo)* 17, 335–339.

Jizba, J., Herout, V., and Sorm, F. (1967). Isolation of ecdysterone (crustecdysone) from *Polypodium vulgare* L. rhizomes. *Tetrahedron Letters* pp. 1869–1891.

Johnson, W. S., Li, T., Faulkner, D. J., and Campbell, S. F. (1968). A highly stereoselective synthesis of the racemic juvenile hormone. *J. Am. Chem. Soc.* 90, 6225–6226.

Kaplanis, J. N., Thompson, M. J., Robbins, W. E., and Bryce, B. M. (1967). Insect hormones: alpha ecdysone and 20-hydroxyecdysone in bracken fern. *Science* 157, 1436–1437.

Karlson, P. (1956). Biochemical studies on insect hormones. *Vitamins Hormones* 14, 227–266.

Karlson, P. (1963). Chemistry and biochemistry of insect hormones. *Angew. Chem. Intern. Ed. English* 2, 175–182.

Karlson, P., and Nachtigall, M. (1961). Ein biologischer Test zur quantitativen Bestimmung der Juvenilhormonaktivität von Insektenextrakten. *J. Insect Physiol.* 7, 210–215.

Karlson, P., Hoffmeister, H., Hummel, H., Hocks, P., and Spiteller, G. (1965). Zur Chemie des Ecdysons. VI. Reaktionen des Ecdysonmoleküls. *Chem. Ber.* 98, 2394–2402.

Kerb, U., Schulz, G., Hocks, P., Wiechert, R., Furlenmeier, A., Fürst, A., Langemann, A., and Waldvogel, G. (1966). Zur Synthese des Ecdysons. IV. Die Synthese des natürlichen Häutungshormons. *Helv. Chim. Acta* 49, 1601–1606.

Kerb, U., Wiechert, R., Furlenmeier, A., and Fürst, A. (1968). Über eine Synthese des Crustecdysons (20-hydroxyecdyson). *Tetrahedron Letters* pp. 4277–4280.

Knipling, E. F. (1955). Possibilities of insect control or eradication through the use of sexually sterile males. *J. Econ. Entomol.* 48, 459–462.

Knipling, E. F. (1959). Sterile male method of population control. *Science* 139, 902–904.

Knipling, E. F. (1962). Potentialities and progress in the development of chemosterilants for insect control. *J. Econ. Entomol.* 55, 782–786.

Kobayashi, M., Takemoto, T., Ogawa, S., and Nishimoto, N. (1967a). The moulting hormone activity of ecdysterone and inokosterone isolated from *Achyranthis radix*. *J. Insect Physiol.* 13, 1395–1399.

Kobayashi, M., Nakanishi, K., and Koreeda, M. (1967b). The moulting activity of ponasterones on *Musca domestica* (Diptera) and *Bombyx mori* (Lepidoptera). *Steroids* 9, 529–536.

Law, J. H., Yuan, C., and Williams, C. M. (1966). Synthesis of a material with high juvenile hormone activity. *Proc. Natl. Acad. Sci. U.S.* 55, 576–578.

Masner, P., Sláma, K., and Landa, V. (1968). Sexually spread insect sterility induced by analogues of juvenile hormone. *Nature* 219, 395–396.

Meyer, A. S., Schneiderman, H. A., Hanzmann, E., and Ko, J. H. (1968). The two juvenile hormones from the Cecropia silk moth. *Proc. Natl. Acad. Sci. U.S.* 60, 853–860.

Mori, H., Shibata, K., Tsuneda, K., and Sawai, M. (1968). Synthesis of ecdysone. *Chem. Pharm. Bull. (Tokyo)* 16, 563–566.

Nakanishi, K. (1968). Conference on insect-plant interactions. *BioScience* 18, 791–799.

Ohtaki, T., and Williams, C. M. (1970). Inactivation of ecdysone and phytoecdysones. In preparation.

Ohtaki, T., Milkman, R. D., and Williams, C. M. (1967). Ecdysone and ecdysone analogues: their assay on the fleshfly *Sarcophaga peregrina*. *Proc. Natl. Acad. Sci. U.S.* 58, 981–984.

Riddiford, L. M. (1970). Effects of orally administered phytoecdysones on the development of the Cecropia silkworm. In preparation.

Riddiford, L. M., and Williams, C. M. (1967). The effects of juvenile hormone analogues on the embryonic development of silkworms. *Proc. Natl. Acad. Sci. U.S.* 57, 595–601.

Robbins, W. E., Kaplanis, J. N., Thompson, M. J., Shortino, T. J., Cohen, C. F., and Joyner, S. C. (1968). Ecdysones and analogs: effects on development and reproduction of insects. *Science* 161, 1158–1160.

Röller, H. (1968). Conference on insect-plant interactions. *BioScience* 18, 791–792.

Röller, H., and Dahm, K. H. (1968). The chemistry and biology of juvenile hormone. *Recent Progr. Hormone Res.* 24, 651–680.

Röller, H., and Bjerke, J. S. (1965). Purification and isolation of juvenile hormone and its action in lepidopteran larvae. *Life Sci.* 4, 1617–1624.

Röller, H., Dahm, K. H., Sweely, C. C., and Trost, B. M. (1967). The structure of the juvenile hormone. *Angew. Chem. Intern. Ed. English* **6**, 179–180.

Romañuk, M., Sláma, K., and Sorm, F. (1967). Constitution of a compound with a pronounced juvenile hormone activity. *Proc. Natl. Acad. Sci. U.S.* **57**, 349–352.

Rosenfeld, A. (1958). The ultimate weapon in an ancient war. *Life* **45**, No. 14, 105–108.

Sato, Y., Sakai, M., Imai, S., and Fujioka, S. (1968). Ecdysone activity of plant originated moulting hormones applied on the body surface of lepidopterous larvae. *Appl. Entomol. Zool.* **3**, 49.

Schmialek, P. (1961). Die Identifizierung zweier in Tenebriokot und in Hefe vorkommender Substanzen mit Juvenilhormonwirkung. *Z. Naturforsch.* **16b**, 461–464.

Schmialek, P. (1963). Metamorphose Hemmung von *Tenebrio molitor* durch Farnesylmethyl äther. *Z. Naturforsch.* **18b**, 513–515.

Siddall, J. B., Cross, A. D., and Fried, J. H. (1966). Steroids. CCXCII. Synthetic studies on insect hormones. II. The synthesis of ecdysone. *J. Am. Chem. Soc.* **88**, 862–863.

Sláma, K., and Williams, C. M. (1966a). The juvenile hormone. V. The sensitivity of the bug, *Pyrrhocoris apterus*, to a hormonally active factor in American paper-pulp. *Biol. Bull.* **130**, 235–246.

Sláma, K., and Williams, C. M. (1966b). "Paper factor" as an inhibitor of the embryonic development of the European bug, *Pyrrhocoris apterus*. *Nature* **210**, 329–330.

Snodgrass, R. E. (1954). Insect metamorphosis. *Smithsonian Inst. Misc. Collections* **122**, No. 9, 1–124.

Spielman, A., and Williams, C. M. (1966). Lethal effects of synthetic juvenile hormone on larvae of the yellow fever mosquito, *Aedes aegypti*. *Science* **154**, 1043–1044.

Staal, G. B. (1967a). Plants as a source of insect hormones. *Koninkl. Ned. Akad. Wetenschap., Proc., Ser. C* **70**, 409–418.

Staal, G. B. (1967b). Insect hormones in plants. *Mededel. Rijksfacult. Landbouwwetenschap. Gent* **32**, 393–400.

Staal, G. B. (1968). Personal communication.

Takemoto, T., Ogawa, S., Nishimoto, N., and Hoffmeister, H. (1967a). Steroide mit Häutungshormonaktivität aus Tieren und Pflanzen. *Z. Naturforsch.* **22b**, 681–682.

Takemoto, T., Ogawa, S., Nishimoto, N., Arihara, S., and Bue, K. (1967b). Insect moulting activity of crude drugs and plants. *Yakugaku Zasshi* **87**, 1414–1418.

Thompson, M. J., Kaplanis, J. N., Robbins, W. E., and Yamamoto, R. T. (1967). 20,26-Dihydroecdysone, a new steroid with moulting hormone activity from the tobacco hornworm, *Manduca sexta* (Johannson). *Chem. Commun.* pp. 650–653.

Vinson, J. W., and Williams, C. M. (1967). Lethal effects of synthetic juvenile hormone on the human body louse. *Proc. Natl. Acad. Sci. U.S.* **58**, 294–297.

Wigglesworth, V. B. (1954). "The Physiology of Insect Metamorphosis," pp. 1–152. Cambridge Univ. Press, London and New York.

Wigglesworth, V. B. (1959). "The Control of Growth and Form." pp. 1–140. Cornell Univ. Press, Ithaca, New York.

Wigglesworth, V. B. (1964). "The Life of Insects," pp. 1–360. World, Cleveland, Ohio.

Williams, C. M. (1952a). Morphogenesis and the metamorphosis of insects. *Harvey Lectures Ser.* **47**, 126–155.

Williams, C. M. (1952b). Physiology of insect diapause. IV. The brain and prothoracic glands as an endocrine system in the Cecropia silkworm. *Biol. Bull.* **103**, 120–138.

Williams, C. M. (1954). Isolation and identification of the prothoracic gland hormone of insects. *Anat. Record* **120**, 743.

Williams, C. M. (1956). The juvenile hormone of insects. *Nature* **178**, 212–213.

Williams, C. M. (1958). Hormonal regulation in insect metamorphosis. *Symp. Chem. Basis Develop., Baltimore. Johns Hopkins Univ. McCollum-Pratt Inst. Contrib.* **234**, 794–806.

Williams, C. M. (1961). Insect metamorphosis: an approach to the study of growth. *In* "Growth in Living Systems" (M. X. Zarrow, ed.), pp. 313–320. Basic Books, New York.

Williams, C. M. (1963a). Differentiation and morphogenesis in insects. *In* "The Nature of Biological Diversity" (J. M. Allen, ed.), pp. 243–260. McGraw-Hill, New York.

Williams, C. M. (1963b). The juvenile hormone. III. Its accumulation and storage in the abdomen of certain male moths. *Biol. Bull.* **124**, 355–367.

Williams, C. M. (1966). Selective control of insects by juvenile hormone analogues. *Science* **152**, 677.

Williams, C. M. (1967). Third-generation pesticides. *Sci. Am.* **217**, 13–17.

Williams, C. M. (1968a). Conference on insect-plant interactions. *BioScience* **18**, 791–799.

Williams, C. M. (1968b). Ecdysone and ecdysone-analogues: their assay and action on diapausing pupae of the Cynthia silkworm. *Biol. Bull.* **134**, 344–355.

Williams, C. M., and Law, J. H. (1965). The juvenile hormone. IV. Its extraction, assay, and purification. *J. Insect Physiol.* **11**, 569–580.

Zurflüh, R., Wall, E. N., Siddall, J. B., and Edwards, J. A. (1968). Synthetic studies on insect hormones. VII. An approach to stereospecific synthesis of juvenile hormone. *J. Am. Chem. Soc.* **90**, 6224–6225.

7

Chemical Communication within Animal Species

EDWARD O. WILSON

I. CLASSIFICATION OF CHEMICAL COMMUNICATION SYSTEMS

A. Pheromones, Hormones, Allomones

Evolutionary inference, together with substantial new experimental evidence from studies in animal behavior and natural products chemistry, leads to the conclusion that chemical communication is the paramount mode of communication in most groups of animals. In the early evolution of animal behavior, chemical releasers, or pheromones as they are now generally called, were probably also the first signals put to service. We know that communication among protozoan cells must have preceded the origin of the metazoans, and this primitive signaling was almost certainly chemical. Consider then the possibility that pheromones are in a special sense the lineal ancestors of hormones. At this state of our knowledge it is still reasonable to speculate with Haldane (1955) that as the metazoan soma was organized in evolution, hormones appeared simply as the intercellular equivalent of the pheromones that mediate behavior among the single-celled organisms.

But whatever their precise evolutionary position, it is at least true that chemical communication systems have now been discovered in most of the principal animal phyla, and they continue to turn up regularly in species where a deliberate search is made for them (Table I). Chemical communication consequently must be regarded as a very general biological phenomenon. In fact, it may not be going too far to say that chemical communication in some form involving cells or organisms or both, is one of the fundamental attributes of life itself.

At this point I feel an obligation to try to define the term communication. The word continues to be a crucial but elusive semantic keystone in descriptions of animal behavior – see for example the recent discussion by Altmann (1967). Biological communication can be defined as action on the part of one organism (or cell) that alters the probability pattern of behavior in another organism (or cell) in an adaptive fashion. By adaptive I mean that the signaling, or the response, or both have been genetically programmed to some extent by natural selection. This broad definition still leaves some kinds of interaction in nature ambiguously classified, particularly those physical actions that come out of encounters between competing species and between predator and prey. Furthermore, it does not incorporate some of the features of communication that are most widespread, such as the fact that the energy of the signal input is usually (but not always) less than the energy of response. But it is sufficient for most purposes and wholly adequate for current descriptions of pheromones in particular, since by definition these substances serve as signals only within species.

Where the definition still fits uneasily is the case of chemical signaling between different species. Let me intrude one more definition. If one or both interacting species have clearly evolved chemical signals, or responses, or both to cope with the other species, the substances involved can be referred to as allomones (after Brown and Eisner, in Brown, 1968). Allomones have arbitrarily been excluded from this chapter. The most useful things that can be said about them in passing is first that we shall probably find that they display approximately the same general design features as pheromones, and second that the similarities will be closest in the case of mutualistic symbiosis, where (just as in pheromone evolution) both signal and response are shaped by the same specific agents of natural selection.

B. Modes of Action of Pheromones

Pheromones are secreted from exocrine glands as liquids, transmitted either as liquids or gases, and smelled or tasted by another animal of the same species. They either evoke an immediate behavioral response, which Wilson and Bossert (1963) termed a releaser effect, or else they have a more subtle influence which we referred to as a primer effect. Primer substances activate the chemoreceptors in such a way as to alter the physiology of the receiving organism, probably in most cases through the mediation of the endocrine system, so that the animal is set to display a different response pattern in the future. This new behavioral repertory

TABLE I
The Occurrence of Chemical Communication Systems in Protistan and Animal Phyla

Taxa	Activity of pheromone	Chemical nature of pheromone	Authority
Protista			
Volvox sp.	Female substance induces gonidia to develop into sperm packets	High molecular weight, over 200,000; probably a protein	Starr (1968)
Paramecium bursaria	Mate recognition, by cilial contact	Apparently a protein	Siegel and Cohen (1962)
Aschelminthes			
Brachionus spp. (rotifer)	Recognition of females by males, followed by breeding	Not a protein; otherwise unknown	Gilbert (1963)
Annelida			
Lumbricus terrestris (earthworm)	Alarm and evasion; secreted in mucus	Unknown	Ressler *et al.* (1968)
Mollusca			
Helisoma spp. and some other aquatic snails	Alarm: self-burying or escape from water	Polypeptides from tissue; mol. wt. about 10,000	Snyder (1967 and personal communication)
Arthropoda			
Decapoda			
Portunus sanguinolentus (crab)	Sex attractant from female urine	Unknown	Ryan (1966)
Cirripedia			
Balanus balanoides and *Elminius modestus* (barnacles)	Aggregation and settlement of larvae, by contact with pheromone on substratum	Protein	Crisp and Meadows (1962)
Arachnida			
Lycosidae (wolf spiders)	Female sex attractant	Unknown	Kaston (1936)
Salticidae (jumping spiders)	Female sex attractant	Unknown	Crane (1949)
Insecta	*Sex attractants.* Female attractants are common and very widespread, having been demonstrated in the following orders: Dictyoptera including Isoptera, Lepidoptera, Coleoptera, Hymenoptera, Diptera. Male attractants and "aphrodisiac" agents are also common and widespread, having been reported from Dictyoptera (Blattaria only), Hemiptera, Mecoptera, Neuroptera, Lepidoptera, Coleoptera, Diptera, Hymenoptera. See reviews by Jacobson (1965), Butler (1964, 1967), and Wilson (1968)		
	Alarm substances, trail substances, recognition odors, etc. Occur in most social insects. Reviews by Wilson (1963, 1965) and Butler (1967)		
Chordata			
Vertebrata	*Sex attractants,* both male and female, are widespread in amphibious reptiles, and mammals, although they are still poorly documented in most groups. See reviews by Wilson and Bossert (1963), Wilson (1968), and Bronson (1969). These pheromones are now known to be common in primates, including even the female rhesus (Michael and Keverne, 1968)		
	Dominance odors and territorial and home-range markers are common in mammals (e.g., Mykytowycz, 1962, 1964; Schultze-Westrum, 1965; Thiessen *et al.*, 1968). Individual odor involved in territorial defense has been reported in fish (Todd *et al.*, 1967)		

will, in turn, be triggered by additional stimuli, which may or may not be pheromonal. Primer effects have been especially well documented in rodents (Bruce, 1966; Whitten, 1966; Bronson, 1969). In the manner of the medical sciences, the different kinds of physiological change are called after their discoverers. I like this practice because it is mnemonic, and so to the better known "effects" listed below I have added the most recently reported as the "Ropartz effect."

1. Bruce Effect

Exposure of a recently impregnated mouse female to a male with an odor sufficiently different from that of her stud results in failure of the implantation and rapid return to estrus. As Bronson (1969) points out, the adaptive advantage to the new male is obvious, but it is less easy to see why it is advantageous to the female and therefore how the response could have been evolved by direct natural selection.

2. Lee-Boot Effect

When about four or more female mice are grouped together in the absence of a male, estrus is suppressed and pseudopregnancies develop in as many as 61% of the individuals. The adaptive significance of the phenomenon is unclear, but it is evidently one of the devices responsible for the well-known reduction of population growth under conditions of high population density.

3. Ropartz Effect

The odor of other mice alone causes the adrenal glands of individual mice to grow heavier and to increase their production of corticosteroids, resulting in a decrease in reproductive capacity of the animal (Ropartz, 1966, 1968). Here we have part but surely not all of the explanation of the well-known stress syndrome. Some ecologists have invoked the syndrome as the explanation of population fluctuation, including the occasional "crashes" of overly dense populations.

4. Whitten Effect

An odorant found in the urine of male mice induces and accelerates the estrus cycle of the female. The effect is most readily observed in females whose cycles have been suppressed by grouping; the introduction of a male then initiates their cycles more or less simultaneously, and estrus follows in 3 or 4 days.

The elucidation of the rodent primer pheromones has important implications for ecology. Here we have concrete evidence for the first time of elementary signals whose summed (and measurable) effects provide feedback to population growth — positive in the case of the Whitten effect, negative in the others. A similar primer system has been found to play a role in the maturation rate and ovarian development of migratory locusts and hence serves as a regulator of population growth in these economically important insects (Norris, 1968).

That both releaser and primer effects can be produced by a single pheromone has been shown by the work of C. G. Butler and his associates at Rothamstead, England, on the "queen substance" of the honey bee (Butler, 1964). This pheromone, 9-ketodecenoic acid, attracts males during the nuptial flight and inhibits royal-cell building by workers in the nest; it also inhibits ovarian development in the workers.

The releaser effects described in a rapidly growing list of animal species, especially those of insects, made an impressively long and varied catalog (Jacobson, 1965; Butler, 1967; Wilson, 1968; Bronson, 1969). Pheromones are used in the assembly of elementary aggregations, in sexual attraction and stimulation, in territorial and home range marking, in the repelling and dispersal of individuals when populations grow overly dense, in the recognition of group, caste, and rank in social aggregations, and in alarm recruitment in other social aggregations. In short, chemical communication is now known to be as fully versatile as acoustical and visual communication.

II. THE CHEMICAL CHANNEL

A. The Signal Is the Active Space

As a pheromone is released from the skin surface or opening of an exocrine gland, it passes into the air either as a gaseous puff or a continuous gaseous stream. In either case, diffusion results in a concentration gradient descending from the point of emission. There is a zone around the point of emission within which the pheromone molecules are at or above the concentration required for a behavioral response (or in the case of primer effects, a physiological response). This zone is itself the signal, and it has been termed the *active space* by Bossert and Wilson (1963). When the pheromone is released from an arboreal station, say the point of a leaf or twig, into still air, its active space is spherical in shape. When it is released into a wind, its space is ellipsoidal with the long axis aligned downwind. If the pheromone is released from a point on the surface of flat ground, the spaces in still air and wind are hemispherical and semiellipsoidal, respectively. Consider for example the active space of a typical insect sex attractant. Working from the data of A. Butenandt, D. Schneider, and others on the production and molecular threshold concentration of the female attractants of moths, Bossert and I constructed a model of the space created by a single female moth releasing her pheromone downwind. As represented in Fig. 1, the space is postulated to form a gigantic semiellipsoid, with a long axis as much as several kilometers long and vertical and lateral axes each over 100 m in extent.

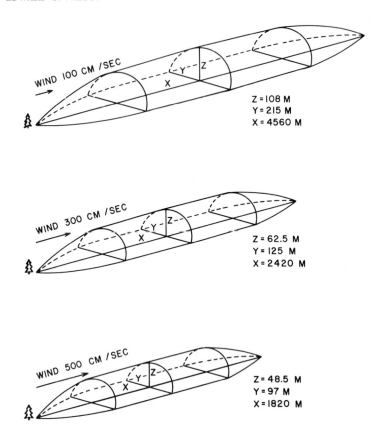

Fig. 1. The postulated active space created by a single female moth releasing sex attractant downwind, in a case where the attractant is as potent as in *Bombyx* and *Porthetria*. So long as the female continues releasing the substance, the space will persist as a semiellipsoid of approximately the same dimensions. Males flying into the active space turn upwind and eventually locate the female at the upwind apex, in this case the pine tree. If wind velocity is increased, the active space shrinks.

Surprisingly, this great signal power can be accomplished with the release of less than 1 μg/sec of sex attractant. This is because the sensitivity of the male to the pheromone molecules is so fantastically acute. In the case of the silkworm moth (*Bombyx mori*), for example, the male begins to react when the molecular density is as low as 100 molecules per cubic centimeter of air. The ecological implications of these findings are considerable. We know that most animals in a given ecosystem are being guided from moment to moment through the mediation, at least in part, of chemical signals. It is difficult for human observers, being exceptionally visual and auditory organisms, to appreciate the predominantly chemical *Umwelt* in which so many other animal species live. Any ordinary terrestrial community of animals contains hundreds or thousands of animal species and scores to hundreds of plant species, each of which produces its own characteristic odor. The odor environment of a given animal is enormously complex. It comprises a kaleidoscopically shifting maze of overlapping active spaces, from which the animal must select those few critical signals that can lead it to a food plant or a prey or a host of the desired kind, to a receptive mate of its own species, and away from certain predator species that specialize in catching it. How is this selection accomplished? In the majority of insect species that have been studied, the chemoreceptors are able to screen stimuli to some extent. They are sensitive, sometimes exceedingly so, to the relevant stimuli and less sensitive to the irrelevant ones. In other words, evolution of the sensory apparatus has enlarged the important active spaces and shrunk the unimportant ones. In some cases, most notably in the vertebrates, the same effect has been achieved by a more subtle form of screening within the central nervous system; the olfactory bulbs receive a wide spectrum of signals and sort them out, often on the basis of learned experience.

B. How to Analyze a Chemical Communication System

Needless to say, our present understanding of the role of chemical signals in the life of most animal species is rudimentary. I would like to recommend the following proposition as being fundamental to the future development of the chemical ecology of animals: *the relation of chemosensory physiology to ecology can be fully elucidated only through the analysis of the active spaces.* This approach depends a great deal, in turn, on the development of a workable methodology. The measurement of active spaces in particular requires the determination of the following parameters of the individual chemical signals: (1) amount of pheromone released by the sender animal; (2) evaporation and diffusion properties of the pheromone; and (3) olfactory efficiency in the receiver animal. With the measurements of these properties, it is possible to employ gas diffusion models to estimate the shape of the active space and the rate at which the space expands and contracts – in short, the basic qualities of the chemical signal itself.

With these considerations in mind, F. E. Regnier and I recently (Regnier and Wilson, 1968) set out to define a complete pheromone system in a way that had not been attempted previously. The phenomenon we chose is alarm communication in the formicine ant species *Acanthomyops claviger*. We were already aware that when a worker of this species is disturbed in the vicinity of the nest, it throws out a mixture of volatile substances that diffuse through the air and are smelled by other workers. Nestmates alerted in this way display a characteristic complex response: they open their mandibles and simultaneously raise, extend, and sweep their antennas laterally through the air in typical osmotactic searching; they then begin to run toward the odor source, growing more excited as they draw near. More workers are thus attracted to the point of disturbance. If the stimulus is sustained, an increasing amount of alarm chemicals are discharged and the entire colony can eventually be mobilized.

Our first step in the analysis of this communication system was to identify the volatile alarm substances and their glandular sources in the worker ants. We started with gas chromatographic analysis of whole ants (see Fig. 2), separation of the components by the same technique, and identification of the components by the aid of mass spectrometry. Then we dissected out exocrine glands and repeated the chromatographic analyses in order to pinpoint the glandular source of each component. Finally we had the information summarized in Fig. 3.

We were surprised to find such a rich medley of substances including, in particular, members of the alkane series. The question was then raised, how good is each of the substances as a pheromone? Or, put more precisely, what is the minimal concentration of each substance that causes the behavioral response?

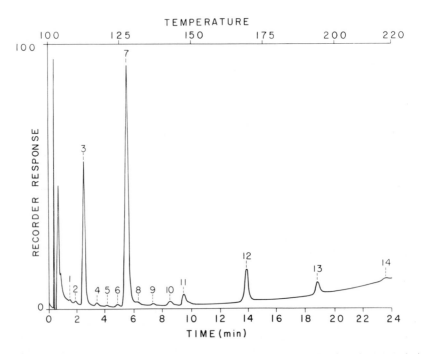

Fig. 2. Gas chromatogram of the total extract of workers of the ant species *Acanthomyops claviger*. This is an example of the first step in the isolation and identification of pheromones, leading to the information presented in Fig. 3.

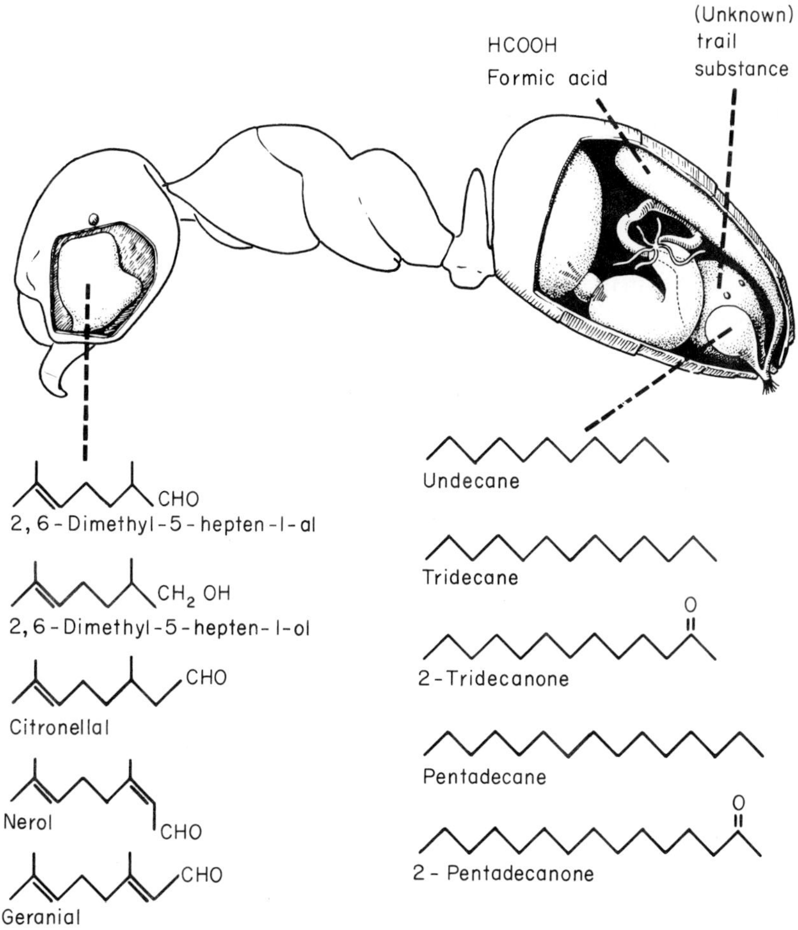

Fig. 3. The structural formulas of volatile substances found in various of the exocrine glands of *Acanthomyops claviger*. Terpenes are located in the mandibular gland in the head, and alkanes and ketones in the Dufour's gland of the abdomen. All but pentadecane and 2-pentadecanone are efficient alarm substances.

This is not an easy measurement to take. The molecular concentrations of diffusing gases are notoriously subject to complex variation affected by many variables. In the course of the *Acanthomyops* work, however, we were able to develop a method which is both relatively simple and very general in applicability. First a droplet of the pure pheromone was inserted into a capillary tube and the subsequent emission rate because of evaporation was measured volumetrically with the aid of a dissecting microscope. The tip of the capillary tube – the point of emission of the evaporating substance – was placed at a predetermined point near a group of insects resting in still air. The time to onset of the response was then measured. From a modification of the diffusion model, the following formula can be derived:

$$K = \frac{Q}{2D\pi r} \text{ erfc } \frac{r}{(4Dt)^{1/2}}$$

where K is the behavioral threshold concentration in molecules/cm³, Q is the emission rate in molecules per second, t is the time to the onset of the response, r is the distance from the emission point to the chemoreceptor of the organism, D is the diffusion coefficient (obtained empirically in an independent procedure), and erfc(x) is the complementary error function. A set of techniques for measuring the parameters and making rapid estimates of K has been provided by Wilson et $al.$ (1969).

Some of the estimates of K of $Acanthomyops$ $claviger$ substances, and their homologs, are given in Fig. 4. It can be seen that the natural $Acan$-

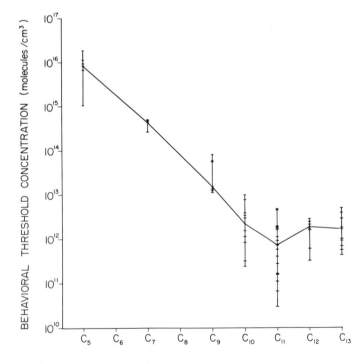

Fig. 4. Estimates of behavioral threshold concentrations (K) of members of the alkane series, from pentane (C_5H_{12}) to tridecane ($C_{13}H_{28}$). Undecane ($C_{11}H_{24}$) and tridecane are the hydrocarbons that occur naturally in the Dufour's gland of $Acanthomyops$ $claviger$ and function as alarm pheromones.

thomyops alkanes, undecane ($C_{11}H_{24}$) and tridecane ($C_{13}H_{28}$) are indeed efficient at eliciting alarm. They are far more efficient, for example, than their lower homologs. It was also determined that the other *Acantho-myops* substances, namely the ketones from Dufour's gland and the terpenoids from the mandibular glands, have approximately the same K values as undecane and tridecane.

By chromatographic measurements of air samples from around disturbed ants, we were able to show that the ants themselves release just enough of the mandibular and Dufour's gland substances, about 1 μg or somewhat less, to broadcast efficiently over the centimeter range. With these "natural Q" values, together with the K values taken earlier, the diffusion model predicts that the active space (i.e., the signal itself) not only should expand until it has a maximum radius of several centimeters, but it should thereafter contract to near zero within a few minutes. These predictions were upheld by direct measurements of the reach and duration of alarm signals around disturbed workers in artificial nests. Thus the pheromones identified by us, and their separately measured physical and behavioral properties, account for all of the observed alarm communication.

The measurements of the *Acanthomyops* signals seem, moreover, to comprise the intuitively optimal set of properties for a chemical alarm system. To see this clearly, consider the Q/K ratios that we have estimated for three widely differing chemical systems (Table II).

The higher the natural Q/K ratio, the greater will be the maximum signal distance and the slower the signal fade-out. Suppose that the alarm system possessed the much lower Q/K value of the fire ant trail system. Then the alarm signal, i.e., the active space, would expand out for a distance of only a few millimeters and it would then fade out a few seconds later. The signal would be consequently ineffective, because any worker ant close enough to perceive it would certainly also be close enough to perceive the cause of the alarm.

Suppose, on the other hand, that the alarm system possessed the extremely high Q/K value of the moth sex attractant. Then if we could somehow allow it to run its course in still air, the alarm signal would theoretically expand outward for a distance of several kilometers, and it would endure for years! This could not happen in a real situation, of course; yet it is undeniably true that periodic signals with such a Q/K value would cause the entire ant colony to be in a perpetual state of turmoil.

C. Differences between Vertebrate and Insect Pheromones

An intriguing generalization that seems to be emerging in the most recent studies of vertebrate pheromones is that these substances tend to occur as complex mixtures. Insect pheromones, in contrast, tend to consist of either a single component or, at most, of a relatively simple mix-

TABLE II

The Q/K Ratios of Three Widely Differing Chemical Communication Systems

Pheromone	$\dfrac{Q}{K} = \dfrac{\text{Natural emission rate in molecules/sec}}{\text{Behavioral threshold conc. in molecules/cm}^3}$
Imported fire ant *S. saevissima* odor trail	1
Acanthomyops sp. alarm substances	$10^3 - 10^5$
Silkworm *Bombyx mori* sex attractant	$10^{10} - 10^{12}$

ture. Examples of the two types are given in Fig. 5. I would like to venture the following guess as to the significance of this difference. The social behavior of most vertebrate species is "personal." It is based on the recognition of individuals—in the maintenance of dominance hierarchies, in leadership, in territorial defense, and in parent–offspring relationships. (Schools of fish form an obvious exception.) Sufficient evidence exists to indicate that where olfaction is involved in many forms of vertebrate social behavior, the animals are able to recognize the pheromones not only of their own species but also of individual members of the species (Leyhausen, 1960; Schultze-Westrum, 1965; Klopfer *et al.*, 1964; Todd *et al.*, 1967; Müller-Schwarze, personal communication). The simplest way to vary odor is to produce an exocrine mixture comprised of many components whose proportions can be varied. Such a device is suggested by the recent work of Dietland Müller-Schwarze and R. M. Silverstein (Müller-Schwarze, personal communication) on the chemistry of the tarsal scent of the black-tailed deer. Insect social behavior, on the other hand, is for the most part impersonal. Perhaps due to the intrinsic limitations of the insect brain, communication operates much more on the principle of elementary stimulus–response. Even the vaunted social insects, the bees, wasps, ants, and termites, organize their colonies primarily in this fashion. Consequently, simple, relatively invariant chemical signals suffice. I am not anxious to push this contrast too far. A notable exception is the nest odors of the social insects which are known to be specific to colonies and variable through time within single colonies. It will be interesting to learn whether the nest odors (which have yet to be characterized chemically) consist of mixtures of comparable complexity to the exocrine secretions of vertebrates.

D. Differences between Airborne and Waterborne Pheromones

The size of pheromone molecules that are transmitted through air can be expected to follow certain rules (Wilson and Bossert, 1963). In general, they should possess a carbon number between 5 and 20 and a molecular weight between 80 and 300. The a priori arguments that lead to this

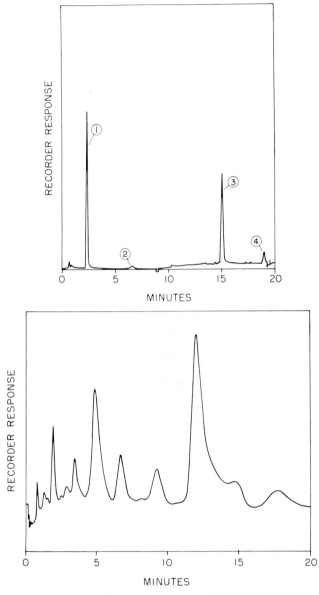

Fig. 5. Two gas chromatograms illustrating the greater complexity that characterizes the exocrine secretions of vertebrates as opposed to those of insects. *Above*: chromatogram of Dufour's gland secretion from a single worker of the ant *Lasius umbratus* (3% OV-17, 6 ft, programmed 100° −200°; F. E. Regnier, unpublished). *Below*: chromatogram of tarsal secretion of a single black-tailed buck (15% Carbowax 20 *M*, 5 ft, isothermal at 200°C; Müller-Schwarze, unpublished). The ant substances are undecane (1), an unknown minor component (2), 2-tridecanone (3), and 2-pentadecanone (4).

prediction are essentially as follows. Below the lower limit, only a relatively small number of kinds of molecules can be readily manufactured and stored by glandular tissue. Above it, molecular diversity increases very rapidly. In at least some insects, and for some homologous series of compounds (for example see Fig. 4), olfactory efficiency also increases steeply. As the upper limit is approached, molecular diversity becomes astronomical so that further increase in molecular sizes confers no further advantage in this regard. The same consideration holds for intrinsic increases in stimulative efficiency, in so far as they are known to exist. On the debit side, large molecules are energetically more expensive to make and to transport, and they tend to be far less volatile. On the other hand, differences in the diffusion coefficient due to reasonable variation in molecular weight do not cause much change in the properties of the active space, contrary to what one might intuitively expect. Wilson and Bossert further predicted that the molecular size of sex pheromones, which generally require a high degree of specificity as well as stimulative efficiency, would prove higher than that of most other classes of pheromones, including, for example, the alarm substances. The empirical rule displayed by insects is that most sex attractants have molecular weights that are between 200 and 300, while most alarm substances fall between 100 and 200. Some of the evidence for this last statement, together with a discussion of the exceptions, has been recently reviewed by Wilson (1968). Since this review was written, the validity of the structure of two of the sex pheromones has been challenged, a serious enough matter to deserve special mention here. The substances reported as the sex attractants, respectively, of the gypsy moth *Porthetria dispar* ("gyptol" and its homolog "gyplure") by Jacobson *et al.* (1960), and of the pink bollworm moth *Pectinophora gossypiella* ("propylure") by Jones *et al.* (1966) were recently synthesized and bioassayed by the German chemists Eiter *et al.* (1967), who found them inactive. Subsequently, Jacobson (1969) reported that only the trans isomer of propylure is active, and that as little as 15% of the cis isomer completely nullifies the activity of the trans isomer. He accounts for the negative result of Eiter *et al.* by the fact that their propylure sample was a mixture of the two isomers. This explanation is plausible but has not yet been supported by molecular threshold measurements.

When we come to pheromones transmitted in water, however, we have a very different situation. The rules concerning diversity of molecular species are, of course, the same; but the rates at which given substances are passed into the medium from films or droplets, as well as the diffusion coefficients, are drastically altered. What kind of molecules might be expected in the aqueous pheromones? Only within the last several years have enough chemical characterizations been made to permit some generalizations (Table III). As far as molecular size is concerned, the substances fall into two distinct classes. On the one hand, are the sex pheromones of fungi and *Lebistes*, along with acrasin, the aggregating

TABLE III
Waterborne Pheromones

Species	Activity	Characterization	Authority
Protistans			
Volvox spp.	Female substance induces gonidia to develop into sperm packets	High molecular weight, 10,000 to over 200,000 in two species studied; probably protein	W.H. Darden and G. D. Kochert, in Starr (1968)
Paramecium bursaria	Mate recognition, by cilial contact	Apparently a protein	Siegel and Cohen (1962)
Fungi			
Allomyces sp. (water mold)	Sperm attractant produced by female gametes; active at $10^{-10}\ M$	Sirenin: oxygenated sesquiterpene with cyclohexane center; $C_{15}H_{24}O_2$	Machlis *et al.* (1966, 1968)
Achlya bisexualis (water mold)	Induction of antheridiol hyphae on the male plant; active at 2×10^{-10} gm/ml	Antheridiol: a steroid $C_{29}H_{42}O_5$	Arsenault *et al.* (1968)
Mucor mucedo	Induction of sexual hyphae in opposite sex	A "gamone": $C_{20}H_{25}O_5$	Plempel (1963)
Dictyostelium discoideum (slime mold)	Attraction and aggregation of ameboid cells	Acrasin: cyclic 3′,5′-adenosine monophosphate	Konijn *et al.* (1967)
Molluscans			
Helisoma duryi	Alarm: burying in substrates	Polypeptides from crushed tissue; mol. wgt. about 10,000	Snyder (1967 and personal communication)
Arthropodans			
Balanus balanoides and *Elminius modestus* (barnacles)	Attraction and settling of larvae	Protein	Crisp and Meadows (1962)
Fish			
Lebistes reticulatus	Increased activity and attraction on part of male	Estrogen, from female ovary	Amouriq (1965a,b)

attractant of the slime mold. These substances are comparable in size with the airborne sex attractants of terrestrial animals. The diffusion coefficients of most water-soluble substances in this range of molecular weights is of the order of 10^{-5} in water and between 10^{-1} and 10^{-2} in air. A thousandfold or more decrease in diffusivity makes a great deal of difference in the properties of the active space. An examination of the example in Table IV will show that at least in the case of discontinuous pheromone release, the maximum radius of the space is the same in water as in air. But the time required to reach the maximum radius and the interval between release of the pheromone and the disappearance of the active space (that is, the "fade-out" time) are approximately 10,000

The Predicted Spread and Fade-Out of a Signal (Active Space) of a Substance
of Low Molecular Weight[a]

Q	Diffusion coefficient		Maximum radius of active space (cm)		Time required to reach maximum radius (sec)		Fade-out time (sec)	
K	Air	Water	Air	Water	Air	Water	Air	Water
1	10^{-1}	10^{-5}	0.6	0.6	0.4	4×10^3	1	10^4
10^2	10^{-1}	10^{-5}	2	2	8	8×10^4	20	2×10^5
10^4	10^{-1}	10^{-5}	10	10	1.5×10^2	1.5×10^5	5×10^2	5×10^6
10^6	10^{-1}	10^{-5}	60	60	4×10^4	4×10^7	10^4	10^8

[a]Diffusion coefficient arbitrarily set at 0.1 cm²/sec.

times greater in water than in air. How then can aquatic and marine
organisms use molecules of this size? A better question is: How can organisms transmit pheromones through water at all? There are in fact
two ways in which the same substance can be employed as efficiently in
water as in the air: (1) the Q/K ratio can be adjusted appropriately; and
(2) the pheromone can be spread more quickly by placing it in natural
currents or creating artificial currents.

By extending the diffusion theory of Bossert and Wilson (1963), I have
examined the possibilities of adjusting the Q/K ratios in aqueous systems with the following results. In order for the same substance to generate about the same intervals to maximum radius and fade-out in water
as on land, it would be necessary for the Q/K ratio to be about a million
times greater in water. In other words, the aquatic or marine species (in
the single-puff model) would have to increase the amount of pheromone
solute a millionfold, or lower its response threshold to a millionth, or
some equivalent combined alteration of the two parameters in order to
achieve the same signal times as a terrestrial species using the same
pheromone in air. This adjustment, incidentally, would result in a hundredfold *increase* in the maximum radius of the active space.

Such a huge increment in Q/K is not as difficult to attain as it might
first seem. The most promising parameter is the emission rate Q. When a
pheromone is emitted as a film, spray, or droplet in air, the emission rate
is, of course, a function of the vapor pressure. In most homologous series,
vapor pressure drops off steeply with increase in molecular weight. In
the alkane series, for example, the emission rate in molecules per second
from a surface of fixed area declines one order of magnitude with every
additional CH_2 group added. Proteins and other macromolecules have
for practical purposes zero vapor pressure and cannot be transmitted by
air unless they are somehow adsorbed onto bubbles or dust particles or
absorbed into droplets of mist. But the same is not true for water transport. The solubility of large polar molecules is moderately high and can
conceivably provide the requisite increase in Q in water as opposed to
air.

This brings us to the proteins, which make up the second class of known waterborne pheromones. In the case of the protistan pheromones and aggregation substance of barnacles (Table III), transport raises no problems since communication is by contact chemoreception or transmission over short distances. In the case of the snail alarm substances, the species have evidently made use of the fact that injured individuals release large quantities of their blood and tissue proteins into turbulent water, involuntarily of course. The ability of the liberated proteins to diffuse is limited but still adequate to generate a large active space. The diffusion coefficients of proteins in water at 20°C range from 0.34×10^{-7} to 1.6×10^{-8} (Edsall, 1953). The long duration of the signal would be in accord with the behavior of the responding snails, who bury themselves or leave the water altogether.

Although the transmission rate to a fixed distance can be increased by enlarging the Q/K ratio, such an adjustment will also increase the time to fade-out (see Table IV). Consequently, in cases where a reasonable short fade-out time is required, we can expect to find additional devices, such as unstable molecular structure or enzymatic deactivation, that cancel the signals. These devices should be more prominently developed in waterborne systems than in airborne systems of similar function.

The foregoing discussion should suffice to point out some of the aspects of analysis that can be undertaken to advance this largely unexplored subject in the behavioral ecology of aquatic and marine organisms.

III. THE INCREASE IN INFORMATION AND ITS UPPER BOUNDARY

Wilson and Bossert (1963) and Wilson (1965, 1968) described the ways in which animal species have evolved to increase the amount of information and the rate of information transfer in chemical signaling. These devices can be summarized as follows.

1. Adjustment of Fading Time

By lowering Q/K, either through reduction of the emission rate Q or raising of the threshold concentration K, the time between the release of the pheromone and final disappearance of the active space can be shortened; the chemical signal can thereby be more sharply pinpointed in space and time. The result is an increase in information per signal and the opportunity for the transmission of more discrete signals. This has been a chief design feature in the evolution of alarm and trail systems. Another way to shorten the life of the signal is to deactivate the pheromone. The only documented instance known to me is the enzymatic deactivation of ingested 9-ketodecenoic acid by worker bees, chiefly through reduction of the pheromone 9-hydroxydecanoic acid and 9-hydroxy-2-decenoic acid (Johnston et al., 1965).

2. Expansion of the Active Space

Information can be increased not only by reducing the duration of a separate signal in time, as just described, but also by enlarging the space within which orientation occurs. Where a pheromone discloses the location of the sending organism, the amount of information transmitted increases as the logarithm of the volume of the active space. We have already seen that the active space is enlarged by increasing Q/K. The insect sex attractants represent an extreme development in this direction. To be sure, the rate of information transfer is kept down in such systems in the sense that signals cannot be rapidly turned on or off; but the total amount of information per signal is increased, since very small targets are pinpointed within very large spaces.

3. Use of Multiple Exocrine Glands

In many insects and mammals, there are multiple glands each of which produces pheromones with a different meaning. The maximum development of this device is seen in the social insects, especially in ants and honeybees where as many as five or more glands are employed by the same individual. The exploration of the fascinating chemical "codes" of the various groups of social insects is still in an early stage.

4. Medleys of Pheromones from the Same Gland

Different responses can be evoked by different pheromones occurring in mixes from the same gland. In the head of the honeybee queen, for example, are to be found at least 32 compounds, including methyl 9-ketodecanoate, methyl 9-keto-2-decenoate, nonanoic acid, decanoic acid, 2-decenoic acid, 9-ketodecanoic acid, 9-hydroxy-2-decenoic acid, 10-hydroxy-2-decenoic acid, 9-keto-2-decenoic acid (Callow et al., 1964). Most or all are present in the mandibular gland secretion. The biological significance of most of these substances is still unknown. Some are undoubtedly precursors to pheromones, but at least two are known pheromones with contrasting effects: the 9-ketodecenoic acid is the "queen substance" already mentioned, and the 9-hydroxydecenoic acid causes clustering and stabilization of worker swarms (Butler et al., 1964).

5. Change of Meaning through Change of Context

Trans-9-keto-2-decenoic acid serves a caste-inhibitory pheromone inside the honey bee nest and as the primary female sex attractant during the nuptial flight. The Dufour's gland secretion of the imported fire ant Solenopsis saevissima is an attractant that is effective on members of all castes during their adult lives. Under different circumstances it serves variously to recruit workers to new food sources, to organize colony emigration, and – in conjunction with a volatile secretion – to cause oriented alarm behavior (Wilson, 1962).

6. Different Responses to Different Concentrations and Durations

Workers of the Florida harvesting ant *Pogonomyrmex badius* react to low concentrations of the mandibular gland pheromone 4-methyl-3-heptanone by simple positive chemotaxis and to higher concentrations of the same substance by aggressive alarm behavior. When exposed to high concentrations for more than a minute or two, many individuals switch from simple alarm to digging behavior (Wilson, 1958; Bossert and Wilson, 1963).

7. New Meanings from Combinations

There are a few examples of pheromones acquiring additional or even different meanings when presented in combination. When released near fire ant workers, cephalic and Dufour's gland secretions cause alarm behavior and attraction, respectively; when expelled simultaneously by a highly excited worker, they cause oriented alarm behavior. Honeybee workers confined closely with queens for hours acquire scents from her which, evidently in combination with their own worker recognition scent, cause them to be attacked by nest mates (Morse and Gary, 1961).

8. Temporal Patterning of Single Pheromones

Frequency and amplitude modulation of single pheromones is a possibility that has been almost wholly neglected by experimentalists. Recently, Bossert (1968) has worked out the theory of the subject with some surprising results (see Fig. 6). He found that under most conceivable conditions, pheromone modulation is not a feasible means of communication. But under two special circumstances, when transmission occurs in still air over a distance of the order of a centimeter or less, or else in a steady, moderate wind, modulation is not only practicable but highly efficient. Under extremely favorable conditions, a perfectly designed system could transmit on the order of 10,000 bits of information a second, an astonishingly high figure considering that only one substance is involved. Under more realistic circumstances, say, for example, in a steady 400-cm/sec wind over a distance of 10 m, the maximum potential rate of information transfer is still quite high — over 100 bits a second, or enough to transmit the equivalent of twenty words of English text per second at 5.5 bits per word. For every pheromone released independently, the same amount of capacity could be added to the channel capacity. We can hardly expect any animal species to achieve more than a minute fraction of the theoretical capacity calculated by Bossert. To do so would require the evolution of a symbolical and syntactical language, something no animal species has done in any other sensory modality. But it is conceivable that modulation has been added somewhere to pheromone communication in order to increase signal specificity, just as a great many visual and acoustical systems have acquired signal modulation in other animal species. To doubt it on the grounds that no examples are yet

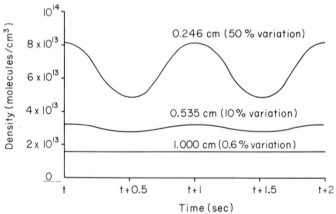

Fig. 6.   In the absence of any known case of pheromone modulation in the real world, two imaginary "caminalcules" are shown using this refinement in chemical communication. (A caminalcule is one of the arbitrarily evolving animal species used in theoretical studies of phylogeny and classification by J. H. Camin and his co-workers at the University of Kansas.) In the inset is an example from Bossert (1968) of the outcome of modulation of an odor under moderately good conditions for information transfer by this means. Following sinusoidal release of the substance in still air at an average of 10^{13} molecules per second and frequency of 1 cps, the density fluctuations are given at three distances from the source.

known is not enough, for human observers are incapable of detecting odor waves, especially under the environmental circumstances Bossert shows to be optimal for the evolution of odor modulation.

References

Altmann, S. A. (1967). The structure of primate social communication. *In* "Social Communication Among Primates" (S. A. Altmann, ed.), pp. 325-362. Univ. of Chicago Press, Chicago, Illinois.

Amouriq, L. (1965a). L'activité et le phénomène social chez *Lebistes reticulatus* (Poeciliidae-Cyprinoidontiformes). *Ann. Sci. Nat. Zool. Biol. Animale* 7, 151-172.

Amouriq, L. (1965b). Origine de la substance dynamogène émise par *Lebistes reticulatus* femelle (Poisson *Poeciliidae*, Cyprinodontiformes). *Compt. Rend.* 260, 2334-2335.

Arsenault, G. P., Biemann, K., Barksdale, A. W., and McMorris, T. C. (1968). The structure of Antheridiol, a sea hormone in *Achlya bisexualis*. *J. Am. Chem. Soc.* 90, 5635-5636.

Bossert, W. H. (1968). Temporal patterning in olfactory communication. *J. Theoret. Biol.* 18, 157-170.

Bossert, W. H., and Wilson, E. O. (1963). The analysis of olfactory communication among animals. *J. Theoret. Biol.* 5, 443-469.

Bronson, F. H. (1969). Pheromonal influences on mammalian reproduction. *In* "Perspectives in Reproduction and Sexual Behavior" (M. Diamond, ed.), cc, Indiana Univ. Press, Bloomington, Indiana.

Brown, W. L. (1968). An hypothesis concerning the function of the metapleural glands in ants. *Am. Naturalist* 102, 188-191.

Bruce, H. M. (1966). Smell as an exteroceptive factor. *J. Animal Sci.* 25, Suppl., 83-89.

Butler, C. G. (1964). Pheromones in sexual processes in insects. *Symp. Roy. Entomol. Soc. London* 2, 66-77.

Butler, C. G. (1967). Insect pheromones. *Biol. Rev. Cambridge Phil. Soc.* 42, 42-87.

Butler, C. G., Callow, R. K., and Chapman, J. R. (1964). 9-Hydroxydec-*trans*-2-enoic acid, a pheromone stabilizing honeybee swarms. *Nature* 201, 733.

Callow, R. K., Chapman, J. R., and Paton, P. N. (1964). Pheromones of the honeybee: chemical studies of the mandibular gland secretion of the queen. *J. Apicult. Res.* 3, 77-89.

Crane, J. (1949). Comparative biology of salticid spiders at Rancho Grande, Venezuela. Part IV. An analysis of display. *Zoologica* 34, 159-214.

Crisp, D. J., and Meadows, P. S. (1962). The chemical basis of gregariousness in cirripedes. *Proc. Roy. Soc.* (B)156, 500-520.

Edsall, J. T. (1953). The size, shape, and hydration of protein molecules. *In* "The Proteins. Chemistry, Biological Activity, and Methods" (H. Neurath and K. Baïley, eds.), Vol. 1B, pp. 549-726. Academic Press, New York.

Eiter, K., Truscheit, E., and Boness, M. (1967). Synthesen von D,L-10-Acetoxy-hexadecen-(7-*cis*)-ol-(1), 12-Acetoxyoctadecen-(9-*cis*)-ol-(1) ("Gyplure") und 1-Acetoxy-10-propyl-tridecadien-(5-*trans*.9). *Ann. Chem.* 709, 29-45.

Gilbert, J. J. (1963). Contact chemoreception, mating behaviour, and sexual isolation in the rotifer genus *Brachionus*. *J. Exptl. Biol.* 40, 625-641.

Haldane, J. B. S. (1955). Animal communication and the origin of human language. *Sci. Progr.* (*London*) 43(171), 385-401.

Jacobson, M. (1965). "Insect Sex Attractants." Wiley (Interscience), New York.

Jacobson, M. (1969). Sex pheromone of the pink bollworm moth: biological masking by its geometrical isomer. *Science* 163, 190-91.

Jacobson, M., Beroza, M., and Jones, W. A. (1960). Isolation, identification and synthesis of the sex attractant of gypsy moth. *Science* 132, 1011-1012.

Johnston, N. C., Law, J. H., and Weaver, N. (1965). Metabolism of 9-ketodec-2-enoic acid by worker honeybees (*Apis mellifera* L.). *Biochemistry* **4**, 1615-1621.

Jones, W. A., Jacobson, M., and Martin, D. F. (1966). Sex attractant of the pink bollworm moth: isolation, identification, and synthesis. *Science* **152**, 1516-1517.

Kaston, B. J. (1936). The senses involved in the courtship of some vagabond spiders. *Entomol. Am.* (N.S.) **16**, 97-167.

Klopfer, P. H., Adams, D. K., and Klopfer, M. S. (1964). Maternal "imprinting" in goats. *Proc. Natl. Acad. Sci. U. S.* **52**, 911-914.

Konijn, T. M., van de Meene, J. G. C., Bonner, J. T., and Barkley, D. S. (1967). The acrasin activity of adenosine-3′,5′-cyclic phosphate. *Proc. Natl. Acad. Sci. U. S.* **58**, 1152-1154.

Leyhausen, P. (1960). Verhaltensstudien an Katzen. Z. *Tierpsychol.* Beiheft **2**, 1-120.

Machlis, L., Nutting, W. H., Williams, M. W., and Rapoport, H. (1966). Production, isolation, and characterization of sirenin. *Biochemistry* **5**, 2147-2159.

Machlis, L., Nutting, W. H., and Rapoport, H. (1968). The structure of sirenin. *J. Am. Chem. Soc.* **90**, 1674-1675.

Michael, R. P., and Keverne, E. B. (1968). Pheromones in the communication of sexual status in primates. *Nature* **218**, 746-749.

Morse, R. A., and Gary, N. E. (1961). Colony response to worker bees confined with queens (*Apis mellifera* L.). *Bee World* **42**, 197-199.

Mykytowycz, R. (1962). Territorial function of chin gland secretion in the rabbit, *Oryctolagus cuniculus* (L.). *Nature* **193**, 797.

Mykytowycz, R. (1964). Territoriality in rabbit populations. *Australian Nat. Hist.* **14**, 326-329.

Norris, M. J. (1968). Some group effects on reproduction in locusts. *Colloq. Intern. Centre Natl. Rech. Sci.* (*Paris*) **173**, 147-159.

Plempel, M. (1963). Die chemischen Grundlagen der Sexualreaktion bei Zygomyceten. *Planta* **59**, 492-508.

Regnier, F. E., and Wilson, E. O. (1968). The alarm-defence system of the ant *Acanthomyops claviger*. *J. Insect Physiol.* **14**, 955-970.

Ressler, R. H., Cialdini, R. B., Ghoca, M. L., and Kleist, S. M. (1968). Alarm pheromone in the earthworm *Lumbricus terrestris*. *Science* **161**, 597-599.

Ropartz, P. (1966). Contribution à l'étude du déterminisme d'un effet de groupe chez les souris. *Compt. Rend.* **262**, 2070-2072.

Ropartz, P. (1968). Role des communications olfactives dans le comportement social des souris males. *Colloq. Intern. Centre Natl. Rech. Sci.* (*Paris*) **173**, 323-339.

Ryan, E. P. (1966). Pheromone: evidence in a decapod crustacean. *Science* **151**, 340-341.

Schultze-Westrum, T. (1965). Innerartliche Verständigung durch Düfte beim Gleitbeutler *Petaurus breviceps papuanus* Thomas (Marsupalia, Phalangeridae). Z. *Vergleich. Physiol.* **50**, 151-220.

Siegel, R. W., and Cohen, L. W. (1962). The intracellular differentiation of cilia. (Abstr.) *Am. Zoologist* **2**, 558.

Snyder, N. (1967). An alarm reaction of aquatic gastropods to intraspecific extract. *Cornell Univ. Agr. Expt. Sta. Mem.* **403**, 1-122.

Starr, R. C. (1968). Cellular differentiation in *Volvox*. *Proc. Natl. Acad. Sci. U.S.* **59**, 1082-1088.

Thiessen, D. D., Friend, H. C., and Lindzey, G. (1968). Androgen control of territorial marking in the Mongolian gerbil. *Science* **160**, 432-433.

Todd, J. H., Atema, J., and Bardach, J. E. (1967). Chemical communication in social behavior of a fish, the yellow bullhead (*Ictalurus natalis*). *Science* **158**, 672-673.

Whitten, W. K. (1966). Pheromones and mammalian reproduction. *Advan. Reprod. Physiol.* **1**, 155-177.

Wilson, E. O. (1958). A chemical releaser of alarm and digging behavior in the ant *Pogonomyrmex badius* (Latreille). *Psyche* **65**, 41-51.

Wilson, E. O. (1962). Chemical communication among workers of the fire ant *Solenopsis saevissima* (Fr. Smith). *Animal Behaviour* **10**, 134-164.

Wilson, E. O. (1963). The social biology of ants. *Ann. Rev. Entomol.* **8**, 345–368.
Wilson, E. O. (1965). Chemical communication in the social insects. *Science* **149**, 1064–1071.
Wilson, E. O. (1968). Chemical systems. *In* "Animal Communication" (T. Sebeok, ed.). Indiana Univ. Press, Bloomington, Indiana.
Wilson, E. O., and Bossert, W. H. (1963). Chemical communication among animals. *Recent Progr. Hormone Res.* **19**, 673–716.
Wilson, E. O., Bossert, W. H., and Regnier, F. E. (1969). A general method for measuring the threshold concentrations of odorant molecules. *J. Insect Physiol.* **15**, 597–610.

8

Chemical Defense against Predation in Arthropods

THOMAS EISNER

I. INTRODUCTION

"Well, what they got their asses up in the air for?" . . .
Hazel turned one of the stink bugs over . . . and the shining black beetle strove madly with floundering legs to get upright again. ". . . why do you think they do it?"
"I think they're praying," said Doc.

John Steinbeck, "Cannery Row"
Viking Press, New York

In nature, there is no turning of the other cheek – survival is based on things more practical than prayer. For beetles of the genus *Eleodes* (Fig. 1A) standing on the head is a defensive maneuver, undertaken by the animals when they are threatened, and signaling the fact that they are about to discharge from the tip of their abdomen an obnoxiously odorous and irritating spray (Figs. 1B and C). The secretion, which contains benzoquinones (Chadha *et al.*, 1961), is the chief line of defense of these beetles, and has been shown to be effectively repellent to a diversity of predators (Eisner, 1966).

Eleodes is by no means an exceptional animal. All organisms interact antagonistically with others in their environment, and the possession of defensive substances, whether for protection against predators, parasites, or competitors, may well be a most general attribute of life itself.

Among terrestrial animals, none is perhaps more diversely endowed with chemical weaponry than the arthropods. Their defenses, and in particular their defenses against predators, have been the subject of considerable recent study. The research has involved the efforts of a contingent of specialists, including biologists, biochemists, and chemists, and has therefore been rather broadly interdisciplinary in character. Although specifically related to arthropods, the findings have relevance to our understanding of chemical ecology as a whole. This chapter is an attempt to summarize some of this work. I have been deliberately and to some extent arbitrarily selective in the treatment of the subject, and only hope that I have done no injustice to any of the arthropods discussed. Priority has been given to work of the most recent years.

II. TYPES OF CHEMICAL DEFENSES

There are two major types of defensive substances in arthropods: those that are elaborated by special exocrine glands and those not strictly of glandular origin that are contained in the blood, gut, or elsewhere in or on the body. Glandular products may in turn be divided into injectable and noninjectable secretions, depending on how they are administered to the enemy. Examples of injectable secretions are the venoms associated with the stings of scorpions and Hymenoptera, mandibles of centipedes, and chelicerae of spiders. These commonly serve not only for defense, but also for incapacitation of prey. Since venom glands are to be reviewed comprehensively in a forthcoming book ["Venomous Animals and Their Venoms" (W. Bücherl, E. Buckley, and V. Deulofeu, eds.) Vol. III, Academic Press; in preparation], they will be ignored here. The only glands to be considered are those whose products are sprayed or otherwise applied topically to an enemy.

A. Glandular Defenses

1. Eversible Glands

The caterpillars of swallowtail and parnassian butterflies (family Papilionidae) possess a defensive gland, the osmeterium, situated dor-

Fig. 1. (A) *Eleodes* (probably *longicollis*), assuming the headstand with which it characteristically responds to disturbance. (B) *Eleodes* sp., dissected, showing the two quinone-secreting glands in rear of abdomen. (C) *Eleodes longicollis* discharging its defensive secretion in response to the pinching of its right hindleg with forceps. The beetle is glued to a tether, and placed on a sheet of filter paper impregnated with a mixture (acidulated KI-starch) that discolors in contact with the quinones present in the secretion.

sally behind the head. Consisting of a two-pronged invagination of the neck membrane, the gland is ordinarily tucked away beneath the integument, but is abruptly everted when the animal is disturbed (Fig. 3A). The extruded "horns" glisten with secretion and are intensely odorous. In the European *Papilio machaon* the secretion contains as its principal components isobutyric (Fig. 2) and 2-methylbutyric acid (Eisner and Meinwald, 1965).

The larva can exercise considerable control over the way in which it employs the gland. It minimizes evaporative loss of secretion by extruding only as much of the osmeterium as is warranted by the conditions of a given attack. Thus, if disturbance is inflicted from the side, the horn of that side is extruded further than its opposite partner. The larva also arches its body and wipes the horns directly against the offending agent (Fig. 3B). Accuracy of application is great, and even small predators such as ants are always contacted and repelled (Eisner and Meinwald, 1965).

In other insects, eversible glands operate on very much the same principle. Staphylinid beetles have a pair of protrusible sacs near the tip of their highly maneuverable abdomen (Jenkins, 1957, and references therein); when attacked, they bend the abdomen in the direction of the disturbance, and attempt to administer secretion on target.

2. Oozing Glands

The large millipede *Narceus gordanus*, which is abundant in certain regions of Florida, possesses a series of glands along each side of the body from which it discharges a secretion containing benzoquinones (Monro *et al.*, 1962). The secretion is emitted as a liquid ooze, which flows from the glands and spreads over the surface of the millipede (Fig. 3E). As a rule, the millipede discharges only from those glands closest to the region of its body subjected to disturbance. Only in response to persistent or generalized stimulation does it call all of its glands into action. The head and the segments immediately behind it lack glands; however, if the front end is disturbed, the millipede coils it ventrally, bringing the head to within the protective range of the first gland-bearing segments, which under these circumstances are the ones to discharge. Sometimes the entire body is coiled in a tight spiral, and this always occurs with the head toward the center. Quinone-secreting glands comparable with those of *Narceus* are found in other millipedes of the orders Spirobolida, Spirostreptida, and Julida (references in Eisner and Meinwald, 1966).

Oozing glands are also present in many other arthropods. An acetylenic acid, 8-*cis*-dihydromatricaria acid (Fig. 2) is contained in a white secretion (Fig. 3D) emitted from paired prothoracic and abdominal glands in the soldier beetle *Chauliognathus lecontei* (Meinwald *et al.*, 1968b).

The secretions are not always irretrievably lost after discharge. In the larva of the cottonwood leaf beetle *Chrysomela scripta*, a fluid contain-

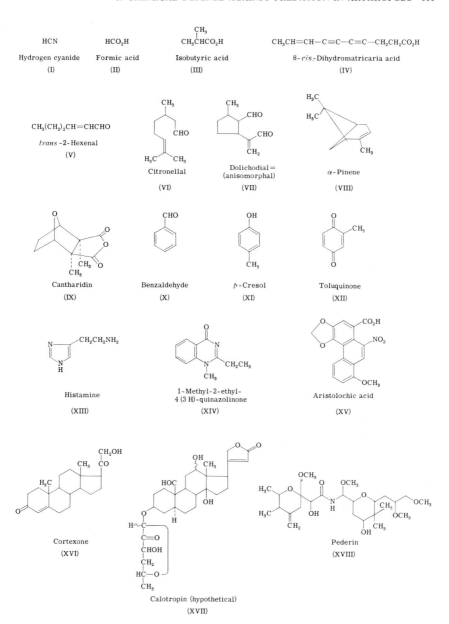

Fig. 2. Representative defensive substances of arthropods. I, II, III, V, VI, VII, VIII, IX, X, XI, XII (references in Eisner and Meinwald, 1966; Jacobson, 1966; Roth and Eisner, 1962; Weatherston, 1967); IV (Meinwald et al., 1968b); XIII (Bisset et al., 1960); XIV (Y. C. Meinwald et al., 1966); XV (von Euw et al., 1969); XVI (Schildknecht et al., 1966a); XVII (Reichstein, 1967); XVIII (Cardani et al., 1965; Matsumoto et al., 1968).

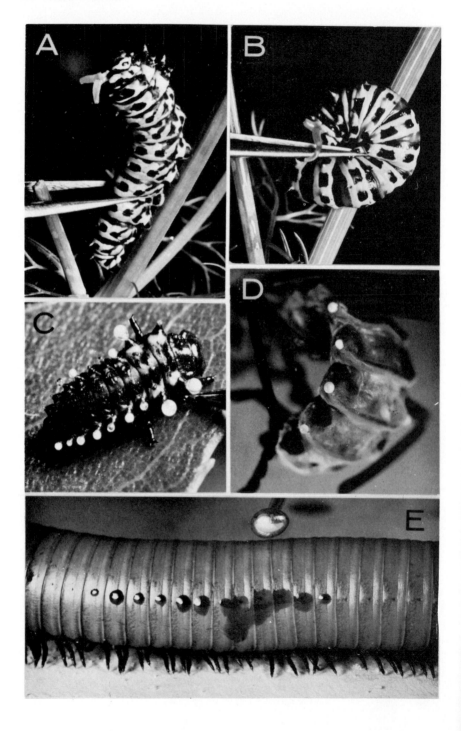

ing salicylaldelyde is emitted from a series of paired prominences on the larva's back (Fig. 3C). The droplets are discharged and "aired" only for the duration of an attack. When the disturbance subsides, much of the secretion is salvaged by being drawn back into the glands. The mechanism is similar to that in other chrysomelids (Garb, 1915).

Secretion is sometimes distributed with the help of the legs. If the front end of *Chauliognathus* is persistently held in forceps, the beetles transfer secretion from the abdominal glands onto the forceps with their hind legs. In adult Hemiptera of the genus *Rhopalus*, the secretion flows from the lateral thoracic gland orifices and collects in a pair of depressions on the body wall. Legs are dipped into the depressions, soaked in secretion, and wiped on the enemy target (Remold, 1962). Nymphs of *Lygaeus saxatilis* may similarly use the legs to transfer secretion from the dorsal abdominal glands (Remold, 1962).

3. Spraying Glands

Many arthropods possess the ability to eject their secretion as a spray, which they aim more or less precisely in different directions (Fig. 4). Some are infallible marksmen.

The carabid beetle, *Galerita janus*, sprays a secretion containing formic acid from two glands that open on the rear of the abdomen beside the anus. The beetle sprays from one gland or the other depending on which side of its body has been stimulated, and it controls the direction of the discharge by flexing the abdominal tip (Figs. 4A and B). Other carabids (e.g., *Chlaenius, Calosoma, Helluomorphoides*) have comparable spray mechanisms (Eisner *et al.*, 1963b,c, 1968), although the secretions in this family differ greatly in chemical composition (Moore and Wallbank, 1968; Schildknecht *et al.*, 1968d).

Aiming may be accomplished in a variety of ways. In Onychophora (Alexander, 1957), the glands of which open on or near the head, the spray is directed by appropriate rotation of the entire front of the animal (Fig. 14A). In the European earwig, *Forficula auricularia*, aiming involves flexion of the abdomen, which bears the glands at its base (Eisner, 1960). The whip scorpion *Mastigoproctus giganteus* (Figs. 4C and D) has glands that open on a veritable gun emplacement—a revolvable knob on the rear of the body. In Hemiptera (Remold, 1962) and certain cockroaches (Waterhouse and Wallbank, 1967), directionality may be controlled through postural adjustments of the body as a whole.

Fig. 3.   (A) Caterpillar of the swallowtail butterfly *Papilio machaon* everting its forked osmeterium in response to pinching with forceps. (B) Same, wiping the osmeterium against the forceps; note that the stimulus is applied on left side, and that it is the left branch of the osmeterium that is everted the furthest. (C) Larva of the chrysomelid beetle *Chrysomela scripta* discharging droplets of salicylaldehyde-containing secretion from its paired dorsolateral glands. (D) The cantharid beetle *Chauliognathus lecontei* emitting droplets of secretion containing 8- *cis*-dihydromatricaria acid from glands in its abdomen. (E) The spiroboloid millipede *Narceus gordanus*, responding to tapping by ejecting quinonoid secretion from its lateral glands.

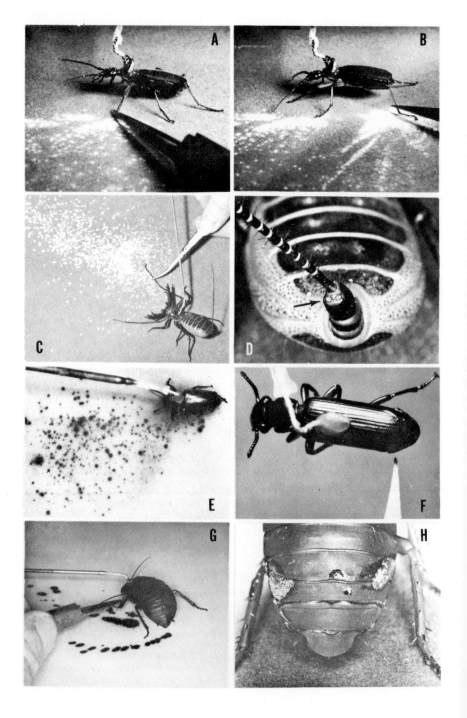

As a rule, arthropods discharge only in response to direct contact stimulation. The range of their spray, which may exceed several feet (Eisner *et al.*, 1961), is therefore indicative only of the force of the ejection and not of the distance over which approaching predators may be reached by a discharge. Actual measurements of ejection force have not been made. For certain Hemiptera, it has been estimated (Remold, 1962) that a pressure of 2500 dyn·cm^{-2} must be generated at the level of the gland orifice in order to expel secretion to a range of 20 cm.

Anisomorpha buprestoides, the large two-striped walking stick (Fig. 16A) from southeastern United States, is exceptional in that it sometimes sprays in response to stimuli generated at a distance. Although contact stimulation is the usual trigger for the discharge, the insects may spray approaching birds before these have actually pecked them (Fig. 8A). Whatever sensory input *Anisomorpha* relies upon in "recognizing" and "getting its bearings" on the approaching bird, it is clear that no crude combination of vibrational and visual cues is involved. Attempts to elicit discharges from the insects by waving objects in their vicinity or by tapping the substrate around them, or by doing both of these things simultaneously, almost always met with failure. A bird is evidently "betrayed" from a distance by peculiar characteristics of its own (Eisner, 1965a).

Secretion is sometimes sprayed from glands that evolved primarily as injection rather than ejection devices. The predaceous reduviid bug from Zanzibar, *Platymeris rhadamantus*, uses its saliva in the usual fashion by injecting it into insect prey, but it also sprays it directionally to a distance of several feet in response to predator attack (Edwards, 1960). The fluid, which has been studied pharmacologically (Edwards, 1961), is said to be similar in several respects to cobra venom, and to serve possibly for defense against monkeys. The European wasps *Vespa germanica* and *V. crabro*, sometimes eject as an aimed spray the venom that they ordinarily inject with the sting. Since the venom contains vola-

---

Fig. 4.   (A), (B) The carabid beetle *Galerita janus*, affixed to a rod and placed on a sheet of filter paper impregnated with alkaline phenolphthalein solution, discharging its aimed formic acid-containing spray in response to pinching of its left foreleg. (A) and left hindleg (B). (C) The whip scorpion *Mastigoproctus giganteus* discharging acid spray (acetic and caprylic acid) in the direction of a pinched appendage. The filter paper used as background is the same as in (A) and (B). (D) Close-up of rear of whip scorpion, showing the revolvable knob that bears the gland openings (arrow). (E) Spray pattern made on filter paper (acidulated KI-starch solution) by the quinonoid discharge of an ozaenine carabid whose left front leg was pinched. (F) Similar beetle, minutes after a discharge, emitting quinone vapors from residual secretion trapped beneath the elytra. A piece of indicator paper, the tip of which is darkening from exposure to the quinones, is being held beside one of the small elytral flanges from under which the vapors make their egress. The leaking vapors provide the beetle with lasting protection. (G) A South African cockroach, *Deropeltis* sp., discharging quinonoid secretion (on filter paper impregnated with KI-starch solution) in response to unilateral stimulation with forceps. (H) Same, showing residual secretion that remains on rear of abdomen after spray ejection. The vapors emanating from this residue may repel predators and prevent them from resuming their attack.

tile "alarm" substances, this behavior undoubtedly serves to alert other wasps to the presence of an enemy that has been topically "labeled" with the poison (Maschwitz, 1964). However, the venom may also have intrinsic deterrent potential, since it contains substances such as kinin and histamine that could be topically irritating to vertebrates, certainly if they should impinge on the eyes or other sensitive surfaces.

Although very little quantitative work has so far been done, it is clear that defensive glands differ considerably in storage capacity, rate of production of secretion, and other functional parameters. Judging from the size of the glands, which in some species may take up a fair share of the body cavity (Figs. 1B and 11A), it would appear that the possession of a high defensive potential is a matter of prime importance to the survival of these animals, and the manufacture of secretion a matter of high metabolic priority. A bombardier beetle, for example, can discharge over twenty times before depleting its resources, and within a day may again be able to eject several times (Eisner, 1958a). Other species cannot discharge as frequently, and replenish their secretion more slowly.

When an arthropod ejects its spray, some secretion inevitably contaminates its own body (Figs. 4G and H). The residual secretion acts as a deterrent to further attack and accounts for the period of partial or complete invulnerability that follows each discharge (Blum, 1961, 1964; Eisner, 1958a; Eisner et al., 1961, 1963b). This is of considerable importance to the animal, particularly as it relates to defense against such predators as ants which may need to be deterred en masse. In some arthropods there exist special adaptations for prolonging the effectiveness of the residual secretion. In adult Hemiptera, the gland openings are surrounded by a region of minutely and elaborately sculptured cuticle. This was originally claimed to serve as a barrier that prevents secretion from entering the insects' own spiracles (Remold, 1962), but is also thought to act as a physical sponge, in which some of the ejected secretion becomes trapped, and from which it evaporates at a retarded rate (Filshie and Waterhouse, 1968a,b). Ozaenine carabids, when they discharge (Fig. 4E), inevitably eject some secretion into the space beneath their folded elytra, and they even have a pair of small elytral flanges that serve to deflect secretion into the subelytral space. Vapors leaking from the trapped secretion protect the beetles for minutes after a discharge (Fig. 4F).

4. "Reactor" Glands

As typically constituted, arthropod defensive glands consist of membranous saclike invaginations of the body wall. The secretory cells are either a part of the wall of the sac itself, or they form a distinct cluster of cells, connected to the sac by way of a duct. The sac may be more or less capacious, and is ordinarily replete with stored secretion.

Certain arthropods have exceptional glands in which the stored products are not the final constituents of the secretion, but merely the chemical precursors thereof. The glands are so constructed that the precursors are mixed at the moment of discharge, with the result that the active principles of the secretion are generated in the ejected fluid. Such glands may appropriately be called "reactor" glands.

In the polydesmoid millipede *Apheloria corrugata* there are 22 glands, arranged in pairs on most body segments, with their openings visible as small pores on the margins of the notal lobes that project above the legs. Each gland consists of two compartments (Fig. 5A). The large saclike inner compartment (reservoir) contains an aqueous emulsion of mandelonitrile, the adduct of benzaldehyde and hydrogen cyanide. The smaller pear-shaped outer compartment (vestibule) contains an enzyme that promotes the dissociation of mandelonitrile. The two compartments are ordinarily sealed from one another by a springlike valve. When the millipede is under attack, it contracts the muscle that opens the valve and squeezes the contents of the reservoir through the vestibule to the outside. The emergent droplet, which now consists of emulsified man-

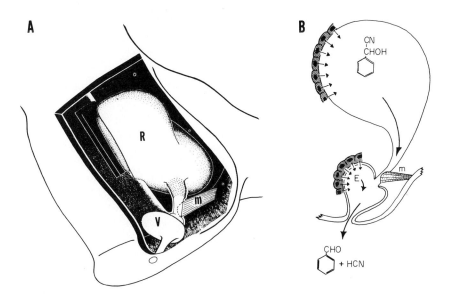

Fig. 5. Two-chambered "reactor" gland of the millipede *Apheloria corrugata*, shown in dissection (A) and in diagrammatic view (B). The inner compartment or reservoir (*R*) stores mandelonitrile, while the smaller vestibule (*V*) contains an enzyme that catalyzes the breakdown of mandelonitrile into hydrogen cyanide and benzaldehyde. A muscle (*m*) operates the valve between the two compartments.

delonitride plus enzyme, remains stuck to the body at the site of discharge. The products of dissociation, hydrogen cyanide and benzaldehyde, are liberated as vapors that enshroud the animal and serve as a protective screen (H. E. Eisner *et al.*, 1963; T. Eisner *et al.*, 1963a). By stimulating millipedes electrically in a closed system in which hydrogen cyanide is trapped for assay, it was shown that the generation of the poison is a gradual and continuing process, which may last for a half hour or longer. This is in line with the finding that a millipede is persistently invulnerable following a discharge. As much as 3.0 mg of mandelonitrile may be stored by a single *Apheloria*. This provides for a generation of about 0.6 mg hydrogen cyanide, which is several times the lethal dose for a mouse. The benzaldehyde also contributes to the effectiveness of the secretion (H. E. Eisner *et al.*, 1967; T. Eisner and Eisner, 1965). Like other aromatic aldehydes, it is powerfully repellent to certain predators (Eisner *et al.*, 1963c).

Reactor glands operating on the same principle as those of *Apheloria* are also present in certain carabid beetles, including the so-called "bombardiers" of the genus *Brachinus*, the members of the *Ozaenini*, and *Metrius* (Aneshansley *et al.*, 1969; Moore and Wallbank, 1968; Schildknecht and Holoubek, 1961). These animals generate benzoquinones by mixing hydroquinones and hydrogen peroxide from one glandular compartment with appropriate enzymes produced in an adjacent compartment. Because of the unusual nature of these glands, whose secretion is ejected at temperatures of up to 100°C, they are discussed separately below (Section III, B).

5. Tracheal Glands

There appear to be two primary mechanisms by which defensive glands expel their contents. One of them, muscular compression, simply involves the contraction of appropriate muscles that surround the storage reservoirs of the glands. This occurs, for example, in carabid beetles and the phasmid *Anisomorpha* (Eisner, 1965a; Eisner *et al.*, 1963b). The other mechanism involves compression of the glands through a rise in hemocele pressure, effected by some sort of contraction of the body as a whole (Eisner *et al.*, 1963b; Fishelson, 1960). This indirect type of compression has actually never been demonstrated, although its occurrence can be safely inferred from the fact that the glands involved lack intrinsic compressor muscles. The two mechanisms are not mutually exclusive; in the cockroach *Eurycotis floridana* the discharge is believed to be effected by a combination of the two (Stay, 1957).

A third and perhaps exceptional mechanism occurs in certain insects that rely on tracheal air pressure to discharge their glands. The glands of these animals have developed as specializations of the tracheal system itself. In the Pacific beetle cockroach *Diploptera punctata*, there are two glands, consisting of dilations of the tracheae leading inward from the second abdominal spiracles. The roaches spray from one or both spiracles depending on whether they have been stimulated unilaterally or bilaterally, and they supposedly effect the discharge by forcing air through the glands (Eisner, 1958b; Roth and Stay, 1958). The large flightless eastern lubber grasshopper *Romalea microptera* emits a froth from the mesothoracic spiracles that consists of a mixture of secretion and tracheal air (Fig. 6A). The secretion is produced by glandular tissue that surrounds the highly coiled trachea (Fig. 6B) associated with the spiracles. The only compound so far isolated from the complex secretion is an allenic sesquiterpenoid (Meinwald *et al.*, 1968a) of unknown defensive merits. Adults of the African grasshopper *Poekilocerus bufonius* also use tracheal air to produce a froth. However, in their case, the secretion, which stems from a gland on the back, runs down over the sides of the body to the second abdominal spiracle where it is mixed with exhaled air (Fishelson, 1960). The discharged fluid contains histamine and cardenolides (von Euw *et al.*, 1967; Reichstein, 1967).

B. Nonglandular Defenses

1. Blood and Other Systemic Factors

Many arthropods possess defensive principles in their blood, and they may have active control over the release of the fluid when they are under attack. In the Mexican bean beetle, *Epilachna varivestis*, blood oozes as droplets from the tibio-femoral joints of the legs. When a single leg is pinched, as an ant might do with its mandibles, a droplet is released from that particular leg (Fig. 6C). A localized stimulus applied to the body itself elicits a response from only the nearest leg. The leg may even be rotated in such a way that the blood-laden knee joint is brought closest to the point of stimulation. The fluid offers effective protection against ants, and these insects may have been the major selective agent that forced the evolution of the mechanism. *Epilachna* can withstand considerable hemorrhage; no noticeable effects result from a loss of one drop per leg (Happ and Eisner, 1961).

Reflex bleeding also occurs in beetles of the family Meloidae, again most commonly from the knee joints (Fig. 6D). The blood of meloids contains cantharidin (Spanish Fly) (Fig. 2, IX), a substance of well-known toxicity to vertebrates (references in Selander, 1960), which may also act as a feeding deterrent to some predaceous insects. As little as 10^{-5} M cantharidin may suffice to render an otherwise acceptable sugar solution unacceptable to ants (Carrel and Eisner, unpublished).

Some species do not hemorrhage spontaneously but nevertheless derive protection from blood that they inevitably emit when wounded. Lycid beetles, which are unacceptable to many predators supposedly because of distasteful factors in their blood, have a rubbery cuticle that is readily subject to injury. They are particularly prone to bleed from the

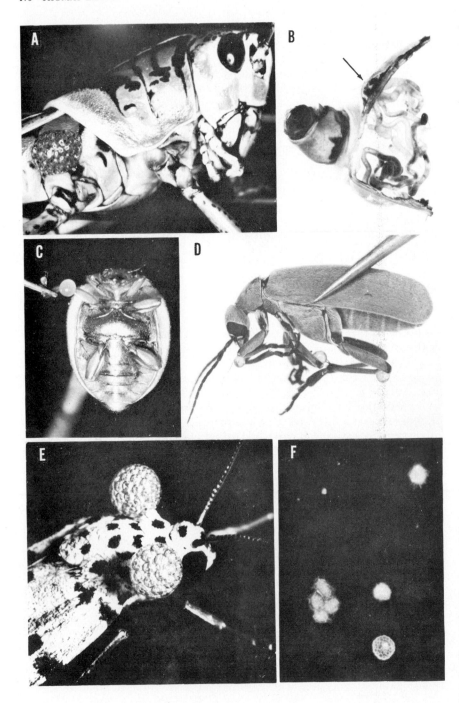

wings, the swollen veins of which are easily ruptured (Darlington, 1938; Linsley *et al.*, 1961). The larvae of *Epilachna* have a soft body beset with erect branched spines that are hollow, brittle, and easily broken when prodded by a predator. The emergent blood offers protection against some insects (Happ and Eisner, 1961).

Insects having protective factors in their blood sometimes have separate defensive glands, and the secretion may contain some of the same substances that are present in the blood. The grasshopper *Poekilocerus bufonius* produces a defensive secretion in an abdominal gland that contains histamine (as the dihydrochloride) and several cardenolides (= cardiac glycosides), including chiefly calactin and calotropin. The blood also contains these compounds, but histamine only in sharply reduced amounts (von Euw *et al.*, 1967; Reichstein, 1967).

Blood may sometimes be discharged together with the products of a gland. In the nymphs of *Poekilocerus*, the ejected defensive secretion is said to contain blood (von Euw *et al.*, 1967), and blood is also sometimes given off with the defensive fluid or froth emitted from the cervical glands of certain arctiid moths (Rothschild and Haskell, 1966) (Figs. 6E and F). The possibility that blood and secretion undergo some sort of chemical interaction as a result of being mixed has not been investigated. Nor has the mechanism been clarified whereby blood is supposedly ejected through a gland.

The relatively few toxic substances that have so far been identified from the blood and tissues of insects include some of the most complex nonproteinaceous poisons known from these animals. Sometimes, as in the case of pederin (Fig. 2, XVIII), a powerful poison extracted from the staphylinid beetle *Paederus fuscipes* (Pavan, 1963), the isolation and identification of the material has in itself been a matter of challenge (Cardani, *et al.*, 1965; Matsumoto *et al.*, 1968). Other substances, such as the cardenolides (Fig. 2, XVII) in certain grasshoppers and danaid butterflies (Reichstein, 1967), the senecio alkaloids in the moth *Callimorpha jacobaeae* (Aplin *et al.*, 1968), and aristolochic acid-I in the papilionid butterfly *Pachlioptera aristolochiae* (Fig. 2, XV) (von Euw *et. al.*, 1969) have been of special interest because they have been shown (or suspected) to be incorporated by the insects from the plants upon which they feed (see Section III, G, below). Some of the pharmacological agents that have been isolated are sometimes present in high concentrations. Histamine (Fig. 2, XIII), which occurs in some arctiid moths, is present

---

Fig. 6. (A) Front end of lubber grasshopper (*Romalea microptera*), showing froth being discharged from anterior thoracic spiracle. (B) Part of thoracic wall of *Romalea* (with base of leg still attached) showing spiracle (arrow), and the coiled glandular tracheae that produce the secretion. (C) Mexican bean beetle (*Epilachna varivestis*) emitting a droplet of blood from the knee of a leg that is being pinched in forceps. (D) Unidentified meloid beetle, reflex-bleeding from its knee joints while held in forceps. (E) Arctiid moth (*Utetheisa bella*) discharging from its cervical glands; the fluid is partly blood, as evidenced by the cells that it contains (F).

in the abdominal tissues of the cinnabar (*Callimorpha jacobaeae*) in the amount of 750 μg/gm (Bisset *et al.*, 1960). It is not unusual for a single species to possess more than one defensive substance in its body. Thus, *C. jacobaeae* contains senecio alkaloids in addition to histamine (Alpin *et al.*, 1968), while *Zygaena lenicerae* has histamine and cyanogenetic principles (Bisset *et al.*, 1960; Jones *et al.*, 1962). There is little doubt that much is yet to be learned about toxic factors of systemic distribution in arthropods. Even in cases studied thus far it is by no means always clear how the poisons are administered to the predator (whether through reflex bleeding, bleeding from injury, or through partial or complete ingestion of the animal) and how precisely the compounds effect their defensive action.

2. Enteric Discharges

Many arthropods, when handled or otherwise disturbed, regurgitate (Figs. 7A and B) or defecate. Although this is known to virtually anyone who has collected these animals in the field, the possibility that these reflexes are essentially defensive in character appears not to be generally accepted.

Best known among insects that regurgitate are grasshoppers. The amount of fluid that they emit may be considerable, and stems, certainly in large measure, from their capacious and usually laden crop. There is evidence that the regurgitate is toxic to mammals. It is a topical irritant to the eyes, may induce vomiting when swallowed, and may cause severe symptoms and even death when injected (Curasson, 1934; Freeman, 1968).

The regurgitate has been shown to be protective against some predators. Grasshoppers (*Brachystola magna*, *Romalea microptera*) that had been glued to tethers and placed beside natural ant colonies (*Pogonomyrmex* spp.) were promptly overrun by ants and induced to regurgitate. The fluid had an instantaneous dispersing effect on the assailants. Those that had been contaminated directly were the most obviously affected, and they became engaged in intensive and persistent cleansing activities. Experiments in which pieces of cut up grasshopper, some treated by addition of regurgitate, were placed in the trail of foraging ants showed that the fluid can render otherwise acceptable food unacceptable; only the untreated pieces were carried away by the ants (Eisner, Kafatos, and Shepherd, unpublished). Tests with captive jays (*Cyanocitta cristata*; *Aphelocoma ultramarina*) have shown that these animals have a rather elegant way of circumventing the regurgitative defenses of grasshoppers. When given *Romalea*, the birds almost invariably pecked at the head first, then pulled the head, together with the crop and its contents, from the body. They subsequently ate parts of the body only, leaving the crop and head behind [Fig. 7c (Eisner, unpublished)]. I have found a grasshopper head, with crop attached, in the

field, suggesting that even in nature predators may occasionally evis-
cerate grasshoppers in this fashion.

Enteric materials can even be put to defensive use in other than liquid
form. The larva of the chrysomelid beetle *Cassida rubiginosa* carries a
tight packet of cast skins and dried feces on a fork held over its back (Fig.
7H). The fork is highly maneuverable and the packet serves as a shield
that protects the larva against attack. No matter from where on its
flanks an attack is initiated, the larva responds by attempting to interpose the shield between itself and the offending agent (Fig. 7I).
Branched spines that project outward from the body of the larva act as
"sensors" that alert the larva to the probings of the predator. Ants are
probably among the larva's chief enemies, and the shield offers good protection against them. The shield is more than a mere mechanical device.
The presence of fresh wet feces near their site of deposition at the base
of the shield adds to the effectiveness of the weapon. Ants that make
casual contact with this pasty material when they first encounter a larva, or those that have the material smeared onto them when the shield is
mobilized against them, flee and clean themselves (T. Eisner *et al.*,
1967).

3. Detachable Outgrowths and Artificial Coverings

The integument of arthropods sometimes bears a more or less dense
covering of hairs or scales which, because they are detachable, can under some circumstances serve effectively for protection. In a sense, the
shedding of such dispensable solid outgrowths can be viewed as a form
of chemical defense.

In moths, scales offer protection against capture by orb-weaving spiders. Whereas smooth-bodied insects fly into webs and remain stuck,
moths simply lose some of their scales to the viscid threads and fly on.
The adhesiveness of scale-covered cuticle is two to six times lower than
that of similar but denuded cuticle. Other insects, such as caddis flies
(Trichoptera), which are covered by loose hair, and white flies
(Hemiptera: Aleurodidae), which bear a "waxy" powder, may be similarly protected against entrapment (Figs. 7D, E, and F) (Eisner *et al.*,
1964a).

Moths can also avoid capture by sundew plants. Tests in which flying
moths were lured at night to illuminated sundew plants placed for observation on the stage of a stereomicroscope showed that the moths merely
collide with the plants and fly on, leaving behind some of their scales
stuck to the secretory droplets on the stalked glands of the leaves (Fig.
7G) (Eisner, 1967; Eisner and Shepherd, 1966).

Thysanura, which have a slippery body covered with scales, also escape from sundew plants (Eisner and Shepherd, 1966). In addition, they
can elude ants. Their natural agility is an asset in itself, but if caught,
they simply slip away, and a "mouthful" of scales is all an ant is likely to

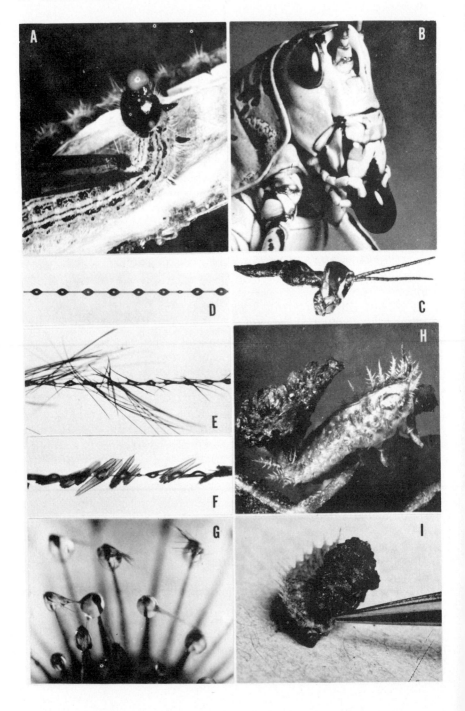

gain from the encounter (Eisner, 1965b). Scales may similarly protect certain lycaenid butterflies whose larvae inhabit ant nests and which as freshly emerged adults must cope with the occasional onslaught of their hosts while they make their exit from the nest. The butterflies are said to have an especially thick covering of scales (references in Hinton, 1951).

A most remarkable defense mechanism is possessed by certain beetle larvae of the family Dermestidae, which bear prominent tufts of segmented, spearheaded setae on their abdomen. When attacked, the larvae respond with directional striking movements of the abdomen, with the result that masses of setae are detached and adhere to the predator. The setae become randomly oriented and interlock, thus effectively repelling or hopelessly entangling small enemies such as ants and beetles (Figs. 8B and C). Toads, lizards, birds, and rodents are undeterred (Nutting and Spangler, 1969).

Protection may also be derived from detachable coverings of an artificial nature. Some larvae of the lace-wing family Chrysopidae carry on their back a loose packet of "trash," consisting of miscellaneous debris, including sometimes the sucked out remains of their insect prey. Although under some circumstances the packet may serve for camouflage, its primary function is undoubtedly defensive. Insects that attempt to bite a larva, find the packet moved like a shield in their direction, and may end up with no more than a load of trash in their mandibles (Fig. 8D). Momentarily distracted as they back away and rid themselves of their worthless load, they may lose track of their intended prey or abandon pursuit altogether. A mature larva may withstand the attack of over a dozen ants before its trash supply is depleted. Once denuded, it is vulnerable. The trash packet may also protect against reduviid bugs, the proboscis of which may not be long enough to reach the larva's body through the shield (Fig. 8E) (Eisner, unpublished).

III. CHEMISTRY AND EFFECTIVENESS OF DEFENSIVE SUBSTANCES

Much has been learned in the last few years about the chemical nature of the defensive substances. Dozens of compounds have been isolated and identified, and several reviews have appeared in which these are

Fig. 7. (A) Notodontid caterpillar (*Ichthyura* sp.) regurgitating in response to pinching with forceps. (B) Lubber grasshopper (*Romalea microptera*) regurgitating. (C) Head and crop of immature *Romalea*, pulled from the body of the grasshopper and discarded by an attacking jay. (D) Enlarged view of viscid thread from web of orb-weaving spider. Moths, when they strike a web, may merely lose some of their scales to the threads and fly on (F); caddis flies may lose some of their hairs (E). (G) Glandular hairs of a sundew plant (*Drosera capillaris*) showing droplets of secretion beset with scales torn from a moth that brushed against them. (H) Larva of the chrysomelid beetle *Cassida rubiginosa* with packet of dried feces and molted skins on its back. The packet is a maneuverable shield which the larva attempts to interpose between itself and any hostile agent (I).

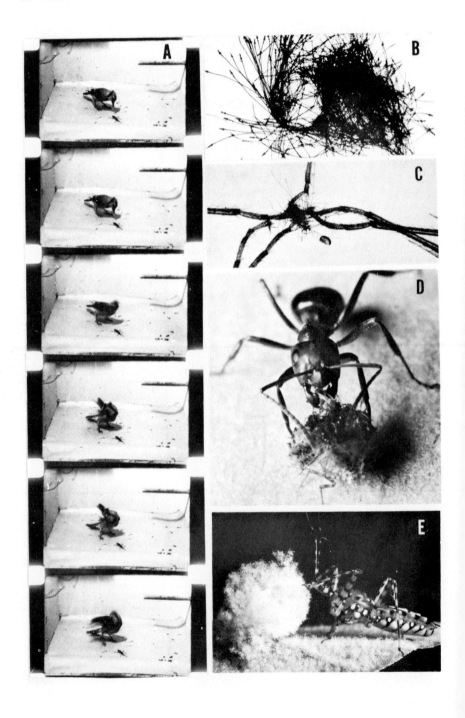

more or less comprehensively discussed and listed (Eisner and Mein-wald, 1966; Jacobson, 1966; Roth and Eisner, 1962; Schildknecht, *et al.*, 1964; Weatherston, 1967). Representatives of some of the principal types of compounds are shown in Fig. 2.

Efforts have also been made to appraise the effectiveness of the defenses. Since field observations of predator–prey encounters are virtually nonexistent, most of what we know is derived from laboratory experiments in which arthropods with chemical weapons were observed while under attack by a variety of their vertebrate and invertebrate enemies. Although it is clear from these tests that the defensive substances are indeed powerfully effective, it is also apparent that effectiveness may be attributable to a number of different causes. Sometimes the substances are distasteful or malodorous, but this need not be the prime basis of their mode of action. They may also act as topical irritants, or as systemic poisons, or – as in cases where they are sticky – as mechanical entangling agents. The effect may even be a thermal one, as in the case of secretions that are ejected at high temperatures. These various mechanisms of action are not mutually exclusive, and they may operate in combination. Some of the principal ones are discussed in the sections that follow.

A good deal of what has been learned about the chemical defenses of arthropods is of more than purely intrinsic interest and relevant to much of ecology and other fields of biology. The last sections of the chapter are designed to point up some aspects of this relevance.

A. Nonspecific Toxicants

The active principles of most defensive secretions are substances of low molecular weight, highly volatile, and as a rule strongly odorous. In fact, it is often the odor given off by the animal when it is collected in the field that provides the first clue to its possession of defensive glands. However, the secretions are not merely odorous in the conventional sense. Their vapors may also in some cases have an irritating effect, particularly on the eyes and other exposed surfaces of the face, and they may be painful upon inhalation. Contact with the actual droplets of se-

---

Fig. 8. (A) Excerpt from a motion picture sequence (downward, 18 frames/sec) showing a blue jay (*Cyanocitta cristata*) being sprayed by the walking stick *Anisomorpha buprestoides*. Note that the bird is hit (frames 2 to 3) before having contacted the insect. (B) Mass of hastisetae detached from larva of a dermestid beetle (*Trogoderma variabile*). The setae may entangle the legs of attacking ants (C). (D) Formicine ant in the process of attacking a trash-carrying chrysopid larva. A "mouthful" of trash is all the ant may gain from the encounter. (E) Unidentified chrysopid larva, hidden beneath the package it constructs from hairs removed from the sycamore leaves on which it lives. A reduviid bug is probing with its proboscis in an attempt (often vain) to reach the larva. [Figs. (B) and (C): courtesy of W. L. Nutting, University of Arizona, Tucson.]

cretion may cause more or less immediate itch or pain, certainly on mucous surfaces, but occasionally even on the skin itself. The substances responsible for these effects are for the most part compounds previously known to chemists from other sources. They include aliphatic acids (e.g., formic acid, caprylic acid, isobutyric acid, 8-*cis*-dihydromatricaria acid), aliphatic aldehydes (e.g., *n*-hexanal, *trans*-2-hexenal, *trans*-2-dodecenal), aromatic compounds (e.g., *m*-cresol, benzaldehyde, salicylaldehyde), quinones (e.g., benzoquinone, toluquinone, ethylquinone, methoxyquinone), terpenes (e.g., dolichodial, iridomyrmecin), and others (references in Weatherston, 1967). They are often present in remarkably high concentrations, and on occasion even in virtually pure form [e.g., *trans*-2-hexenal in the spray of the cockroach *Eurycotis floridana* (Roth, *et al.*, 1956)]. Sometimes the mixtures are, for all intents and purposes, natural imitations of histological fixatives, as for instance the spray of a whip scorpion, which contains 84% acetic acid (Eisner, *et al.*, 1961), and that of certain cockroaches, which contains 95% ethyl acrolein (Waterhouse and Wallbank, 1967). Such secretions are clearly poisons of a very general nature, capable of interfering with any number of cellular events, and they might appropriately be classed as nonspecific toxicants (Loomis, 1968).

The effect of such secretions on vertebrate predators is usually immediate. Startled by the secretory discharges of their intended prey, they desist from the attack, retreat, and then immediately clean themselves. Rodents rub their muzzle by wiping it with the paws or by plowing it in the soil (Eisner *et al.*, 1961, 1963b,c). Frogs and toads spit out the prey (Fig. 9B) and may then scratch their tongue or other affected regions with their feet (Eisner, 1960; Eisner *et al.*, 1959). Birds rub their head in the body plumage and wipe their eyeballs with the nyctitating membranes (Fig. 9C) (Eisner, 1965a; Eisner *et al.*, 1961, 1963c). Lizards scrape the sides of their mouth against the ground (Eisner *et al.*, 1961).

Fig. ˜ (A) Two cockroaches (*Diploptera punctata*) placed side by side within range of attack by ants (*Pogonomyrmex badius*). The one on left has had its glands removed and is under persistent attack; the other was attacked, but sprayed its quinonoid secretion (pattern is shown on acidulated KI-starch indicator paper) and repelled the assailants. (B) *Bufo* spitting out an individual of the cyanogenetic millipede *Apheloria corrugata*. (C) Bluejay (*Cyanocitta cristata*) jumping back after having been sprayed by the walking stick *Anisomorpha buprestoides*. The "veiled" appearance of the bird's eye is due to the nictitans which, already in action, is shown drawn across the eyeball. (D) The carabid beetle, *Calosoma prominens*, responding to the defensive chemical discharge of insect prey by dragging its mouthparts in the substrate. (E) Formicine ant, biting the rear end of a mealworm larva (*Tenebrio molitor*), while simultaneously flexing its gaster and discharging its acid secretion toward the site of the bite. (F) and (G) Two droplets, both consisting of aqueous acetic acid (84%), but one (G) containing also 5% caprylic acid, are placed on insect cuticle, and shown in a timed sequence of photographs. Whereas the first droplet merely shrinks away by evaporation, the second spreads widely and achieves a broad area of contact. The defensive spray of the whip scorpion, *Mastigoproctus giganteus* (Fig. 4C) has the constitution of the caprylic acid-containing drop.

Arthropod predators show similar reactions. Spiders clean their pedi-palps (Eisner, 1958a), while ants and carabid beetles (Fig. 9D) drag their mouthparts in the soil and cleanse their antennae with the front legs (Blum, 1961; Blum and Crain, 1961; Eisner, 1958a, 1960; Eisner and Eisner, 1965; Eisner et al., 1961). Bioassays have been developed which employ such cleansing reflexes as a basis for determining the "irritant effectiveness" of secretions and their components. Decapitated cock-roaches, for example, perform highly stereotyped scratch reflexes in re-sponse to topical application of these substances. The delay to onset of scratching following application of material is a useful measure of the effectiveness of the sample (Eisner et al., 1961, 1963b).

Bioassays of this sort have served to show that certain secretions are rendered especially effective by virtue of the fact that they are mixtures rather than single components. For example, the defensive secretion of certain carabid beetles (Helluomorphoides spp.) contains two compo-nents, of which one, formic acid, is a notable irritant. The other compo-nent, n-nonyl acetate, is innocuous in itself, but fulfills the important function of being a penetration-promoting agent. Without the ester, the secretion would not be nearly as effective in penetrating the skeletal shield of an arthropod or the skin of a mammal. Sprayed on human skin, an aqueous solution of formic acid may induce itch and pain, but the effect is usually not immediate, and if application is sparse may not be felt at all. If the mixture contains n-nonyl acetate, the reaction has ear-lier onset, is more intense, and may be more persistent (Eisner et al., 1968). Other secretions are comparably formulated. The spray of the whip scorpion Mastigoproctus giganteus contains 84% acetic acid, 11% water, and 5% caprylic acid. Despite its low concentration, caprylic acid effectively accelerates penetration, while at the same time acting as a wetting agent that promotes the spread of the spray droplets over the integument of the enemy target (Figs. 9F and G) (Eisner et al., 1961). "Additives" suspected of serving as spreading and penetration-promot-ing agents are also present in the secretions of certain Hemiptera and tenebrionid beetles (Blum and Crain, 1961; Gilby and Waterhouse, 1965; Remold, 1962). In Hemiptera, the presence of hydrocarbons is said to hasten the infiltration of the secretion into the respiratory trachea of in-sect predators (Remold, 1962).

Some arthropods resort to mechanical rather than chemical means to insure proper penetration of their defensive secretion. Formicine ants, which spray an aqueous solution of formic acid, usually (but not always; see Fig. 15F) eject their spray while simultaneously biting with their mandibles, and they direct the spray forward toward the site of the bite (Fig. 9E). When the enemy is an arthropod, the bite need not perforate the integument in order to be effective. All that is necessary is that the outermost layer of the cuticle, the epicuticle, be abraded. This wax-laden layer is the chief barrier to the penetration of the acid (Ghent, 1961).

Despite the broad effectiveness of these general toxicants, occasional predators are known which, for one reason or another, appear to be undeterred by them (Fig. 10B). Such predators need not be chemically insensitive to the secretions, but may simply minimize their exposure to the fluids by attacking the prey in specialized fashion. Grasshopper mice feed on *Eleodes* and other beetles with caudal glands by holding them head up while forcing their rear ends into the ground. The secretion is thus ineffectually discharged into the soil (Fig. 10A) (Eisner, 1966; Eisner *et al.*, 1963b). The mouse opossum *Marmosa demararae* can overcome the stick insect *Anisomorpha buprestoides*. It initiates its attack by grasping the insect in its jaws, and is invariably sprayed in return (Fig. 10C). It then scurries about in obvious discomfort while wiping its muzzle on the ground, but it continues to hold the insect with a front paw (Fig. 10D). Squeezed in this fashion, the insect continues spraying, but its discharges are now aimed at the relatively insensitive paw of the opossum and miss the sensitive eyes and snout. When the secretion is finally exhausted, the insect is eaten (Eisner, 1965a).

The duration of the effect of a discharge on a predator varies, and depends on the nature of the secretion, the type of predator, the extent to which the predator has been contaminated with secretion, the sensitivity of the sites affected, and other factors. As a rule, not enough secretion is administered to a predator to induce lasting, or at least noticeable lasting ill-effects. The cleansing reactions subside in a matter of seconds or minutes, and eventual recovery appears in most cases to be complete. Whether or not secretion penetrates the integument of the predator in amounts sufficient to induce tissue damage or other more or less persistent covert effects is a question that has generally been ignored. In the case of formicine ants, which may inject their acid spray into open wounds inflicted with their mandibles, the secretion may have lethal effects and is in fact used to kill prey. But severe toxic effects have otherwise been induced only under conditions that may not always be representative of those that prevail in nature. Thus, hemipteran secretion can cause paralysis when administered in high topical dosages to insects (Remold, 1962) and death may result from prolonged confinement of animals with the vapors of certain toxicants (e.g., lizards and ants with vapors of benzoquinones).

Very little is known about the sensory basis of irritant perception in animals, and it is therefore not clear how nonspecific toxicants exert the itch and pain that characteristically follow their application and on which their defensive effectiveness probably very largely depends. Quite possibly the receptors involved are the so-called "free" nerve endings which, acting in the manner of nonspecific chemoreceptors, mediate what has been called the "general" or "common" chemical sense (Parker, 1922). But the precise manner in which the toxicants effect these receptors remains a matter of speculation. They might act upon

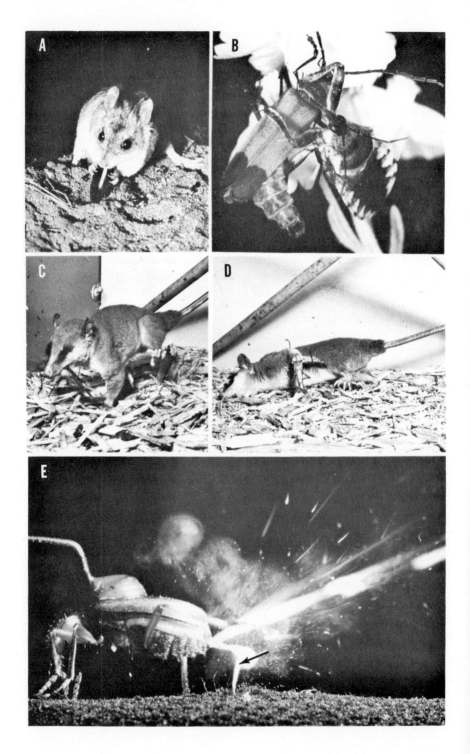

the receptors directly, or they might do so indirectly, through induction of biochemical changes in nonnervous cells and tissue fluids around the nerve endings (Keele and Armstrong, 1964). Either way, the sensitivity of the system is likely to be much the same in animals of very diverse kinds, which may be why the secretions are so generally effective against predators. It would be interesting to know whether those analgesics that abolish topical sensitivity to irritants in man also do so in other animals, including nonvertebrates. Besides affecting the general chemical sense, nonspecific toxicants probably act also via the ordinary senses of smell and taste. Defensive secretions are for the most part obnoxiously odorous and distasteful, certainly to man, but not only to him. It is perhaps significant in this connection that the vapors emanating from discharged secretion may be powerfully repellent to some predators, even from a distance (Blum, 1961; Eisner, 1958a; Eisner *et al.*, 1961, 1963b).

Nonspecific toxicants are always the products of glands and appear to be absent from nonglandular defensive fluids such as blood. This probably reflects the fact that these substances cannot be tolerated systemically at higher concentrations even by the animals that produce them. In certain Hemiptera, for example, it has been shown that death follows quickly if some of their secretion is injected into them (Remold, 1962). But if arthropods are sensitive to their own secretion, how do they withstand their own discharges? And how are the secretions stored in the glands, or manufactured by the gland cells, without the animals poisoning themselves?

It is possible that the possession of an especially impervious integument protects some arthropods against the topical effects of their discharged secretions. Remold (1962) has shown that the cuticle of Hemiptera is generally impermeable to hemipteran secretion (or to an artificial mixture chemically resembling it). Moreover, abrasion of the surface of the cuticle renders it permeable, suggesting that impenetrability is normally attributable to the epicuticle or to a component thereof. The only Hemiptera found to have permeable cuticle were species that also lack defensive glands or have reduced glands of unknown function, as well as freshly molted individuals whose epicuticle had not yet fully formed. Some arthropods may only be partially insensitive to the topical action of their secretion. The carabid beetle, *Calosoma prominens*,

Fig. 10. (A) Grasshopper mouse (*Onychomys torridus*) eating the tenebrionid beetle *Eleodes longicollis* (Fig. 1A); the beetle is held in such a way that its secretion is ejected into the soil. (B) Reduviid bug eating the cantharid beetle *Chauliognathus lecontei* (Fig. 3D); this predator is undeterred by the beetle's secretion. (C) and (D) Two consecutive stages in the attack of a mouse opossum (*Marmosa demararae*) on a walking stick (*Anisomorpha buprestoides*); in (C) the opossum has been sprayed in the eye after the initial seizure of the insect; in (D) it is running about wiping itself, but without releasing the prey, which is eventually eaten. (E) Bombardier beetle, discharging its hot secretion. The photograph was taken with an electronic flash unit, triggered by a thermocouple (arrow) placed in the path of the spray. (The beetle has been glued to a wire to prevent its escape.)

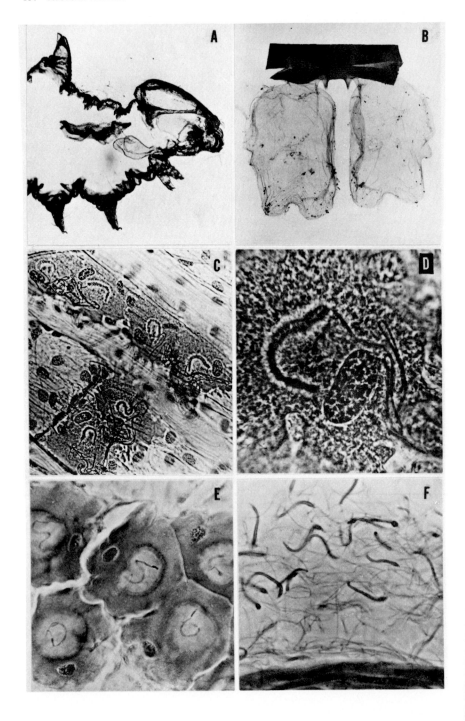

which sprays a secretion containing salicylaldehyde, shows no effect from application of the aldehyde to its body and legs, but cleans itself promptly if application is made to the mouth. However, the mouth of the beetle is ordinarily not accessible to the spray, since ejection occurs from the rear of the abdomen. Nevertheless, beetles might still be sprayed on the mouth when attacking one another, and it may be that oral sensitivity in *Calosoma* is an adaptive safeguard against cannibalism (Eisner *et al.*, 1963c).

In the glands themselves the secretions are stored harmlessly, out of contact with living tissue. Being invaginations of the body wall, the glands are lined internally with a cuticular membrane (Figs. 11A and B), and it is presumably this membrane that provides the insulation necessary to prevent the toxicants from seeping into the body cavity. Actual studies of the permeability of these membranes have not been made.

A hypothesis has recently been advanced, designed to explain how the gland cells might produce poisons without poisoning themselves. The hypothesis takes into account a basic characteristic of many arthropod gland cells, namely their possession of certain more or less elaborate cuticular chambers and ducts (Fig. 11, C–F), and postulates that the synthesis of toxicants takes place within the lumen of these cuticular organelles rather than in the cytoplasm of the living gland cells associated with them (Eisner *et al.*, 1964b; Eisner and Meinwald, 1966). Support for this hypothesis has now been provided by the work of Happ (1968), who by means of chromatographic and histochemical techniques investigated the mechanism of quinone production in the defensive glands of certain tenebrionid beetles. Previous extensive work on these and other quinone-producing glands (Brunet and Kent, 1955; Eisner *et al.*, 1964b; Hurst *et al.*, 1964; Kent and Brunet, 1959; Roth and Stay, 1958) had led to the supposition that the overall pathway for quinone formation proceeds from phenol glucoside, to free diphenol, and then to quinone. Happ gives evidence for the presence in various parts of the glandular tissue of all the essential enzymes and substrates that such a pathway would presuppose. Most important is the finding that the final step in the sequence, the formation of the toxic quinones themselves, appears to occur inside the cuticular organelles, and therefore presumably in isolation, or relative isolation, from the living cells. Whether the hypothesis is of broad or

Fig. 11.   (A) Front end of notodontid caterpillar (*Schizura leptinoides*), showing defensive gland projecting inward from ventral neck region. The preparation has been treated with KOH, and the gland consists only of its cuticular lining. (B) KOH-treated abdominal defensive glands of a cockroach (*Deropeltis* sp.; Fig. 4G); only the membranous cuticular lining remains of each sac. (C) Quinone-secreting glandular epithelium of *Deropeltis*. Note the lengthy cuticular ducts leading away from the enlarged terminal organelles within which the poisons are presumably synthesized. One of these intracellular "reaction chambers" is shown (beside a nucleus) in (D). (E) Secretory cells associated with the defensive glands of *Eleodes longicollis* (Fig. 1A). Note the large vesicles with their tubular cuticular organelles. (F) Cuticular organelles and ducts from the quinone-secreting defensive glands of the cockroach *Diploptera punctata* (Fig. 9A), isolated by treatment with KOH.

limited applicability remains to be seen. It cannot apply to all defensive glands, since there is at least one type—the osmeterium of papilionid caterpillars—which in some species produces general toxicants but lacks cuticular organelles (Crossley and Waterhouse, 1969).

B. Hot Secretions

Over 100 years ago, the claim was made that certain South American representatives of the widely distributed beetle genus *Brachinus* (Coleoptera: Carabidae), "on being seized . . . immediately . . . play off their artillery, burning . . . the flesh to such a degree, that only few specimens (can) be captured with the naked hand" (Westwood, 1839). The species of *Brachinus* have long been a favorite of naturalists, to whom they are known as "bombardier beetles." Their "artillery" is a defensive spray, ejected from a pair of glands that open at the tip of the abdomen. Proof has now been obtained that the secretion is indeed hot when ejected (Aneshansley *et al.*, 1969).

The glands of *Brachinus* are essentially reactor glands (see Section II, A, 4 above), which in their general structure and mode of operation resemble the cyanogenetic glands of the millipede *Apheloria* (Fig. 5). Each gland is a two-compartmented apparatus (Fig. 12). The inner compartment (reservoir) contains an aqueous solution of hydroquinones (hydroquinone and methylhydroquinone) and hydrogen peroxide, while the outer compartment (vestibule) contains a mixture of catalases and peroxidases (Schildknecht and Holoubek, 1961; Schildknecht *et al.*, 1968b). In order to effect a discharge, the beetle squeezes some reservoir fluid into the vestibule, thereby triggering what is essentially an instantaneous and explosive set of events: the catalases promote the decomposition of hydrogen peroxide, while the peroxidases force the oxidation of the hydroquinones to their respective quinones. Under pressure of the free oxygen, the mixture "pops" out (Fig. 12C).

Given the known concentration of the reactants (hydrogen peroxide = 25%; the two hydroquinones = 10%), as well as the thermodynamic properties of the reaction they undergo, it was predicted that the secretion should be ejected with a heat content of about 0.19 cal/mg, an amount sufficient to bring the spray to the boiling point (100°C) and to vaporize about one-fifth of it. Actual measurements made with a microcalorimeter showed the heat content to be about 0.22 cal/mg (Fig. 12B). Temperature, measured with appropriate thermocouples and thermistors, was consistently recorded at 100°C. Photographs taken of the spray showed it to consist of droplets as well as vapor (Fig. 10E) (Aneshansley *et al.*, 1969).

Brachinus can eject its spray in virtually all directions (Eisner, 1958a), and it invariably aims the discharge in such a way as to hit an attacker "full blast" (Fig. 13). There can be no doubt that the thermal properties of the spray contribute to its effectiveness. Quinones are repel-

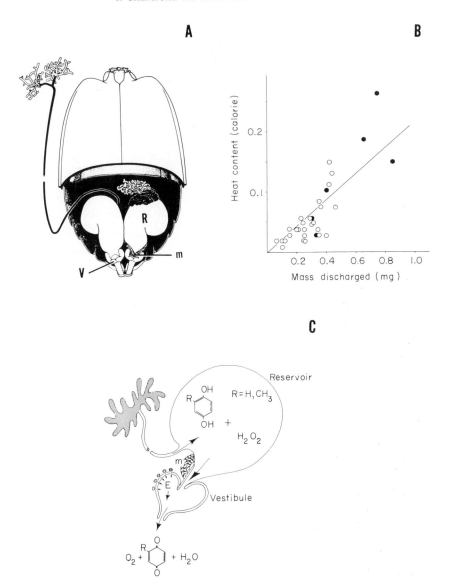

Fig. 12. Glandular defense mechanism of bombardier beetle (*Brachinus* sp.) (A) Abdomen, showing glands in place. Each gland is a two-chambered organ, consisting of an inner reservoir (*R*) and a small outer vestibule (*V*). A muscle (*m*; removed from left gland) operates the valve between the two compartments. The secretion in the reservoir is produced by a glandular tissue drained by a highly coiled duct (the duct has been unraveled in the left gland). (B) Heat content of spray, plotted as a function of the mass discharged (open circles = single discharges; closed circles = double or triple discharges). The secretion is discharged at 100°C (after Aneshansley *et al.*, 1969). (C) Diagram of the chemical basis of the discharge mechanism. Details in text, Section III, B (after Schildknecht and Holoubek, 1961).

Fig. 13. Consecutive stages in the attack of a bombardier beetle by an ant. After approaching (A), and poising itself for the attack (B), the ant bites the beetle's left rear leg, and is accurately sprayed in return (C).

lent even when cold, but their immediate irritant effect is usually re-stricted (at least in man) to the more permeable surfaces of the integu-ment, such as the mucous surfaces. By being hot, the beetle can make itself felt thermally, even where its chemical "message" cannot get through.

Remarkable as it is, the defense mechanism of *Brachinus* is not unique. Comparable reactor glands, also producing hot quinones by oxi-dation of hydroquinones, are found in certain other carabids (members of the tribes Ozaeniinae and Metriinae) not directly related to *Brachinus* (Aneshansley *et al.*, 1969; Eisner and Chalmers, unpublished).

C. Sticky, Slimy, Supercooled, and Resinous Fluids

Defensive secretions may take their effect by other than purely chemi-cal means. The discharged materials are sometimes sticky or slimy, and as a result may serve to entangle, restrain, or otherwise hinder the preda-tor mechanically.

Species of the primitive arthropod-related phylum Onychophora (including *Peripatus* and its relatives) spray a viscous fluid (Fig. 14A) from a pair of voluminous glands that open on the head (Cuénot, 1949). The odorless fluid, which may be ejected to a distance of a foot or more, hardens to a rubbery consistency on exposure to air (Fig. 14B) (Alexander, 1957; Lawrence, 1950; Manton and Heatley, 1937). It pro-vides effective defense against centipedes (Alexander, 1957), as well as against spiders and ants (Eisner and Carrel, unpublished). The dis-charge is aimed with precision, and even single ants, biting the onycho-phoran anywhere on its body, may be hit and "pinned down" by secre-tion. It may take minutes for the ants to free themselves. Some never do, and these eventually die (Fig. 14C) (Eisner and Carrel, unpublished). Comparable defense mechanisms occur in some lithobiid centipedes, which discharge sticky threads from their posterior legs to entangle ants, lycosid spiders, and other potential enemies (Verhoeff, 1925), and in cer-tain scytodid spiders, whose sticky cheliceral discharge is also used against arthropods (McAlister, 1960). Some of these secretions may also serve offensively for prey capture. This is said to be the case in Onycho-phora (Alexander, 1957), and is certainly true for scytodid spiders (Bristowe, 1958), which may only secondarily employ the spray for de-fense.

Sticky secretions need not always be sprayed, but may be present as a persistent coating on part of the body. Certain cockroaches possess a layer of slime on the dorsal rear of the abdomen (Roth and Stahl, 1956), and this has been shown to offer protection against carabid beetles, ants, and centipedes (Nayler, 1964). One might mention, incidentally, that the slimy integumental coating of earthworms, slugs, and land planarians may be one of the chief adaptive assets that enables these soft-bodied and relatively helpless animals to survive in coexistence with the many

arthropod predators that share their habitat. The slime gums up the mouthparts, antennae, and legs of arthropods, and acts as an effective feeding deterrent (Fig. 14E) (Eisner and Carrel, unpublished).

Relatively little work has been done on the chemistry of most sticky secretions. Some, as one might expect, are proteinaceous in nature (Manton and Heatley, 1937; Roth and Stahl, 1956), but the possibility remains that they are heterogeneous mixtures containing additional factors of secondary function. Such a possibility would not be without precedent. The sticky proteinaceous droplets discharged from mid-dorsal glandular pores in the millipede *Glomeris marginata* (Fig. 14D) contain two quinazolinones chemically related to certain plant alkaloids (Fig. 2, XIV) (Y. C. Meinwald *et al.*, 1966; Schildknecht *et al.*, 1967c). These compounds are bitter and toxic (Schildknecht *et al.*, 1966b, 1967c), and thereby contribute to the overall effectiveness of the secretion (see Section III, D, below).

A remarkable mechanism has recently been described by Edwards (1966), involving the droplets of fluid that ooze from the tips of the two peglike processes (cornicles) on the abdomen of aphids (Fig. 14F). The fluid apparently consists of droplets of wax in an aqueous vehicle. The wax is liquid when discharged, but solidifies promptly on contact with extraneous surfaces. It is suggested that the wax is initially in a stable liquid crystalline state (i.e. supercooled) and changes to the solid crystal phase on contact with a seeding nucleus. The melting point of the wax (37.5°-48°C depending on species) is higher than normal summer temperatures, indicating that crystallization by seeding can occur under natural conditions. The discharge of cornicle wax occurs when aphids are disturbed, and there is little doubt that the mechanism is defensive. Edwards himself reports finding aphids in the field with the shriveled bodies of hymenopteran parasitoids fixed to their back by plaques of cornicle wax.

Sticky fluids also serve for defense in termite soldiers. In the primitive species *Mastotermes darwiniensis*, a highly odorous material, rich in *p*-benzoquinone (and with traces of toluquinone), is discharged from the buccal cavity. The liquid is initially mobile, but soon sets to a dark, rubberlike material that supposedly serves to immobilize the foe. Hardening

---

Fig. 14. (A) Unidentified onychophoran from Panama, discharging its slimy secretion in response to mild pinching with forceps. The secretion assumes the consistency of a rubbery glue on exposure to air, and it may have a powerful sticking action, even on human fingers (B). (C) Dead formicine ant, pinned down on previous day by the sticky discharge of an onychophoran. (D) An oniscomorph millipede (*Glomeris marginata*), coiled into a tight sphere in response to disturbance, has discharged three droplets of secretion from glands opening on the dorsal midline; the sticky secretion can be drawn into fine threads, as is here being done with the tip of a needle (see also Fig. 15, C–E). (E) An Australian bull ant (*Myrmecia* sp.), its mouthparts gummed up with slime, backing away from a slug it has just bitten. (F) An aphid, seen in rear-end view, with spherical droplets freshly discharged from its two cornicles. The left droplet is still liquid; the one on right has solidified (see text, Section III, C).

may be a result of the tanning action of the quinones, which may combine with proteins from the saliva (Moore, 1968). The precise source of the quinones, which because of their powerful repellent action (Eisner and Meinwald, 1966) undoubtedly contribute significantly to the effectiveness of the secretion, remains unknown.

In other termite soldiers the defenses are different. In *Coptotermes lacteus* (Rhinotermitidae) a glandular material is emitted from a cephalic gland consisting of a milklike fluid that soon dries to form a colorless resilient film. The fluid comprises a suspension of lipids in an aqueous mucopolysaccharide vehicle (Moore, 1968). In the family Termitidae, the soldiers produce cephalic secretions containing terpenes (Fig. 2, VIII) (e.g., α-pinene, β-pinene, limonene, terpineole, etc.), and they sometimes eject these in a sticky resinous mixture (Moore, 1968). The nature of the resin remains unknown. In *Nasutitermes*, the soldiers have a pointed cephalic nozzle, from which secretion is ejected as a fine thread that hardens rapidly in air. The material quickly incapacitates and sometimes even kills insects. Although its action is primarily mechanical and attributable to the resinous base (Ernst, 1959), it is not known to what extent the terpenes themselves convey deterrent or toxic properties upon the secretion. The terpenes do contribute indirectly to effectiveness, by acting as alarm pheromones that induce other termites to converge upon a site where chemical combat is in progress (see Section III, E, below).

Blood, by virtue of the fact that it clots on exposure to air, may also have a mechanical defensive effect. In the Mexican bean beetle (*Epilachna varivestis*), which reflex bleeds in the adult, and bleeds from ruptured spines in the larva, the clotting droplets gum up the mouthparts and appendages of attacking ants. Ants may even become stuck in groups, if they contact one another after having been contaminated with blood (Happ and Eisner, 1961).

Sticky and other mechanically acting defensive fluids are probably of significance primarily as they are used against arthropod predators. Vertebrates are probably unaffected, except to the extent that they may be sensitive to such repellent or toxic compounds as may also be present in the materials.

D. Poisons of Delayed Effect (Emetics, Vesicants, and Others)

Among the more intriguing defensive substances produced by arthropods are those that may be classed as truly poisonous in the conventional sense of the term. Broadly speaking, these substances differ from the general toxicants discussed earlier in that they may be active in relatively low concentrations, and in that their primary action may involve a more or less delayed and usually systemic effect, rather than a topical effect of immediate onset. They also differ from general toxicants in that they are frequently tolerated systemically by the organisms that produce them, and may therefore be present in general or more or less general distribution through blood and tissues, rather than only within the insulated confines of special integumental glands.

Some of the poisons recently isolated from arthropods are similar chemically to drugs long known from medicinal plants. Cardiac glycosides (= cardenolides) have been identified from body extracts of the familiar danaine butterfly known as the monarch (*Danaus plexippus*). The four principal glycosides in the butterfly are calotropagenin, calotoxin, calotropin (Fig. 2, XVII), and calactin (Reichstein *et al.*, 1968). Like other cardiac glycosides, these compounds are potentially capable of inducing a number of systemic effects, but their defensive action is probably attributable chiefly to their emetic properties. Birds are among the principal enemies of the monarch, and both starlings and jays have been shown to vomit within several minutes after ingestion of monarchs or their extracts (Figs. 15A, and B) (Brower *et al.*, 1968; Parsons, 1965; Reichstein *et al.*, 1968). Cardiac glycosides are also produced by grasshoppers of the genus *Poekilocerus* and *Phymatus* (von Euw *et al.*, 1967; Fishelson, 1960; Reichstein, 1967).

Steroids, but of other types, are also produced by aquatic beetles of the family Dytiscidae. These insects have a pair of glands in the neck region from which they discharge a fluid containing, in several cases, C_{21} corticosteroids (Schildknecht and Hotz, 1967; Schildknecht *et al.*, 1966a, 1967b,d), and in one case, the C_{17} androstene, testosterone (Schildknecht *et al.*, 1967a). The amount of steroid produced per beetle may be extraordinarily high, certainly as compared with the quantities of these substances normally produced as hormones by vertebrates. Thus, as much as 0.4 mg deoxycorticosterone (= cortexone) (Fig. 2, XVI) may be stored by a single *Dytiscus marginalis* (Schildknecht *et al.*, 1966a). Blunck (1917), in a pioneer paper dealing with extensive experimentation with the secretion of this and related dytiscids, shows the material to be primarily toxic to fish and Amphibia, the most likely vertebrate enemies of the beetles. The secretion has a gradual, protracted, and usually reversible narcotizing effect, such as is said to be elicitable also with some of the steroidal principles themselves (Schildknecht *et al.*, 1966a, 1967b). Ingestion of live beetles by frogs and toads has been said to result in their being regurgitated (Blunck, 1917), sometimes live and covered with bloody slime (Schildknecht *et al.*, 1967a). However, this need not be attributable to the steroids alone, since dytiscids have powerful mandibles and in addition have other exocrine glands, such as the pygidial glands (Casper, 1913) which, despite the unconvincing claim that they are antimicrobial (Maschwitz, 1967; Schildknecht and Weis, 1962), may well also be protective against predators.

The production by dytiscid beetles of defensive substances such as these, which are essentially replicates of vertebrate steroid hormones, brings to mind the similar, although not strictly comparable case of the

use of insect hormones by plants. Both ecdysonelike and juvenile hormonelike substances are known to be produced by a diversity of plants, which presumably employ them as means for interfering with the development of their insect enemies (e.g., Bowers *et al.*, 1966; Nakanishi *et al.*, 1966; Sláma and Williams, 1966; Williams, 1967).

Arthropods have also been found to contain substances belonging to certain categories of poisonous plant alkaloids. The moth *Callimorpha jacobaeae* contains senecio alkaloids (Aplin *et al.*, 1968). The insect is unacceptable to a wide range of predators, but this may not be attributable to the alkaloids alone, since the animal also possesses other chemical defenses. The European millipede *Glomeris marginata* produces two closely related quinazolinones (Fig. 2, XIV) [1-methyl-4(3H)-quinazolinone and 1,2-dimethyl-4(3H)-quinazolinone] as part of a sticky proteinaceous secretion discharged from a row of dorsal glands (Y. C. Meinwald *et al.*, 1966; Schildknecht *et al.*, 1967c). Ingestion of *Glomeris* is said to cause delayed general symptoms and sometimes even death in mice, and behavioral effects in birds (Schildknecht *et al.*, 1967c). We have found the quinazolinones to have a dramatic effect on lycosid spiders. For minutes or even hours after an attack (Fig. 15C) upon a *Glomeris* (which the *Glomeris* may not survive), the spider may present a fully normal appearance. However, it may then gradually develop symptoms of motor impairment, and eventually may become totally motionless. Placed on its back, it will thus remain without righting itself (Fig. 15D). Recovery may not occur until hours later. In nature, such prolonged paralysis is likely to have fatal consequences. The spider might itself fall victim to predation (Fig. 15E), or it may prove unable to withstand the heat of day outside its usual diurnal shelter.

Two other poisons of considerable interest are cantharidin and pederin (Fig. 2, IX and XVIII). The former, long known as Spanish Fly, and traditionally (but incorrectly) reported to have aphrodisiac properties, is the toxic principle in the blood and tissues of beetles of the family Meloidae. The latter has only recently been identified (Cardani *et al.*, 1965; Matsumoto *et al.*, 1968), and stems from certain staphylinid beetles of the genus *Paederus*. Both substances have vesicating properties when applied

---

Fig. 15. (A) A bluejay (*Cyanocitta cristata*) eating a monarch butterfly. Minutes later, it vomits (B). (C)-(E) Fate of a lycosid spider that attacked and partly devoured (C) an oniscomorph millipede (*Glomeris marginata*). The quinazolinones in the secretion of the millipede have a protracted immobilizing effect on the spider (D), which is rendered defenseless when subsequently attacked by ants (E). (F) Mound of the formicine ant *Formica rufa,* being tapped by hand, showing the collective response of hundreds of worker ants, which squirt their acid spray into the air. (G) Nasute termite soldiers (*Rhynchotermes perarmatus* from Panama), lined up as guards along the sides of the foraging column shown in rear. When provoked, the soldiers eject sticky odorous secretion from their pointed cephalic "nozzles." [Photos (A) and (B): courtesy of Lincoln P. Brower, Amherst College; Photo (F): taken in collaboration with Mario Pavan, Universita di Pavia.]

to human skin, and both are capable of inducing severe systemic effects when ingested. Cantharidin causes gastroenteritis, as well as marked irritation of the urogenital tract in the course of its excretion. Vertebrates vary in their sensitivity to cantharidin, man being among the more sensitive (a lethal dose of 0.5 mg/kg has been reported) (Kaiser and Michl, 1958). The amounts of cantharidin present in meloids (0.2 to 2.3% of body weight) (Kaiser and Michl, 1958) should suffice for even single ingested beetles to cause toxic effects when swallowed by some vertebrates. Pederin is a powerful cytotoxin, capable of inhibiting growth of cultured cells at concentrations of the order of 1.5 ng/ml (Brega *et al.*, 1968). Individual beetles contain about 1 μg of material (Pavan, 1963). The precise way in which cantharidin and pederin affect predators in nature is not understood. Cantharidin need not necessarily act as a true poison in all cases. As indicated earlier (see Section II, B, 1, above), there is evidence that insect predators (ants, carabid beetles) may discriminate against this substance on the basis of taste alone (Carrel and Eisner, unpublished).

There are a number of ecological and evolutionary questions that come to mind regarding the mode of action of poisons of this general sort. Given the fact that their principal noxious effects may be of delayed rather than immediate onset, one wonders how accurately a predator "reasons in retrospect" to associate the ill-effects with the particular type of animal eaten that brought them about. Only if such an association is made can the predator be expected to learn to discriminate against the causative agent. In the case of poisons that have an emetic effect, it is conceivable that the predator associates the taste signals present in the food as it is regurgitated with the noxious experience of the emesis syndrome. Conditioned in this way, it may subsequently discriminate against prey on the basis of its flavor when first caught, and through further association may then recognize the prey by appearance and discriminate against it on the basis of sight alone. Such a mechanism has been suggested for the interaction of birds and cardenolide-containing insects such as the monarch butterfly (Brower, 1969).

But might not a poison be adaptively justified even if the affected predators never learn to identify and ignore the carrier? Might not the poison have a chronic debilitating effect on the predators of an area, or result in their impaired fecundity (as might well be the case after ingestion of cantharidin), and might not this be of benefit to the species that produces the poison? If the poison is lethal, and if the predator evolves no counter-measures, one could even imagine the eventual transformation of a predator into a species from which the tendency to capture the lethal prey has been selected out altogether. The extent to which selective forces of this sort might have been instrumental in forging the evolutionary restriction of food habits is by no means entirely clear. Similar questions can be raised about herbivores, since plants produce comparable poisons.

One wonders about the evolution of the poisons themselves. If they have delayed action, and if in order to exert their effect they must be ingested with the carrier organism, then a mechanism for their evolution must be postulated which justifies the sacrifice of the ingested individuals in adaptive terms. One possibility is that the sacrifice involves true "altruism," and that the surviving individuals that profit from the sacrifice are the close kin of those sacrificed. It would be interesting to determine whether the dense and often highly localized and relatively static aggregations that characterize some poisonous arthropods, including for example many meloid beetles, are in fact assemblages of relatively uniform genetic constitution.

E. Collective, Parental, and Intraspecific Defenses

In social insects, defense is often achieved through group action. In the mound-building formicine ant, *Formica rufa*, a tapping of the nest immediately induces hundreds of workers to eject their acid spray toward the source of the disturbance (Fig. 15F). Similarly, the soldiers of nasute termites instantly converge upon any instrument used to prod open their nest, discharging upon it and coating it with their resinous and odorous secretion. Some nasute termites make daylight forays above ground in which the columns of workers are flanked on each side by rows of soldiers. Oriented with their glandular *nasus* pointed outward, the soldiers offer effective guard against the trespass of arthropod enemies (Fig. 15G).

Group defense in social insects may be regulated to a greater or lesser extent through the action of alarm pheromones. By definition, these chemical messengers serve to alert the society to a state of emergency, causing its soldiers, or such other members as are usually enlisted for defense, to respond to the disturbance and to converge upon its site. Sometimes, as in the case with formicine ants (Maschwitz, 1964) and nasute termites (Ernst, 1959), the defensive secretions may in themselves have an alarming action. But in other social insects, glands may be present that serve exclusively or primarily for the dissemination of alarm substances. The topic has been the subject of considerable study and of excellent reviews (Blum, 1969; Butler, 1967; Cavill and Robertson, 1965; Maschwitz, 1964; Regnier and Law, 1968; Wilson, 1965; Wilson and Bossert, 1963). Defensive secretions may also serve secondarily as alarm pheromones in subsocial rather than truly social insects. In the bug *Dysdercus intermedius*, exposure to secretion is said to cause dispersal of the aggregations in which the species normally lives. Thus, individuals attacked by predators may, through their discharges, "forewarn" others of impending danger and cause them to escape (Calam and Youdeowei, 1968). For a species that has no collaborative way of opposing a predator, an avoidance reaction is clearly in order.

Some insects may, while paired sexually, profit from the pooling of their defensive resources. In the large Southern walking stick, *Anisomorpha buprestoides*, the situation is a special one. The male of the species is considerably smaller than the female, and its supply of defensive secretion is correspondingly reduced. The male is usually found astride the female, even while the two are not mating (Fig. 16A), and pair formation may occur already in the nymphal stages (Eisner, 1965a). The partnership may well be a defensive one, certainly as it involves the immatures.

Provision is sometimes made by parents for the defenses of their offspring. Such provision may consist simply of endowing the eggs with enough reserves to enable the first-instar young to hatch with functional and replete glands — *Anisomorpha*, for example, can spray and repel ants immediately after eclosion (Eisner, 1965a) — but in other cases it may involve equipping an egg with a chemical defense of its own. The eggs of the mosquito *Culex pipiens*, which are laid in rafts on water, bear on their posterior pole a small droplet of lipoidal fluid. The liquid, which is of maternal origin, has been shown to offer protection against certain ants that are natural enemies of the eggs (Hinton, 1968). The Neuroptera of the family Chrysopidae typically lay stalked eggs. The stalks are usually naked, but in two species (*Chrysopa claveri; Nodita floridana*) they bear droplets, arranged on the stalks like beads on a string (Fig. 16B). Ants that make contact with the droplets back away and clean themselves (McLeod, Carrel, and Eisner, unpublished).

Chemical factors may also serve to protect an individual from others of the same species, but little is known about this general subject. In indiscriminately predaceous arthropods, cannibalism can be a potential threat, and chemical factors such as distastefulness may well operate against it in some species. Aphrodisiacs or other sex pheromones, produced by aggressive species during courtship, might serve to prevent mating from becoming a cannibalistic feast [as it may become, for example, in mantids (Roeder, 1967)]. Whether the extrusible glands of male mantispids (Eltringham, 1932) have a protective function of this sort, remains to be demonstrated. An unusual intraspecific defense is suggested by the observation of Roth (1967) that in certain cockroaches large amounts of uric acid, voided from the accessory sex glands of males, are poured over the spermatophore during copulation. It is argued

Fig. 16. (A) Male and female of the walking stick *Anisomorpha buprestoides*, paired as they are usually found, even when not mating (see text, Section III, E). (B) Egg of the neuropteran, *Chrysopa claveri*, the stalk of which bears droplets of a defensive fluid. (C) Two freshly killed cockroaches placed within range of attacking ants. The one at right has been treated with a droplet of catnip (nepetalactone), which is repelling the ants. (D) A group of lycid beetles (*Lycus loripes*), closely clustered, as they are often found within their aggregations. (E) The mimetic cerambycid beetle *Elytroleptus ignitus*, eating the lycid beetle (*Lycus loripes*) that serves as its model. (F) and (G) Two glandless beetles that mimic the headstand of *Eleodes longicollis* (Fig. 1A): *Megasida obliterata* (F); *Moneilema appressum* (G).

that the acid coating may serve (or may have served in the evolutionary past) to protect the spermatophore from being eaten by the female herself, or by other insects.

Intraspecific chemical antagonism may also function for other purposes than protection from cannibalism. The female of the mosquito, *Aedes aegypti*, may copulate repeatedly, but is inseminated only once. Subsequent insemination is inhibited by a substance ("matrone") from the male accessory gland that is transferred to the female in seminal fluid at the initial mating (Craig, 1967; Fuchs *et al.*, 1968; Spielman *et al.*, 1967). Virgin females injected with an extract of male glands are rendered sterile for life (Craig, 1967). Chemical factors such as these, which regulate sperm distribution in populations, are likely to be of more general occurrence than previously thought, and their investigation should prove rewarding from both basic and applied points of view.

F. *Entspannungsschwimmen*

A most remarkable defense mechanism, involving escape over the surface of water, is possessed by certain staphylinid beetles of the genus *Stenus*, and water striders of the genus *Velia* (Hemiptera: Veliidae). *Stenus*, like other staphylinids, has a pair of eversible glands on the tip of its abdomen. The beetles are normally terrestrial, but they occasionally forage on water or may be wind-blown on water. It is under these conditions that they may rely on the surface tension-depressant properties of their secretion to propel themselves over the surface of the liquid. By touching their everted glands to the water, they weaken the surface tension behind them, and are then carried forward by the "contracting" surface as it is withdrawn before them. The velocity that they may achieve through such *Entspannungsschwimmen* is considerable: up to 15 m may be covered in one stretch at 40–75 cm/sec. The beetles employ this form of locomotion only in emergencies, as perhaps when threatened by gerrids. Ordinarily they simply paddle along with their feet. In *Velia* the mechanism is comparable, but the propellant is apparently saliva, which is discharged posteriorly from the beak (Linsenmair, 1963; Linsenmair and Jander, 1963).

G. The Parallel with Plants

The defensive substances of arthropods are, with relatively few exceptions, compounds that had been previously known from plants. Their presence in plants, and particularly the question of their adaptive justification, has to this day remained a matter of controversy. Some feel that the name of "secondary plant substances" that is usually given these compounds is a truly descriptive one, since the substances do not play a role in the fundamental biochemical processes of the living plant and would hence appear to be superfluous. But there are others who consider the substances to be defensive (e.g., Ehrlich and Raven, 1965; Fraenkel, 1959). To me, the very fact that plants should possess the same materials that in other organisms are *known* to be defensive, may in itself be considered to be circumstantial evidence in support of the latter view.

The parallel is a truly striking one. As is evident by comparing the lists of arthropod defensive substances compiled by Weatherston (1967) with those available of secondary substances of plants (Karrer, 1958), virtually all major categories of compounds produced by the former group are also represented among the latter. Some of the substances in arthropods, including such compounds as *trans*-2-hexenal, benzaldehyde, salicylaldehyde, citral, and citronellal (Fig. 2), are in fact widely distributed among plants. Compounds such as pederin (Fig. 2, XVIII), which have no close counterpart among plants, are rare. Even the mechanism of release of the defensive principles may be similar in the two groups of organisms. In plants, hydrogen cyanide is usually generated by hydrolysis of cyanohydrin glycosides. In the larvae of certain chrysomelid beetles of the tribe Paropsini, whose cyanogenetic secretion contains both benzaldehyde and glucose, a similar hydrolytic mechanism is probably at play (Moore, 1967). The reactor glands of the millipede *Apheloria corrugata* (Fig. 5) also operate on this principle, except that the cyanogenetic precursor is a cyanohydrin (mandelonitrile) rather than one of its glycosides (Eisner *et al.*, 1963a).

Most secondary plant substances have not been screened systematically for repellency, toxicity, or other defensive properties. In cases where these substances resemble known defensive factors of arthropods, one might think that their specific mode of defensive action could be similar in plants. In at least one case, a prediction of this sort has received some support. The cyclopentanoid monoterpene of plant origin known as catnip (nepetalactone) had long been known for its excitatory effect on cats. Monoterpenes similar to catnip, but produced by insects, (e.g., Fig. 2, VII), were known to be powerfully repellent or toxic to predators, including other insects (Eisner, 1965a; Pavan, 1952, 1959). Catnip, when tested, was also found to be repellent to insects (Fig. 16C), and this, one might argue, is indicative of its true adaptive justification in plants (Eisner, 1964).

H. Intrinsic and Extrinsic Origin of Defensive Substances

There are two principal ways in which an arthropod may come upon the possession of a defensive substance. It may synthesize the material itself, or it may sequester it from an exogenous source.

Production through intrinsic synthesis has been demonstrated conclusively in a few species, using radiotracer incorporation techniques. The walking stick *Anisomorpha buprestoides* and the ant *Acanthomyops claviger* produce their monoterpenes (Chadha *et al.*, 1962; Meinwald *et al.*, 1962) from acetate, through mevalonate, according to the usual terpenoid biosynthetic scheme (Happ and Meinwald, 1965; Meinwald *et al.*, 1966a). Similarly, the hemipteran *Nezara viridula* incorporates acetate into both the carbonyl and hydrocarbon fractions of its secretion (Gordon *et al.*, 1963), and the millipede *Glomeris marginata* synthesizes quinazolinone from anthranilic acid (Schildknecht and Wenneis, 1967). In the tenebrionid beetle *Eleodes longicollis*, two pathways are apparently involved in the production of quinones: *p*-benzoquinone is synthesized from preformed aromatic precursors (phenylalanine, tyrosine), whereas the alkylated quinones (toluquinone, ethylquinone) are produced from acetate (Meinwald *et al.*, 1966b). An investigation of cantharidin biosynthesis in the meloid beetle *Lytta vesicatoria* led to the surprising finding that, although both adult sexes contain the substance, only the adult male produces it (from acetate and mevalonate, by other than a tail to tail linkage of two isoprene units). Since biosynthesis occurs in larvae, it is presumed that the adult females obtain their cantharidin by manufacturing it as immatures (Meyer *et al.*, 1968; Schlatter *et al.*, 1968). The significance of this biochemical sexual dimorphism remains obscure, although it raises interesting questions regarding the origin and possible shift of function of cantharidin during the evolutionary history of meloids.

Most of what we know about defensive substances of *extrinsic* origin relates to factors that are incorporated from plants. The evidence for incorporation may only be indirect or circumstantial, but it may be strongly suggestive just the same. The papilionid butterfly, *Pachlioptera aristolochiae*, contains aristolochic acid (Fig. 2, XV). The caterpillar feeds exclusively on plants of the family Aristolochiaceae and presumably ingests the acid, which is absorbed systemically and carried through the pupa to the adult insect (von Euw *et al.*, 1969). The cinnabar moth *Callimorpha jacobaeae* contains senecio alkaloids, just as do the composite plants of the genus *Senecio* upon which it feeds (Aplin *et al.*, 1968). The cantharid beetle *Chauliognathus lecontei* produces a defensive secretion containing 8-*cis*-dihydromatricaria acid (Fig. 2, IV). Similar acetylenic compounds are found in some composite plants, and the beetles are known to aggregate at times on species of this family (Meinwald *et al.*, 1968b). Incorporation mechanisms such as these, in which toxic or repellent secondary plant substances that evolved initially for the protection of a plant have become appropriated secondarily for their own defense by herbivores that evolved means of coping with the plant, may be more widespread than suspected. Excellent discussions of the coevolution of herbivores and plants, and of the chemical factors that are likely to have influenced their reciprocal evolutionary relationships, are given by Brower and Van Zandt Brower (1964), Dethier (1954), and Ehrlich and Raven (1965). An interesting case about which much has been learned recently concerns milkweed plants (Asclepiadaceae), their cardenolide toxins, and the insects that sequester these poisons. Since cardenolides are steroids, and since insects have only a limited ability to synthesize steroids from nonsteroidal precursors (Clayton, 1964), it was suspected from the outset that the cardenolides found in insects stem from the plants ingested. The evidence, both pharmacological and chemical, strongly suggests that they do. It has been shown, for instance, that a jay will eat a specimen of the grasshopper *Poekilocerus bufonius* when this has been fed on dandelion, but will vomit after ingestion of an individual fed on *Asclepias* (Rothschild, 1966). Similarly, when monarch butterflies, which ordinarily feed on Asclepiadaceae, are reared on cabbage, the larvae, prepupae, and adults are palatable to bluejays. Rearing the butterflies on Asclepiadaceae can cause them to be toxic, but toxicity is variable and may reflect the variability in the cardenolide content of the plants themselves. Thus, the monarchs reared on *Calotropis procera* and *Asclepias curassavica* are, respectively, about six and five times as emetic as those fed a species of *Gomphocarpus*, while those raised on an asclepiad that lacks cardenolides (*Gonolobus rostratus*) are not emetic at all (Brower *et al.*, 1967, 1968). Given a certain diversity in its larval food habits, the monarch in nature is thus likely to be a species of variable palatability.

Secondary plant substances need not be incorporated systemically in order to afford protection. It may suffice for them to be taken into the gut, from where they may then be regurgitated and employed defensively in that fashion. The regurgitate of the grasshopper *Romalea microptera* varies in repellency depending on what the insect ate. When given two of its natural food plants (*Eupatorium capillifolium; Salix nigra*), the regurgitate was strongly repellent to ants. An unnatural diet of lettuce or of *Myrica cerifera*, on the other hand, produced innocuous regurgitates, despite the fact that *Myrica* is an aromatic plant of high intrinsic repellency. The evidence suggests that the repellent principles of *Myrica* are inactivated enterically by the grasshopper. It is tempting to conclude from this that in its selection of a natural foodplant, an animal such as *Romalea* is guided not only by its ability to cope with the plant, but also by its ability to withstand in unaltered form the defensive principles of that plant, since these are potentially employable for the defense of the grasshopper itself (Eisner, Kafatos, and Shepherd, unpublished). The enteric preservation of an ingested plant poison may thus be thought of as being adaptively more meritorious than its detoxification, even in animals that have not as yet hit upon the evolutionary expedient of absorbing the chemical, storing it throughout their body, and transmitting it from one instar to the other.

It has been suggested that an insect may derive an advantage by mere virtue of its residence on an unpalatable plant. Absence of large browsers may prevent its eggs and young from being incidentally ingested

(Reichstein *et al.*, 1968), and in the case of insects that subsist on highly aromatic plants, protection might be obtained from living in the repellent chemical vapors that emanate from their host (might entomophagous parasites be among those avoiding such plants?). These possibilities are clearly open to testing.

Even plant materials that are merely carried topically may serve for defense. The loose "trash packets" carried for protective purposes (see Section II, B, 3, above) by many chrysopid larvae sometimes consist of vegetable matter, as is the case with the unidentified species from Arizona shown in Fig. 8E, which covers itself with the hairs from the underside of sycamore leaves. A remarkable group of large flightless weevils has recently been described (Gressit *et al.*, 1965, 1966), from the high altitude moss forests of New Guinea, which have a dense dorsal covering of living plant growth. A secretion in depressions on the beetles' backs appears to foster this flora, which consists of variously mixed groups of fungi, algae, mosses, lichens, and liverworts. Camouflage need not be the only or even principal function of this vegetal covering. Field observations (Gressit *et al.*, 1968) suggest that predation of the weevils is relatively uncommon even when these are released in situations where they are conspicuous. It is therefore possible that the animals are distasteful, and that the plants are partly or wholly responsible for this.

Very little is known about the extent to which defensive substances might be transmitted from animal to animal in a food chain, rather than from plant to animal as discussed so far. There can be no doubt that such transmission must occur, since many animals subsist on others that are potentially toxic or repellent and may, as a result, incorporate the noxious principles themselves. It is conceivable, therefore, that arthropods feeding on danaid butterflies (e.g., *Argiope* on *Danaus gilippus berenice*, as I have seen in Florida) become toxic as a result. Similarly, it has been suggested that the mimetic cerambycid beetles of the genus *Elytroleptus*, which feed on the lycid beetles that serve as their models (Fig. 16E), must inevitably acquire some of the unpalatability of the latter, at least for the period immediately following the meal (Eisner *et al.*, 1962). An interesting occurrence has been recorded in the older literature, involving the transmission of an insect poison, through an insect predator, to man himself. It seems that French soldiers in Algiers, after eating the meat of frogs locally captured, sometimes developed urogenital symptoms, including "érections douloureuses et prolongées," such as were known to doctors of the time to be characteristic of cantharidin poisoning. The frogs were shown to come from areas where meloid beetles were abundant, and examination of their stomach contents showed them to be feeding on the beetles (Vézien, 1861; Meynier, 1893). Substances incorporated by an animal may also have an effect on its taste to man. It is well known, certainly to some Europeans, that the "fishy" flavor of a fish-fed goose may spoil the Christmas dinner.

I. Chemotaxonomic Considerations

It is clear already from what we know about the chemistry of arthropod defenses, that this body of information, like any other about a group of organisms, lends itself to interpretation in phyletic terms. It is equally clear, however, that this interpretation cannot be carried out in disregard of other character systems. The pitfalls of single-character analyses are well known to evolutionists, and the dangers inherent in their taxonomic misuse have been pointed out repeatedly, even as they apply to chemical systems (Brown, 1967). Thus, to propose, for example, that geophilid centipedes and polydesmoid millipedes are directly related to one another simply because both have cyanogenetic glands (Schildknecht *et al.*, 1968a) is clearly unwarranted. Arthropods vary in their defensive chemistry as they vary in anything else. A similarity in chemical product need, therefore, be no more indicative of close phyletic affinity than a chemical disparity is of a major phyletic gap. Nevertheless, providing other characters are also taken into account, the distinction between homologous and analogous defenses can often be made with safety.

Several instances are known in which arthropods of very diverse kinds produce the same or closely similar compounds, and there can be no doubt in their case that the similarities evolved in parallel. Cyanogenesis, for example, occurs not only in polydesmoid millipedes (references in Eisner and Meinwald, 1966) and geophilid centipedes (Schildknecht *et al.*, 1968a), but also in some chrysomelid larvae (Moore, 1967) and moths (Jones *et al.*, 1962). The first three groups discharge cyanogenetic secretions (from glands that are clearly not homologous), while the latter release their hydrogen cyanide from injured tissues. Formic acid has also evolved as a defensive substance independently in several groups. It is ejected as part of the spray of formicine ants (Wray, 1670), certain notodontid caterpillars (Poulton, 1888; Roth and Eisner, 1962; Schildknecht and Schmidt, 1963) and some carabid beetles (Moore and Wallbank, 1968; Schildknecht *et al.*, 1968d). Among the most widespread glandular defenses are the *p*-benzoquinones (e.g., Fig. 2, XII). These powerful oxidizing agents tan human skin and are responsible for the characteristic darkening of the fingers that results from collecting arthropods that produce them. Quinonoid secretions occur in opilionids (Fieser and Ardao, 1956), millipedes (references in Jacobson, 1966; Roth and Eisner, 1962; Weatherston, 1967), termites (Moore, 1968), cockroaches (Eisner, unpublished, and see Fig. 4G; Roth and Stay, 1958), earwigs (Eisner and Blumberg, 1959; Schildknecht and Weis, 1960), tenebrionid beetles (references in Jacobson, 1966; Roth and Eisner, 1962; Weatherston, 1967), and carabid beetles (Aneshansley *et al.*, 1969; Moore and Wallbank, 1968; Schildknecht *et al.*, 1968d). The glands that produce the quinones are morphologically distinct entities in the various groups, and they must have evolved separately in each. But *de novo* evolution need not be invoked as it applies to quinone biosynthesis itself. Arthropods as a group utilize quinones (albeit o-quinones) for the tanning of their skeleton (Gilmour, 1965), and in order to employ such substances for defense, they need only have acquired the ability to manufacture them *en masse* and in the form of the more stable and volatile (and hence presumably defensively more effective) p-quinones. Comparable arguments probably apply also to other defensive substances which might similarly be synthesized by preexisting biosynthetic pathways of largely unspecialized nature. There is certainly no need to postulate elaborate and novel biosynthetic schemes in order to account for the formation of aliphatic acids, ketones, aldehydes, hydrocarbons, phenols, and such other relatively simple compounds as arthropod glands usually produce.

Instances are also known in which homologous glands of related organisms produce substances of very different chemical nature. In some of these cases, the evolutionary trends implied by the chemical diversification are in accord with accepted phyletic schemes previously established by systematists. Thus, among the millipedes, quinone production occurs in the orders Julida, Spirostreptida, and Spirobolida, cyanogenesis appears to be confined to the Polydesmida, the single species of Chordeumida studied secretes a phenol, and quinazolinones are produced by Glomerida (Y. C. Meinwald *et al.*, 1966; Schildknecht *et al.*, 1967c; references in Eisner and Meinwald, 1966). The glands of Glomerida, which may be altogether lacking in some species (Eisner and Davis, 1967), may not be homologous to those in other orders (Eisner, unpublished).

Within individual phyletic groups, the extent of chemical diversification undergone by the secretory systems varies. Conservatism appears to have been the trend in some cases. Thus, in nine bugs of the superfamily Coreoidea that were examined, the secretions contain similar major components (Waterhouse and Gilby, 1964), and in several species of the beetle family Tenebrionidae, the secretions are quinonoid (references in Eisner and Meinwald, 1966). However, as pointed out by Waterhouse and Gilby (1964), examination of a few species does not prove a rule. A coreid with exceptional secretion is already known (Blum *et al.*, 1961), and a tenebrionid has now been described in which one pair of glands produces phenols instead of quinones (Tschinkel, 1969). The suggestion (Schildknecht *et al.*, 1964) that the Tenebrionidae might be called "quinone beetles" was never appropriate, since in two of the three subfamilies (Tentyriinae and Asidinae) the glands are frequently and perhaps always lacking.

The most comprehensive studies done so far on the defensive chemistry of any one group of arthropods are two excellent and largely parallel investigations carried out in Australia (Moore and Wallbank, 1968) and in Germany (Schildknecht *et al.*, 1968c,d) on beetles of the family Carabidae, which discharge secretions of varying composition from a pair of abdominal glands that are (at least in their basic features) homologous throughout the family. The organic components of the secretions fall into eight main classes: hydrocarbons, formic acid, higher saturated acids, unsaturated acids, simple phenols, salicylaldehyde, and quinones. The secretion of any one beetle may combine components from more than one class. As pointed out by Moore and Wallbank (1968), considerable parallel evolution in these chemical systems is evident at the subfamily level, but below this level, at the tribal and generic levels, the chemical character can be usefully employed as an additional criterion for the recognition of natural phyletic lines. Major variability between closely related species is not the rule. In chemotaxonomic studies of this sort, the criteria used for the establishment of evolutionary relationships should be carefully chosen. For example, it seems doubtful to me that one need assume that chemically less effective secretory formulations are evolutionarily older than more effective ones, and particularly when the criteria for effectiveness (cytotoxicity, antimicrobial action) (Schildknecht et al., 1968c) are not of proven applicability to the chemical defenses in question.

Chemophyletic considerations of broad ecological implications also emerge from a consideration of phytophagus insects and their food plants. In a masterful treatment of the subject, Ehrlich and Raven (1965) analyze the stepwise reciprocal selective pressures that are likely to have governed the evolutionary interaction of plants and butterflies. Among other things, they emphasize how the explosive adaptive radiation of various angiosperm subgroups is likely to have been triggered by the acquisition of novel chemical defenses, and how certain groups of butterflies, after having "crashed" through these biochemical barriers and incorporated the compounds systemically for their own defense, followed suit and underwent diversification of their own. The existence among butterflies of certain phyletic units that have become specialized for the exploitation of certain complexes of related species of plants can be explained in this fashion. The assumption that the defensive chemistry of such butterflies matches that of the plants upon which they feed, is gaining support from a growing body of evidence (Brower et al., 1968; von Euw et al., 1969; Reichstein, 1967; Reichstein et al., 1968).

J. Aposematism and Mimicry

Many arthropods with chemical defenses are aposematically or "warningly" colored. Some, moreover, have the habit of living in dense and often conspicuous aggregations (Cott, 1957). Rather than spacing themselves more or less evenly throughout what is seemingly a uniformly favorable habitat, they occur in distinct, sporadically distributed clusters. Many meloid, cantharid, and lycid beetles, as well as a variety of Hemiptera, among others, are known for this habit. It has now been shown in one species, the lycid beetle Lycus loripes, that aggregation is maintained through the action of a volatile pheromone, released by the males, and serving to attract both males and females (Eisner and Kafatos, 1962). Comparable pheromones may eventually also be found in other insects that "pool" their aposematic resources in collective displays.

Aposematism can also be achieved through contrast alone, without resort to the use of color. Many desert insects are primarily nocturnal, and their principal exposure to visually oriented predators may occur at dawn and dusk. Those that live at ground level, such as many tenebrionid beetles (Fig. 1A), are often black and highly conspicuous against the sandy soil. Since true color is only minimally discernable in a crepuscular setting, and since some of the principal enemies (e.g., rodents) of these beetles are likely to be totally or relatively color blind, it makes sense, perhaps, that they should display themselves in black.

Aposematism, as the reinforcing stimulus that it is, may also involve the use of acoustical instead of visual cues. Many chemically protected arthropods emit sounds when disturbed (Haskell, 1961) and it seems likely, certainly in some cases (Rothschild and Haskell, 1966), that this represents a way of warning a predator against repetition of its past "mistakes." In insects such as many Hemiptera, which may also be visually aposematic, sound production (Leston and Pringle, 1963) may serve primarily as an aposematic signal in the night. In this connection it is of interest that moths of the family Arctiidae, which include many unpalatable forms (Bisset et al., 1960), possess tympanic organs capable of detecting the echolocating chirps of bats (Haskell and Belton, 1956). The moths may respond to these chirps by generating acoustical signals of their own, which are rich in ultrasound and have been shown to cause foraging bats to turn away in midair (Blest et al., 1963; Dunning and Roeder, 1965). It appears therefore that the calls are aposematic, and that bats may know that "a prospective meal that answers back is likely to taste bad" (Roeder, 1967).

Unpalatability is perhaps the chief basis of mimicry in nature, and many arthropods with chemical defenses do indeed figure as elements in mimetic assemblages of one kind or another. The mimicry may involve behavioral as well as visual imitation, as is the case with the beetles *Megasida obliterata* (Fig. 16F) (Eisner, 1966) and *Moneilema appressum* (Raske, 1967) (Fig. 16G) which accurately imitate the headstand of *Eleodes* (Fig. 1A). Neither *Megasida* (a tenebrionid) nor *Moneilema* (a cerambycid) have defensive glands, and they are therefore presumed to be Batesian mimics of *Eleodes*.

The distinction between Batesian and Müllerian mimicry need not always be a rigorous one. In the case of certain mimetic cerambycids (*Elytroleptus* spp.), which have the peculiar habit of feeding on the distasteful lycids (*Lycus* spp.) that serve as their models (Figs. 16D, 16E) (Eisner et al., 1962; Selander et al., 1963), it has been suggested (although not proven) that the cerambycids might be alternatively Batesian or Müllerian depending on how recently they last fed on a lycid (Eisner et al., 1962).

Surprisingly little is known about the extent to which distasteful arthropods imitate one another, or are imitated by palatable species, on the basis of nonvisual characteristics. Future chemical and behavioral work should attempt to determine whether unpalatable species are recognizable by body odors of constant emission, and whether such "aposematic odors" are imitated by others. That chemical mimicry of this sort must occur has been suggested (Rothschild, 1961). Acoustical mimicry (e.g., Lane and Rothschild, 1965) must also be more widespread than so far demonstrated. One is tempted to predict that palatable moths will eventually be found which "fool" bats by "faking" the calls of arctiids.

A new category of mimicry — automimicry — has recently been erected (Brower et al., 1967), applicable to situations in which the members of a species include both unpalatable and palatable individuals, with the latter acting essentially as mimics of the former. The term was coined as a result of the finding that monarch butterflies (Danaus plexippus) may

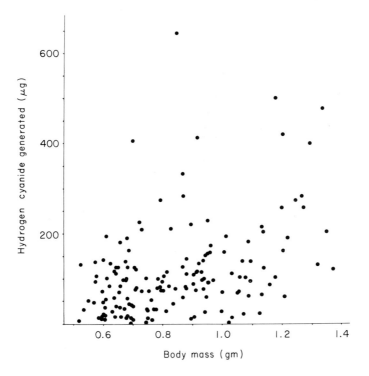

Fig. 17. Cyanogenetic yield in the millipede Apheloria corrugata. Each point gives the total output of a single individual, assayed shortly after capture. The variability in output depicted on this chart is a true indication of the variability in defensive potential that prevails in a natural population of this animal (from H. E. Eisner et al., 1967).

be variously palatable, depending on the amount of cardenolide supplied to them by the particular species of milkweed that they ate as larvae (Brower *et al.*, 1968).

The concept is undoubtedly widely applicable, since palatability may be subject to intraspecific variability more often than it is not. In species that have glands, for example, automimicry may also prevail, since natural populations of these animals are always likely to include individuals whose glands have become partly or wholly depleted as a result of predator attacks. It is these relatively vulnerable individuals that profit, as do any mimics, from their resemblance to those whose defenses are intact. Unfortunately we know very little about the relative states of depletion of the defensive glands of individuals in nature. The only assays made, on certain cyanogenetic millipedes (H. E. Eisner *et al.*, 1967), show that defensive potential is indeed a highly variable parameter (Fig. 17). It remains to be seen whether this sort of variability is attributable to intrinsic diversity in the potential for secretory production, or whether it is — as one would obviously very much like to know — a true reflection of the natural exposure of a species to predator attack. There is clearly a good deal yet to be learned about the arthropods' own approach to "better living through chemistry."

Acknowledgments

Rosalind and David Alsop, James E. Carrel, Maria Eisner, William Thompson, Walter Tschinkel, and Cynthia Walters were helpful in the preparation of the manuscript. Messrs. Alsop and Carrel gave permission for the use of some of their unpublished data and illustrations. The collaboration of my friend, Jerrold Meinwald, is gratefully acknowledged. Dr. Lincoln Brower (Amherst College) and Dr. William L. Nutting (University of Arizona, Tucson) provided Figs. 15A, 15B, and 8B, 8C, respectively. The Honorable Miriam Rothschild (Ashton, Peterborough, England) and Dr. Douglas Waterhouse (CSIRO, Canberra) kindly provided advance copies of their papers still in the press. My research on this general subject has been supported by the National Institutes of Health (Grant AI-02908) and by funds from the Bache Fund of the National Academy of Sciences, S. C. Johnson & Son, Inc., and the Upjohn Company.

References

Alexander, A. J. (1957). Notes on onychophoran behavior. *Ann. Natal Museum* 14, 35–43.
Aneshansley, D., Eisner, T., Widom, J. M., and Widom, B. (1969). Biochemistry at 100°C: the explosive discharge of bombardier beetles (*Brachinus*). *Science* 165, 61–63.
Aplin, R. T., Benn, M. H., and Rothschild, M. (1968). Poisonous alkaloids in the body tissues of the cinnabar moth (*Callimorpha jacobaeae* L.). *Nature* 219, 747–748.
Bisset, G. W., Frazer, J. F. D., Rothschild, M., and Schachter, M. (1960). A pharmacologically active choline ester and other substances in the garden tiger moth, *Arctia caja* (L.). *Proc. Roy. Soc. (London)* B152, 255–262.
Blest, A. D., Collett, T. S., and Pye, J. D. (1963). The generation of ultrasonic signals by a New World arctiid moth. *Proc. Roy. Soc. (London)* B158, 196–207.
Blum, M. S. (1961). The presence of 2-hexenal in the scent gland of the pentatomid *Brochymena quadripustulata. Ann. Entomol. Soc. Am.* 54, 410–412.

Blum, M. S. (1964). Insect defensive secretions: hex-2-enal-1 in *Pelmatosilpha coriacea* (Blattaria) and its repellent value under natural conditions. *Ann. Entomol. Soc. Am.* **57**, 600-602.

Blum, M. S. (1969). Alarm pheromones. *Ann. Rev. Entomol.* **14**, 57-80.

Blum, M. S., and Crain, R. D. (1961). The occurrence of *para*-quinones in the abdominal secretion of *Eleodes hispilabris* (Coleoptera: Tenebrionidae). *Ann. Entomol. Soc. Am.* **54**, 474-477.

Blum, M. S., Crain, R. D., and Chidester, J. B. (1961). *Trans*-2-hexenal in the scent gland of the hemipteran *Acanthocephala femorata*. *Nature* **189**, 245-246.

Blunck, H. (1917). Die Schreckdrüsen des *Dytiscus* und ihr Sekret. *Z. Wiss. Zool.* **117**, 205-256.

Bowers, W. S., Fales, H. M., Thompson, M. J., and Uebel, E. C. (1966). Juvenile hormone: identification of an active compound from balsam fir. *Science* **154**, 1020-1021.

Brega, A., Falaschi, A., de Carli, L., and Pavan, M. (1968). Studies on the mechanism of action of pederine. *J. Cell Biol.* **36**, 485-496.

Bristowe, W. S. (1958). "The World of Spiders." New Naturalist, Collins, London.

Brower, L. P. (1969). Ecological chemistry. *Sci. Am.* **220**(2), 22-29.

Brower, L. P., and Van Zandt Brower, J., (1964). Birds, butterflies, and plant poisons: A study in ecological chemistry. *Zoologica* **49**, 137-159.

Brower, L. P., Van Zandt Brower, J., and Corvino, J. M. (1967). Plant poisons in a terrestrial food chain. *Proc. Natl. Acad. Sci. U.S.* **57**, 893-898.

Brower, L. P., Ryerson, W. N., Coppinger, L. L., and Glazier, S. C. (1968). Ecological chemistry and the palatability spectrum. *Science* **161**, 1349-1351.

Brown, K. S. (1967). Chemotaxonomy and chemomimicry: The case of 3-hydroxykynurenine. *Systematic Zool.* **16**, 213-216.

Brunet, P. C. J., and Kent, P. W. (1955). Observations on the mechanism of a tanning reaction in *Periplaneta* and *Blatta*. *Proc. Roy. Soc.* **B144**, 259-274.

Butler, C. G. (1967). Insect pheromones. *Biol. Rev. Cambridge Phil. Soc.* **42**, 42-87.

Calam, D. H., and Youdeowei, A. (1968). Identification and functions of secretion from the posterior scent gland of fifth instar larva of the bug *Dysdercus intermedius*. *J. Insect Physiol.* **14**, 1147-1158.

Cardani, C., Ghiringhelli, D., Mondelli, R., and Quilico, A. (1965). The structure of pederin. *Tetrahedron Letters* **29**, 2537-2545.

Casper, A. (1913). Die Körperdecke und die Drüsen von *Dytiscus marginalis* L. Ein Beitrag zum feineren Bau des Insektenkörpers. *Z. Wiss. Zool.* **107**, 387-508.

Cavill, G. W. K., and Robertson, P. L. (1965). Ant venoms, attractants, and repellents. *Science* **149**, 1337-1345.

Chadha, M. S., Eisner, T., and Meinwald, J. (1961). Defence mechanisms of arthropods. IV. *Para*-benzoquinones in the secretion of *Eleodes longicollis* Lec. (Coleoptera: Tenebrionidae). *J. Insect Physiol.* **7**, 46-50.

Chadha, M. S., Eisner, T., Monro, A., and Meinwald, J. (1962). Defence mechanisms of arthropods. VII. Citronellal and citral in the mandibular gland secretion of the ant *Acanthomyops claviger* (Roger). *J. Insect Physiol.* **8**, 175-179.

Clayton, R. B. (1964). The utilization of sterols by insects. *J. Lipid Res.* **5**, 3-19.

Cott, H. B. (1957). "Adaptive Coloration in Animals." Methuen, London.

Craig, G. B., Jr. (1967). Mosquitoes: Female monogamy induced by male accessory gland substance. *Science* **156**, 1499.

Crossley, A. C., and Waterhouse, D. F. (1969). The ultrastructure of the osmeterium and the nature of its secretion in *Papilio* larvae (Lepidoptera). *Tissue Cell* **1**(3), 525-554.

Cuénot, L. (1949). Les onychophores. *Traité Zool.* **6**, 3-37.

Curasson, G. (1934). Sur la toxicité de la sécrétion buccale des Sauterelles. *Bull. Acad. Vet. France* **7**, 377-382.

Darlington, P. J. (1938). Experiments on mimicry in Cuba, with suggestions for future study. *Trans. Roy. Entomol. Soc. London* **87**, 681-695.

Dethier, V. G. (1954). Evolution of feeding preferences in phytophagous insects. *Evolution* **8**, 33-54.

Dunning, D. C., and Roeder, K. D. (1965). Moth sounds and the insect-catching behavior of bats. *Science* **147**, 173–174.

Edwards, J. S. (1960). Spitting as a defensive mechanism in a predatory reduviid. *Intern. Kongr. Entomol., Verhandl., 11 Kongr., Vienna* **3**, 259–263.

Edwards, J. S. (1961). The action and composition of the saliva of an assassin bug *Platymeris rhadamanthus* Gaerst. (Hemiptera, Reduviidae). *J. Exptl. Biol.* **38**, 61–77.

Edwards, J. S. (1966). Defence by smear: super-cooling in the cornicle wax of aphids. *Nature* **211**, 73–74.

Ehrlich, P. R., and Raven, P. H. (1965). Butterflies and plants: A study in coevolution. *Evolution* **18**, 586–608.

Eisner, H. E., Eisner, T., and Hurst, J. J. (1963). Hydrogen cyanide and benzaldehyde produced by millipedes. *Chem. Ind. (London)* pp. 124–125.

Eisner, H. E., Alsop, D. W., and Eisner, T. (1967). Defense mechanisms of arthropods. XX. Quantitative assessment of hydrogen cyanide production in two species of millipedes. *Psyche* **74**, 107–117.

Eisner, T. (1958a). The protective role of the spray mechanism of the bombardier beetle, *Brachynus ballistarius* Lec. *J. Insect Physiol.* **2**, 215–220.

Eisner, T. (1958b). Spray mechanism of the cockroach *Diploptera punctata*. *Science* **128**, 148–149.

Eisner, T. (1960). Defense mechanisms of arthropods. II. The chemical and mechanical weapons of an earwig. *Psyche* **67**, 62–70.

Eisner, T. (1964). Catnip: its raison d'être. *Science* **146**, 1318–1320.

Eisner, T. (1965a). Defensive spray of a phasmid insect. *Science* **148**, 966–968.

Eisner, T. (1965b). Insect's scales are asset in defense. *Nat. Hist.* **74**(6), 26–31.

Eisner, T. (1966). Beetle's spray discourages predators. *Nat. Hist.* **75**(2), 42–47.

Eisner, T. (1967). Life on the sticky sundew. *Nat. Hist.* **76**(6), 32–35.

Eisner, T., and Blumberg, D. (1959). Quinone secretion: A widespread defensive mechanism of arthropods. *Anat. Record* **134**, 558–559.

Eisner, T., and Davis, J. A. (1967). Mongoose throwing and smashing millipedes. *Science* **155**, 577–579.

Eisner, T., and Eisner, H. E. (1965). Mystery of a millipede. *Nat. Hist.* **74**(3), 30–37.

Eisner, T., and Kafatos, F. C. (1962). Defense mechanisms of arthropods. X. A pheromone promoting aggregation in an aposematic distasteful insect. *Psyche* **69**, 53–61.

Eisner, T., and Meinwald, J. (1966). Defensive secretions of arthropods. *Science* **153**, 1341–1350.

Eisner, T., and Meinwald, Y. C. (1965). Defensive secretion of a caterpillar (*Papilio*). *Science* **150**, 1733–1735.

Eisner, T., and Shepherd, J. (1966). Defense mechanisms of arthropods. XIX. Inability of sundew plants to capture insects with detachable integumental outgrowths. *Ann. Entomol. Soc. Am.* **59**, 868–870.

Eisner, T., McKittrick, F., and Payne, R. (1959). Defense sprays of roaches. *Pest Control* **27**, 11–12, 44–45.

Eisner, T., Meinwald, J., Monro, A., and Ghent, R. (1961). Defence mechanisms of arthropods. I. The composition and function of the spray of the whipscorpion, *Mastigoproctus giganteus* (Lucas) (Arachnida, Pedipalpida). *J. Insect. Physiol.* **6**, 272–298.

Eisner, T., Kafatos, F. C., and Linsley, E. G. (1962). Lycid predation by mimetic adult Cerambycidae (Coleoptera). *Evolution* **16**, 316–324.

Eisner, T., Eisner, H. E., Hurst, J. J., Kafatos, F. C. and Meinwald, J. (1963a). Cyanogenic glandular apparatus of a millipede. *Science* **139**, 1218–1220.

Eisner, T., Hurst, J. J. and Meinwald, J. (1963b). Defense mechanisms of arthropods. XI. The structure function, and phenolic secretions of the glands of a chordeumoid millipede and a carabid beetle. *Psyche* **70**, 94–116.

Eisner, T., Swithenbank, C., and Meinwald, J. (1963c). Defense mechanisms of arthropods. VIII. Secretion of salicylaldehyde by a carabid beetle. *Ann. Entomol. Soc. Am.* **56**, 37–41.

Eisner, T., Alsop, R., and Ettershank, G. (1964a). Adhesiveness of spider silk. *Science* **146**, 1058–1061.

Eisner, T., McHenry, F., and Salpeter, M. M. (1964b). Defense mechanisms of arthropods. XV. Morphology of the quinone-producing glands of a tenebrionid beetle (*Eleodes longicollis* Lec.). *J. Morphol.* **115**, 355–399.

Eisner, T., van Tassell, E., and Carrel, J. E. (1967). Defensive use of a "fecal shield" by a beetle larva. *Science* **158**, 1471–1473.

Eisner, T., Meinwald, Y. C., Alsop, D. W., and Carrel, J. E. (1968). Defense mechanisms of arthropods. XXI. Formic acid and *n*-nonyl acetate in the defensive spray of two species of *Helluomorphoides*. *Ann. Entomol. Soc. Am.* **61**, 610–613.

Eltringham, H. (1932). On an extrusible glandular structure in the abdomen of *Mantispa styriaca*, Poda (Neuroptera). *Trans. Roy. Entomol. Soc. London* **80**, 103–105.

Ernst, E. (1959). Beobachtungen beim Spritzakt der *Nasutitermes*-Soldaten. *Rev. Suisse Zool.* **66**, 289–295.

Fieser, L. F., and Ardao, M. I. (1956). Investigation of the chemical nature of gonyleptidine. *J. Am. Chem. Soc.* **78**, 774–781.

Filshie, B. K., and Waterhouse, D. F. (1968a). The fine structure of the lateral scent glands of the green vegetable bug, *Nezara viridula* (Hemiptera, Pentatomidae). *J. Microscopie* **7**, 231–244.

Filshie, B. K., and Waterhouse, D. F. (1968b). The structure and development of a surface pattern on the cuticle of the green vegetable bug *Nezara viridula*. *Tissue Cell* **1**(2), 273–293.

Fishelson, L. (1960). The biology and behaviour of *Poekilocerus bufonius* Klug, with special reference to the repellent gland. *Eos (Madrid)* **36**, 41–62.

Fraenkel, G. S. (1959). The *raison d'être* of secondary plant substances. *Science* **129**, 1466–1470.

Freeman, M. A. (1968). Pharmacological properties of the regurgitated crop fluid of the African migratory locust, *Locusta migratoria* L. *Comp. Biochem. Physiol.* **26**, 1041–1049.

Fuchs, M. S., Craig, G. B., Jr., and Hiss, E. A. (1968). The biochemical basis of female monogamy in mosquitoes. I. Extraction of the active principle from *Aedes aegypti*. *Life Sci.* **7**, 835–839.

Garb, G. (1915). The eversible glands of a chrysomelid larva, *Melasoma lapponica*. *J. Entomol. Zool.* **7**, 88–97.

Ghent, R. L. (1961). Adaptive refinements in the chemical defense mechanisms of certain Formicinae. Ph.D. Thesis, Cornell Univ., Ithaca, New York.

Gilby, A. R., and Waterhouse, D. F. (1965). The composition of the scent of the green vegetable bug, *Nezara viridula*. *Proc. Roy. Soc. (London)* **B162**, 105–120.

Gilmour, D. (1965). "The Metabolism of Insects." Freeman, San Francisco, California.

Gordon, H. T., Waterhouse, D. F., and Gilby, A. R. (1963). Incorporation of ^{14}C-acetate into scent constituents by the green vegetable bug. *Nature* **197**, 818.

Gressitt, J. L., Sedlacek, J., and Szent-Ivany, J. J. H. (1965). Flora and fauna on backs of large Papuan moss-forest weevils. *Science* **150**, 1833–1835.

Gressitt, J. L., Aoki, J., and Samuelson, G. A. (1966). Epizoic symbiosis. *Pacific Insects* **8**, 221–297.

Gressitt, J. L., Samuelson, G. A., and Vitt, D. H. (1968). Moss growing on living Papuan moss-forest weevils. *Nature* **217**, 765–767.

Happ, G. M. (1968). Quinone and hydrocarbon production in the defensive glands of *Eleodes longicollis* and *Tribolium castaneum* (Coleoptera: Tenebrionidae). *J. Insect Physiol.* **14**, 1821–1837.

Happ, G. M., and Eisner, T. (1961). Hemorrhage in a coccinellid beetle and its repellent effect on ants. *Science* **134**, 329–331.

Happ, G. M., and Meinwald, J. (1965). Biosynthesis of arthropod secretions. I. Monoterpene synthesis in an ant (*Acanthomyops claviger*). *J. Am. Chem. Soc.* **87**, 2507.

Haskell, P. T. (1961). "Insect Sounds." Quadrangle, Chicago, Illinois.

Haskell, P. T., and Belton, P. (1956). Electrical responses of certain lepidopterous tympanal organs. *Nature* **177**, 139–140.

Hinton, H. E. (1951). Myrmecophilous Lycaenidae and other Lepidoptera – a summary. *Proc. S. London Entomol. Nat. Hist. Soc.* 1949–1950, 111–175.

Hinton, H. E. (1968). Structure and protective devices of the egg of the mosquito *Culex pipiens. J. Insect Physiol.* 14, 145-161.

Hurst, J. J., Meinwald, J., and Eisner, T. (1964). Defense mechanisms of arthropods. XII. Glucose and hydrocarbons in the quinone-containing secretion of *Eleodes longicollis. Ann. Entomol. Soc. Am.* 57, 44-46.

Jacobson, M. (1966). Chemical insect attractants and repellents. *Ann. Rev. Entomol.* 11, 403-422.

Jenkins, M. F. (1957). The morphology and anatomy of the pygidial glands of *Dianous coerulescens* Gyllenhal (Coleoptera: Staphylinidae). *Proc. Roy. Entomol. Soc. London* A32, 159-167.

Jones, D. A., Parsons, J., and Rothschild, M. (1962). Release of hydrocyanic acid from crushed tissues of all stages in the life-cycle of species of the Zygaeninae (Lepidoptera). *Nature* 193, 52-53.

Kaiser, E., and Michl, H. (1958). "Die Biochemie der Tierischen Gifte." Deuticke, Vienna.

Karrer, W. (1958). Konstitution und Vorkommen der organischen Pflanzenstoffe (exclusive Alkaloide). Birkhäuser, Basel.

Keele, C. A., and Armstrong, D. (1964). "Substances Producing Pain and Itch." Arnold, London.

Kent, P. W., and Brunet, P. C. J. (1959). The occurrence of protocatechuic acid and its 4-O-β-D-glucoside in *Blatta* and *Periplaneta. Tetrahedron* 7, 252-256.

Lane, C., and Rothschild, M. (1965). A case of Müllerian mimicry of sound. *Proc. Roy. Entomol. Soc. London* A40, 156-158.

Lawrence, D. F. (1950). *Peripatus*: a living museum of antiquites. *African Wild Life* 4, 112-120.

Leston, D., and Pringle, J. W. S. (1963). Acoustical behaviour of Hemiptera. *In* "Acoustical Behaviour of Animals" (R. G. Busnel, ed.), pp. 391-410. Elsevier, Amsterdam.

Linsenmair, K. E. (1963). Das Entspannungsschwimmen. *Kosmos* 59, 331-334.

Linsenmair, K. E., and Jander, R. (1963). Das "Entspannungsschwimmen" von *Velia* und *Stenus, Naturwissenschaften* 50, 231.

Linsley, E. G., Eisner, T., and Klots, A. B. (1961). Mimetic assemblages of sibling species of lycid beetles. *Evolution* 15, 15-29.

Loomis, T. A. (1968). "Essentials of Toxicology." Lea & Febiger, Philadelphia, Pennsylvania.

McAlister, W. H. (1960). The spitting habit in the spider *Scytodes intricata* Banks (Family Scytodidae). *Texas J. Sci.* 12, 17-20.

Manton, S. M., and Heatley, N. G. (1937). Studies on the Onychophora. II. The feeding, digestion, excretion, and food storage of *Peripatopsis. Phil. Trans. Roy. Soc. London* B227, 411-464.

Maschwitz, U. (1964). Gefahrenalarmstoffe und Gefahrenalarmierung bei sozialen Hymenopteren. *Z. Vergleich. Physiol.* 47, 596-655.

Maschwitz, U. (1967). Eine neuartige Form der Abwehr von Mikroorganismen bei Insekten. *Naturwissenschaften* 54, 649.

Matsumoto, T., Yanagiya, M., Maeno, S., and Yasuda, S. (1968). A revised structure of pederin. *Tetrahedron Letters* 60, 6297-6300.

Meinwald, J., Chadha, M. S., Hurst, J. J., and Eisner, T. (1962). Defense mechanisms of arthropods. IX. Anisomorphal, the secretion of a phasmid insect. *Tetrahedron Letters* 1, 29-33.

Meinwald, J., Happ, G. M., Labows, J., and Eisner, T. (1966a). Cyclopentanoid terpene biosynthesis in a phasmid insect and in catmint. *Science* 151, 79-80.

Meinwald, J., Koch, K. F., Rogers, J. E., and Eisner, T. (1966b). Biosynthesis of arthropod secretions. III. Synthesis of simple *p*-benzoquinones in a beetle (*Eleodes longicollis*). *J. Am. Chem. Soc.* 88, 1590-1592.

Meinwald, J., Erickson, K., Hartshorn, M., Meinwald, Y. C., and Eisner, T. (1968a). Defensive mechanisms of arthropods. XXIII. An allenic sesquiterpenoid from the grasshopper *Romalea microptera. Tetrahedron Letters* 25, 2959-2962.

Meinwald, J., Meinwald, Y. C., Chalmers, A. M., and Eisner, T. (1968b). Dihydromatricaria acid: acetylenic acid secreted by soldier beetle. *Science* 160, 890-892.

Meinwald, Y. C., Meinwald, J., and Eisner, T. (1966). 1,2-dialkyl-4(3H)-quinazolinones in the defensive secretion of a millipede (*Glomeris marginata*). *Science* **154**, 390–391.

Meyer, D., Schlatter, C., Schlatter-Lanz, I., Schmid, H., and Bovey, P. (1968). Die Zucht von *Lytta vesicatoria* im Laboratorium und Nachweis der Cantharidinsynthese in Larven. *Experientia* **24**, 995–998.

Meynier, J. (1893). Empoisonnement par la chair de grenouilles infectées par des insectes du genre *Mylabris* de la famille des Méloides. *Arch. Med. Pharm. Mil.* **22**, 54–56.

Monro, A., Chadha, M. S., Meinwald, J., and Eisner, T. (1962). Defense mechanisms of arthropods. VI. *Para*-benzoquinones in the secretion of five species of millipedes. *Ann. Entomol. Soc. Am.* **55**, 261–262.

Moore, B. P. (1967). Hydrogen cyanide in the defensive secretions of larval Paropsini (Coleoptera: Chrysomelidae). *J. Australian Entomol. Soc.* **6**, 36–38.

Moore, B. P. (1968). Studies on the chemical composition and function of the cephalic gland secretion in Australian termites. *J. Insect Physiol.* **14**, 33–39.

Moore, B. P., and Wallbank, B. E. (1968). Chemical composition of the defensive secretion in carabid beetles and its importance as a taxonomic character. *Proc. Roy. Entomol. Soc. London* **B37**, 62–72.

Nakanishi, K., Koreeda, M., Sasaki, S., Chang, M. L., and Hsu, H. Y. (1966). Insect Hormones. I. The structure of ponasterone A, an insect moulting hormone from the leaves of *Podocarpus nakaii* Hay. *Chem. Commun.* pp. 915–917.

Nayler, L. S. (1964). The structure and function of the posterior abdominal glands of the cockroach *Pseudoderopeltis bicolor* (Thunb.). *J. Entomol. Soc. S. Africa* **27**, 62–66.

Nutting, W. L., and Spangler, H. G. (1969). The hastate setae of certain dermestid larvae: an entangling defense mechanism. *Ann. Entomol. Soc. Am.* **62**, 763–769.

Parker, G. H. (1922). "Smell, Taste, and Allied Senses in Vertebrates." Lippincott, Philadelphia, Pennsylvania.

Parsons, J. A. (1965). A digitalis-like toxin in the monarch butterfly, *Danaus plexippus* L. *J. Physiol.* (*London*) **178**, 290–304.

Pavan, M. (1952). "Iridomyrmecin" as insecticide. *Symp. 9th Intern. Congr. Entomol., 1951* **1**, 321–327.

Pavan, M. (1959). Biochemical aspects of insect poisons. *Proc. 4th Intern. Congr. Biochem., Vienna, 1958* **12**, 15–36.

Pavan, M. (1963). Ricerche biologiche e mediche su pederina e su estratti purificati di *Paederus fuscipes* Curt. (Coleoptera Staphylinidae). 93 pp. Industrie Lito-Tipografiche M. Ponzio, Pavia.

Poulton, E. B. (1888). The secretion of pure aqueous formic acid by lepidopterous larvae for the purpose of defence. *Brit. Assoc. Advan. Sci. Rept.* **5**, 765–766.

Raske, A. G. (1967). Morphological and behavioral mimicry among beetles of the genus *Moneilema* (Coleoptera: Cerambycidae). *Pan-Pacific Entomologist* **43**, 239–244.

Regnier, F. E., and Law, J. H. (1968). Insect pheromones. *J. Lipid Res.* **9**, 541–551.

Reichstein, T. (1967). Cardenolide (herzwirksame Glykoside) als Abwehrstoffe bei Insekten. *Naturw. Rundschau* **20**, 499–511.

Reichstein, T., von Euw, J., Parsons, J. A., and Rothschild, M. (1968). Heart poisons in the monarch butterfly. *Science* **161**, 861–866.

Remold, H. (1962). Über die biologische Bedeutung der Duftdrüsen bei den Landwanzen (Geocorisae). *Z. Vergleich. Physiol.* **45**, 636–694.

Roeder, K. D. (1967). "Nerve Cells and Insect Behavior." Harvard Univ. Press, Cambridge, Massachusetts.

Roth, L. M. (1967). Uricose glands in the accessory sex gland complex of male Blattaria. *Ann. Entomol. Soc. Am.* **60**, 1203–1211.

Roth, L. M., and Eisner, T. (1962). Chemical defenses of arthropods. *Ann. Rev. Entomol.* **7**, 107–136.

Roth, L. M., and Stahl, W. H. (1956). Tergal and cercal secretion of *Blatta orientalis* L. *Science* **123**, 798–799.

Roth, L. M., and Stay, B. (1958). The occurrence of *para*-quinones in some arthropods, with emphasis on the quinone-secreting tracheal glands of *Diploptera punctata* (Blattaria). *J. Insect Physiol.* **1**, 305–318.

Roth, L. M., Niegisch, W. D., and Stahl, W. H. (1956). Occurrence of 2-hexenal in the cockroach *Eurycotis floridana*. *Science* 123, 670–671.

Rothschild, M. (1961). Defensive odours and Müllerian mimicry among insects. *Trans. Roy. Entomol. Soc. London* 113, 101–121.

Rothschild, M. (1966). Experiments with captive predators and the poisonous grasshopper *Poekilocerus bufonius* (Klug). *Proc. Roy. Entomol. Soc. London* C31, 32.

Rothschild, M., and Haskell, P. T. (1966). Stridulation of the garden tiger moth, *Arctia caja* L., audible to the human ear. *Proc. Roy. Entomol. Soc. London* A41, 167–170.

Schildknecht, H., and Holoubek, K. (1961). Die Bombardierkäfer und ihre Explosionschemie. V. Mitteilung über Insekten-Abwehrstoffe. *Angew. Chem.* 73, 1–7.

Schildknecht, H., and Hotz, D. (1967). Identification of the subsidiary steroids from the prothoracic protective gland system of *Dytiscus marginalis*. *Angew. Chem. Intern. Ed. English* 6, 881.

Schildknecht, H., and Schmidt, H. (1963). Die chemische Zusammensetzung des Wehrsekretes von *Dicranura vinula*. *Z. Naturforsch.* 18b, 585–587.

Schildknecht, H., and Weis, K. H. (1960). Zur Kenntniss des Pygidialdrüsensekretes vom gemeinen Ohrwurm, *Forficula auricularia*, VI. Mitteilung über Insekten-Abwehrstoffe. *Z. Naturforsch.* 15b, 755–757.

Schildknecht, H., and Weis, K. H. (1962). Zur Kenntniss der Pygidialblasensubstanzen vom Gelbrandkäfer (*Dytiscus marginalis* L.) XIII. Mitteilung über Insektenabwehrstoffe. *Z. Naturforsch.* 17b, 448–452.

Schildknecht, H., and Wenneis, W. F. (1967). Über Arthropoden-Abwehrstoffe XXV. Anthranilsäure als Precursor der Arthropoden-Alkaloide Glomerin und Homoglomerin. *Tetrahedron Letters* 19, 1815–1818.

Schildknecht, H., Holoubek, K., Weis, K. H., and Krämer, H. (1964). Defensive substances of the arthropods, their isolation and identification. *Angew. Chem. Intern. Ed. English* 3, 73–82.

Schildknecht, H., Siewerdt, R., and Maschwitz, U. (1966a). A vertebrate hormone as defensive substance of the water beetle (*Dytiscus marginalis*). *Angew. Chem. Intern. Ed. English* 5, 421.

Schildknecht, H., Wenneis, W. F., Weis, K. H., and Maschwitz, U. (1966b). Glomerin, ein neues Arthropoden-Alkaloid. *Z. Naturforsch.* 21b, 121–127.

Schildknecht, H., Birringer, H., and Maschwitz, U. (1967a). Testosterone as protective agent of the water beetle *Ilybius*. *Angew. Chem. Intern. Ed. English* 6, 558.

Schildknecht, H., Hotz, D., and Maschwitz, U. (1967b). Über Arthropoden-Abwehrstoffe. XXVII. Die C_{21}-Steroide der Prothorakalwehrdrüsen von *Acilius sulcatus*. *Z. Naturforsch.* 22b, 938–944.

Schildknecht, H., Maschwitz, U., and Wenneis, W. F. (1967c). Neue Stoffe aus dem Wehrsekret der Diplopodengattung *Glomeris*. Über Arthropoden-Abwehrstoffe. XXIV. *Naturwissenschaften* 54, 196–197.

Schildknecht, H., Siewerdt, R., and Maschwitz, U. (1967d). Über Arthropoden-Abwehrstoffe, XXIII. Cybisteron, ein neues Arthropoden-Steroid. *Ann. Chem* 703, 182–189.

Schildknecht, H., Maschwitz, U., and Krauss, D. (1968a). Blausäure im Wehrsekret des Erdläufers *Pachymerium ferrugineum*. *Naturwissenschaften* 55, 230.

Schildknecht, H., Maschwitz, E., and Maschwitz, U. (1968b). Die Explosionschemie der Bombardierkäfer (Coleoptera, Carabidae) III. Mitt.: Isolierung und Charakterisierung der Explosionskatalysatoren. *Z. Naturforsch.* 23b, 1213–1218.

Schildknecht, H., Maschwitz, U., and Winkler, H. (1968c). Zur Evolution der Carabiden-Wehrdrüsensekrete. Über Arthropoden-Abwehrstoffe XXXII. *Naturwissenschaften* 55, 112–117.

Schildknecht, H., Winkler, H., and Maschwitz, U. (1968d). Über Arthropoden-Abwehrstoffe XXXI. Vergleichend chemische Untersuchungen der Inhaltsstoffe der Pygidialwehrblasen von Carabiden. *Z. Naturforsch.* 23b, 637–644.

Schlatter, C. Waldner, E. E., and Schmid, H. (1968). Zur Biosynthese des Cantharidins. I. *Experientia* 24, 994–995.

Selander, R. B. (1960). Bionomics, systematics, and phylogeny of *Lytta*, a genus of blister beetles (Coleoptera, Meloidae). *Illinois Biol. Monographs* **28**, 1–295.

Selander, R. B., Miller, J. L., and Mathieu, J. M. (1963). Mimetic associations of lycid and cerambycid beetles (Coleoptera) in Coahuila, Mexico. *J. Kansas Entomol. Soc.* **36**, 45–52.

Sláma, K., and Williams, C. M. 1966. The juvenile hormone. V. The sensitivity of the bug, *Pyrrhocoris apterus*, to a hormonally active factor in American paper-pulp. *Biol. Bull.* **130**, 235–246.

Spielman, A., Leahy, M. G., and Skaff, V. (1967). Seminal loss in repeatedly mated female *Aedes aegypti. Biol. Bull.* **132**, 404–412.

Stay, B. (1957). The sternal scent gland of *Eurycotis floridana* (Blattaria: Blattidae). *Ann. Entomol. Soc. Am.* **50**, 514–519.

Tschinkel, W. R. (1969). Phenols and quinones from the defensive secretions of the tenebrionid beetle, *Zophobas rugipes. J. Insect Physiol.* **15**, 191–200.

Verhoeff, K. W. (1925). Chilopoda. *In* "Klassen und Ordnungen des Tierreiches" (H. G. Bronn, ed.), Vol. 5, pp. 351–365. Akademische Verlagsgesellschaft, Leipzig.

Vézien, M. (1861). Note sur la cystite cantharidienne causée par l'ingestion de grenouilles qui se sont nourries de coléoptères vésicants. *Rec. Mem. Med. Chir. Pharm. Mil.* **4**, 457–460.

von Euw, J., Fishelson, L., Parsons, J. A., Reichstein, T., and Rothschild, M. (1967). Cardenolides (heart poisons) in a grasshopper feeding on milkweeds. *Nature* **214**, 35–39.

von Euw, J., Reichstein, T., and Rothschild, M. (1969). Aristolochic acid-1 in the swallowtail butterfly *Pachlioptera aristolochiae* (Fabr.) (Papilionidae). *Israel J. Chem.* **6**, 659–670.

Waterhouse, D. F., and Gilby, A. R. (1964). The adult scent glands and scent of nine bugs of the superfamily Coreoidea. *J. Insect Physiol.* **10**, 977–987.

Waterhouse, D. F., and Wallbank, B. E. (1967). 2-methylene butanal and related compounds in the defensive scent of *Platyzosteria* cockroaches (Blattidae: Polyzosteriinae). *J. Insect Physiol.* **13**, 1657–1669.

Weatherston, J. (1967). The chemistry of arthropod defensive substances. *Quart. Rev. (London)* **21**, 287–313.

Westwood, J. O. (1839). "An Introduction to the Modern Classification of Insects." Longmans, London.

Williams, C. M. (1967). Third-generation pesticides. *Sci. Am.* **217**(1), 13–17.

Wilson, E. O. (1965). Chemical communication in the social insects. *Science* **149**, 1064–1071.

Wilson, E. O., and Bossert, W. H. (1963). Chemical communication among animals. *Recent Progr. Hormone Res.* **19**, 673–716.

Wray, J. (1670). Some uncommon observations and experiments made with an acid juyce to be found in ants. *Phil. Trans. Roy. Soc. London* pp. 2063–2069.

9

Chemical Ecology of Fish

ARTHUR D. HASLER

I. INTRODUCTION

More species of fish exist today than all other species of land and aquatic vertebrates combined. Perhaps this has happened because as the oldest of the higher vertebrates they have had a longer geological period in which to evolve. Moreover, their aquatic habitat occupies four-fifths of the area of the world, hence it provided vast spaces and innumerable ecological niches in which new species of fish could thrive, compete, or cooperate.

In considering the ecology of fish one observes that their external similarity in shape hides the huge diversity of function found within their bodies and organ systems. It is therefore dangerous to make the generalization that organ physiology is the same from species to species just because fish look so much alike externally. In a comparative estimation, the geological time period between the evolution of salmon and bass might very well exceed that between rodents and primates, hence differences in physiology between salmon and bass might be expected to be at least as great.

Fish have been, over geological time, resourceful in evolving physiological systems for meeting ecological demands and have succeeded in adapting to or invading many inhospitable and difficult habitats.

The most successful adaptive evolutionary "invention" that enabled fish to exploit the aquatic environment was the swimbladder because it freed fish from the force of gravity and enabled them to have access to the enormous food supplies of the open sea. Moreover, to have devised a means for living in the bathypelagic, lightless depths of the ocean (below 600 m) is a noteworthy accomplishment. Surprisingly, it covers 50% of the world's surface and is only now being explored, hence untold wonders in chemical ecology await discovery and exploration.

Adaptations to broad osmotic environments stretch from hard waters of high ionic content to soft, almost distilled water lakes of the high alps, and from slightly brackish estuaries to saline waters, such as the Salton and Caspian Seas, which exceed sea water in salt content. An initial example of osmotic fortitude is the African cichlid *Tilapia grahami*, an endemic species that lives in saline and alkaline hot springs around the shores of Lake Magadi in Kenya (salinity of 20 to 30 parts per thousand, pH 10.5, and temperature 39°C). The alkalinity is due to sodium carbonate. Beadle (1968) kept this species in slowly evaporating water in which they finally succumbed at 80 parts per thousand; however, there was no evidence that the high salinity was actually the cause of death. Nothing is known about the nature of the regulating mechanism which must deal with excess hydroxyl ions. This fish has been introduced into highly alkaline Lake Nakura in Kenya, and shows potential for becoming an important food fish.

Not only is salinity a problem to contend with in osmoregulation, but important ions may be lacking in many types of freshwater. Calcium and magnesium are low in soft water lakes and rivers such as the ion-impoverished pre-Cambrian Shield lakes and the "black" waters of the Rio Negro river of the Amazon system.

In waters influenced by man, poisonous chemicals, from which fish cannot escape, reduce the quality of the environment. Arsenic and copper salts used in aquatic weed and algae control are hazards to fish life. Other heavy metals, pesticides, some industrial wastes, and agricultural runoff make life difficult for fish. Calcium and strontium from radioactive fallout are concentrated in the hard tissues of fish, the consequences of which should be better understood by chemical ecologists.

Several species of fish conduct life processes and grow well in arctic and antarctic seas at −1.7°C. What marvelous biochemical adjustment makes this possible?

Tropical marshes are occupied by air-breathing fishes. In fact, the lungfish inhabiting ephemeral African lakes can pass into a state of suspended animation within a cocoon imbedded in a cake of dry lake mud. It can thus survive a drought for many months. Another amazing adaptation to drought is the East African cyprinodont *Notobranchus taeniopygus* which produces drought-resistant eggs that hatch soon after intermittent rains refill their habitat.

Through a modification of its muscles, the electric eel can produce enough electrical energy to stun its prey in water of low ionic content; torpedo-rays can do it even in highly conductive sea water, obviously with electric organs differently adapted. Other electric fish use weak charges of electricity to defend territories, to find food, and for orientation. Their chemical ecology and physiology needs further elucidation (Marshall, 1966).

Because of its high latent heat, water does not change temperature as rapidly as air, yet it is a more complex and a more variable chemical medium in which to live. Since it is also a capricious environment in which the margins of respiratory safety are less than in air, fish are constantly in danger of asphyxiation. Available to terrestrial arborial animals are 200 ml of oxygen in each liter of air, while a trout in water has only 8.8 ml of dissolved oxygen per liter at 20°C. To make matters worse, a pollutant using as little as one-half of this amount could bring about respiratory distress in a trout; many species asphyxiate at 3 mg/liter.

Not only do fish live in lightless depths, but even in the clearest water vision cannot be very good because water selectively absorbs sunlight reducing its intensity and limiting its quality as it penetrates deeper. Moreover, water may be turbid from silt, or tea-colored by lignious colloids, especially in bog lakes and brownwater rivers, thus restricting underwater visibility. In such a medium it might be suspected that fish would have applied their sense of smell to detect organic substances and use them as guide posts to find their way about. Indeed they have, because odors are relied upon for homing in salmon (Hasler and Wisby, 1951), for predator avoidance (von Frisch, 1941), and in sex and care of young (Kühme, 1963). Taste is used for food location (Bardach *et al.*, 1967) and aggregation of individuals of the same species (Hemmings, 1966). In fact, the sensitivity of fish to olfactory stimuli is astonishing. For example, an eel, by conditioned response training (Teichmann, 1957) can detect β-phenylethyl alcohol in concentrations of 3×10^{-18} parts, and minnows can easily discriminate by olfaction α-chlorophenol from p-chlorophenol at dilutions greater than can man (Hasler and Wisby, 1950).

The above chemical problems associated with fish and their environment are only a selected few, but even these could be the subject matter for a thick book. I chose therefore to discuss only a few examples.

II. THE CHEMISTRY OF THE SWIMBLADDER

And the fish suspending themselves so curiously
below there – and the beautiful curious liquid.
Walt Whitman, "Assimilation"

The evolution of the swimbladder and the physicochemical wonders that provided its hydrostatic functions were the miracle "inventions" of prehistoric physiology and evolution, for it gave fish an energy-saving device that freed them from gravitation and thus enabled them to invade and benefit from the food-rich pelagic zone and the pitch dark bathypelagic regions below 600 m.

The swimbladders in deep-dwelling coregonids in Lake Constance Central Europe and in the Great Lakes of North America have astonishingly high amounts of nitrogen (99% by volume the total gases). How can a biologically inert gas accumulate to such high levels? We will return to this question shortly.

Some species of fish have a duct connecting the swimbladder to the gullet, hence gas can be released readily if the pressure decreases as the fish rises. These types, found mostly in freshwater, are grouped anatomically as physostoma. Other species, including the bathypelagic fish, have no such connection (physoclisti), and must relieve the pressure by an absorptive mechanism, a very vascular oval organ in the swimbladder. In addition, they have a more highly developed gas gland and a rich capillary, countercurrent structure, the rete mirabile; these, in combined action, "pump" up the pressure of the gases (principally O_2, CO_2, and N_2) to make the fish neutrally or positively buoyant.

Bathypelagic fish live at depths of 200–10,000 m (i.e., below the photosynthetic or algae-producing zone of the ocean), but they are quite small (only 50 mm in length). Yet in studies of the deep scattering layer on which they are recorded with echo sounders, they ascend and descend 300 m in a single day (Backus et al., 1968). As might be deduced from the correlation of anatomy and function, their gas glands and rete mirabile are huge when compared with shallower living physoclisti (Marshall, 1966).

> It now seems that one function of the gas gland is to produce a substance that releases these additional gases from the outgoing blood in the retia. The substance may well be lactic acid, which on entering the outgoing capillaries will have two effects on the blood. It will reduce the solubility of all gases and release some of the oxygen bound to haemoglobin. The pressure of each gas in the outgoing blood will thus become greater than that in the ingoing blood. As long as this extra pressure is maintained, gas will diffuse from outgoing to ingoing capillaries and so be carried towards the gas gland, where it will be concentrated and multiplied . . . Oxygen, nitrogen and even the rare gases such as argon are evidently secreted in this way.
>
> From Marshall, "The Life of Fishes," p. 75

When Hüfner (1892) found nitrogen to be 99% (by volume) of the gas in the swimbladder of deepwater whitefish (Coregonus) and at a hydrostatic pressure of 8 atm, physiologists doubted the accuracy of his determination until it was confirmed in a North American species by Scholander et al. (1956). Just how this high percentage is achieved and maintained at enormous pressures is yet to be explained.

The swimbladder amounts to only 7% by volume of freshwater fish and 5% in marine species, yet in those species in which oxygen is the principal gas, it is not used for auxiliary respiration. In fact, there is never enough oxygen to supply the respiratory needs for more than 5 min, even if it could be absorbed into the bloodstream in time to be useful. If gas is withdrawn from the swimbladder of a perch by a hypodermic needle, carbon dioxide is the principal gas used to refill it, but oxygen later becomes the major constituent.

When a fish swims in water containing only the dissolved nuclid of oxygen ^{18}O, the gas in the swimbladder becomes fully labeled and is deposited as molecular oxygen without a preliminary split in oxygen atoms (Wittenberg, 1961). One can assume therefore that there is a direct transport of oxygen from the gills to the swimbladder.

Carbonic anhydrase is high in the gas gland, indicating its probable role in releasing carbon dioxide; and more recently, high energy phosphate compounds have been found in the blood leaving the gas gland (Scholander, personal communication), therefore lending circumstantial evidence to the view that secretory work is performed there.

III. SEX HORMONES

Gonad-stimulating hormones of the pituitary constitute a fascinating chapter in the biochemistry of fish (Pickford and Atz, 1957). Spawning, initiated quite suddenly from ovaries whose eggs are already ripe, appears to be influenced by several environmental factors that affect the release of hormones from anterior pituitary glands and initiate the release of eggs.

In trout, muskellunge, and carp, release of eggs appears to be induced by the changing length of day coupled with appropriate temperature; however, in the equatorial South American tropics, where day length and temperature change very little, it is the rains which come suddenly after a long drought that induce spawning. The monsoons appear to be the stimulus for Indian carp. These environmental stimuli can be substituted artificially by injections of fish pituitaries (Hasler et al., 1939). Transferring pituitary-injected gray mullet from fresh- to marine water seems to augment hormonal action for a reason as yet unexplained (Yashouv, 1968).

Herein lies a host of opportunities for chemists interested in ecology. First, the active hormones have not been isolated, nor are the intermediary events well understood. Moreover, the ecological factors that initiate the hormonal sequence need further elucidation.

IV. ECOLOGY OF OLFACTION

My principal research interest in recent years is to try to determine the cues that fish use for orientation and homing (Hasler, 1966). Two decades ago I postulated that adult salmon locate their parent stream in a coastal river system by recognizing the specific aroma that distinguishes its home tributary from any other river in the system. Every stream has a different odor (Hasler and Wisby, 1951). Apparently the soils and vegetation of a drainage basin lend a fragrance which is distinctive for every stream.

Field and laboratory experiments by us and others have largely confirmed this hypothesis. Fish detect these fragrances in fantastically low concentrations, yet these diluted natural odors appear to imprint their chemical identity upon the "memory" of young fish before they leave for the sea to grow to maturity. Since early conditioning to an odor specific to their home stream is retained to adulthood, the sexually mature fish return and spawn in the stream that smells like home after passing up one stream after another that does not. We are ignorant of the chemical nature of these odors and of the physiochemical events that transpire in the nerve and conducting tissues of fish.

Minnows can discriminate between aquatic plants in their environment by odor differences (Walker and Hasler, 1949). During spawning, red fin shiners recognize the nest of the green sunfish by the odor of the eggs and milt of that species, and are induced to deposit their eggs on the nest of the host species for care during incubation (Hunter and Hasler, 1965). We do not know the chemical nature of this attractant.

Mouthbreeding cichlid fishes can distinguish the odor of their own young from the body odor of the young of another female of the same species (Kühme, 1963). Hence, a plethora of smell-related reactions influence the lives of fish.

Certain species of cyprinid fishes are endowed with both repellent *Schreckstoff* (von Frisch, 1941) and attractant odors that are used in their chemical ecology. Individuals of the fish *Rutilus rutilus* L. are repelled by inflowing water containing filtered extract of ground *Rutilus* skin, but are strongly attracted to water in which uninjured individuals of this species have been previously kept (Hemmings, 1966). Although the strength of response of the fish to the attractant odor undergoes both short-term and long-term declines, it still remains significant even after 180 days. In fact, the attractant odor is as important as a visual cue (presentation of an individual of the same species) in keeping members of this cyprinid species together. In light of these behavioral data, Hemmings suggests that the attractant odor plays an important role in maintaining aggregation or schools of fish, especially at night.

V. OXYGEN AND CARBON DIOXIDE

Dissolved oxygen may reach nearly zero in shallow ponds at night; it is low to absent in marshes, in the deeper water of bog lakes, and under the snow-covered ice of shallow lakes in late winter. Moreover, the depths (hypolimnion) of eutrophic lakes lying in nutrient-rich drainage basins lose oxygen by midsummer and again in late winter. In fact, the great and fantastically deep tropical Lake Tanganyika (a meromictic lake) has oxygen only to a depth of 25 m, a very small part of its 1750-m depth. The depths of the Black and Dead Seas and even some deep parts of the ocean are also anoxic. These widely occurring anoxic waters are obviously fish deserts. As the deep water of a eutrophic lake in the temperate zone loses oxygen in midsummer, the fish are forced to rise to shallower zones, hence seasonal migrations to livable water are forced upon them. In meromictic lakes the anoxic areas are permanent barriers to the depths.

The brief terrestrial occupancy of grunion (*Leuresthes tenuis*) during spawning is one of the most unique illustrations of chemical ecology involving oxygen. During spawning, this atherinid fish is out of the water for as long as several minutes and exerts considerable muscular activity. It escapes asphyxial difficulties under these conditions by developing physiological responses similar to those used by many diving and nondiving vertebrates. When the fish is removed from the water (asphyxial period), it shows a marked bradycardia, a sharp increase in muscle lactate, and virtually no rise in blood lactate (Scholander *et al.*, 1962). When the fish is returned to the water, it resumes a normal rate of heartbeat and shows a gradual decline in muscle lactate, but a sharp initial increase in blood lactate. These observations indicate that the circulation to the muscles is shut off during the asphyxial period, but opened for recovery hyperemia when the fish returns to the water.

Two Canadians, Idler and Clemens (1959), have developed a fascinating biochemical program for migrating salmon, in which they have attempted to evaluate the effect of hydroelectric dams in sapping the strength of salmon migrating upstream to their spawning grounds. Delays at the base of each dam place stresses upon the energy reserves of the fish that had ceased feeding when they entered the river from the sea.

Female sockeye salmon migrating up the Fraser River and swimming from Soda Creek to Fort Hames (421 km and 275 m altitude) consume 26.2% of their fat reserves and 8.5% of their protein. This amounts to a utilization of 2.3 cal/m. Hence, a stock of salmon can easily be destroyed through exhaustion by delaying the migratory run or creating obstructions, all of which tax the limited energy reserves beyond recovery. Spurts of speed to negotiate a waterfall create problems in oxidizing lactic acid for an animal with an inefficient circulatory system such as a fish. Stresses of stream velocity encountered normally and by man-made barriers take their toll. Brett (1964) has measured these in the laboratory showing how very real these hardships are (Fig. 1).

Goldfish use less oxygen per gram of tissue when they are in groups than when isolated (Schlaifer, 1939). Hence social ecology must be considered in metabolic studies.

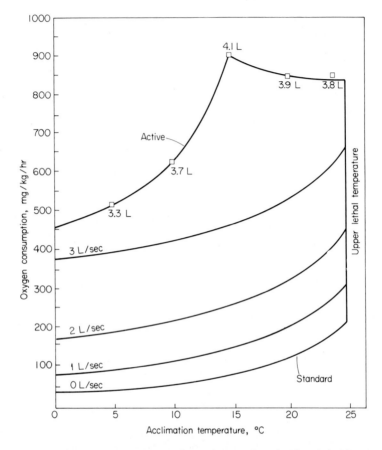

Fig. 1. Relation between oxygen consumption and temperature at various swimming speeds for yearling 50-gm sockeye salmon (*Oncorhynchus nerka*). Active oxygen consumption at each acclimation temperature is given with maximum sustained swimming speed. The relation at the standard level reflects maintenance expenditure. From ''The Respiratory Metabolism and Swimming Performance of Young Sockeye Salmon'' by J. R. Brett (1964), used with permission of *J. Fisheries Res. Board Can.*

VI. TEMPERATURE

Temperature has a marked effect on the distribution of fish (Uda, 1957) (Fig. 2).

Except for the tunas, fish as poikilothermic (cold blooded) animals are incapable of raising their temperature above the water in which they live. As a consequence, all metabolic processes change in rate as the temperature varies. Massive fish kills have been observed on occasions when sudden decreases in temperature occur as a cold front chills the shallow coastal water (Wells *et al.*, 1961). For an unanswered reason, fish are less tolerant of a rapid decrease in the ambient temperature than of a rapid increase. Perhaps because adaptation to low temperatures in fish proceeds at a slower rate than at higher temperatures, mortality may occur following a rapid drop.

A big-eye tuna, *Thunnus obesus*, and the yellow fin, *T. albacares*, can build up temperatures in their swimming muscles that are 3°C to 12°C higher than the ambient water (Carey and Teal, 1966). Carey and Teal believe this is accomplished by a countercurrent system in the blood capillaries similar to that in the swimbladder (i.e., by means of a retia of parallel vessels). Tunas are powerful fish, achieving speeds of up to 70 km/hr, and appear to have the ability to keep their muscles warm. This countercurrent, heat-exchange system provides a thermal barrier to heat loss at the gills and allows the chemical processes in the muscles to proceed presumably at more efficient levels, hence approaching the capabilities of warm-blooded animals.

Temperate zone, arctic, and antarctic fishes have adapted interestingly to a medium of low temperature. Moreover, in deep, oligotrophic

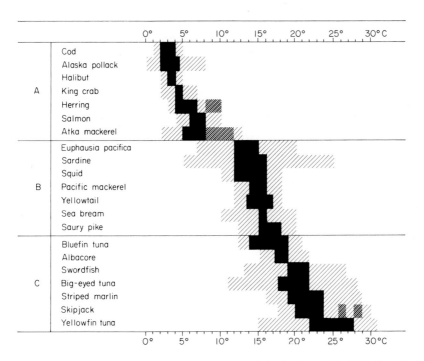

Fig. 2 Optimum water temperature spectra of important fishes in Japan (Uda, 1957). From *Ecology and Resource Management* by K. Watt (1968), used with permission of McGraw-Hill Book Company.

temperate lakes, cold-loving species such as lake trout and cisco conduct life's processes at temperatures of 2° to 6°C., and even spawn at these temperatures; their eggs incubate at 1°–2°C and hatch in the spring. The metabolic rates of certain polar fishes at temperatures close to 0°C would be consistently 0.6 to 0.7 logarithmic units (4–5 times) higher than the average rates for temperate fishes extrapolated to about 0°C (Wohlschlag, 1963). An antarctic fish *Trematomus bernacchi* living in water never exceeding −1.9°C grows as fast as many temperate fishes. What biochemical evolution makes capabilities such as this possible? Interestingly, the arctic fjord cod (*Gadus ognac*) has an antifreeze component in the nonprotein nitrogen fraction of its blood which enables it to survive with no signs of freezing for periods of several hours, even when in continuous contact with ice (Gordon *et al.*, 1959).

Heat pollution will increase as more nuclear-powered generators are built. The heat regime, and therefore the chemistry, of an entire lake or estuary will be altered, e.g., Lake Cayuga (Eipper *et al.*, 1968).

VII. EFFECT OF CHANGING THE CHEMICAL ENVIRONMENT SYNTHETICALLY

Medium to hard water lakes (50–150 ppm, $CaCO_3$) have long been associated in nature with high productivity for plankton and fish, whereas soft water lakes and brownwater bog lakes are poor producers of fish foods. To test this observation experimentally, we divided an hourglass-shaped, soft water, bog lake (Peter and Paul Lakes) of pH 5.9 and an alkalinity of 5 mg/liter/$CaCO_3$ with an earthen barrier. We added to one of these lakes (Peter) sufficient commercial hydrated lime, $Ca(OH)_2$ + $Mg(OH)_2$, to raise the pH to 7.3 (Johnson and Hasler, 1954). The lime, which coagulates the suspended colloids in the treated water, makes it more transparent and increases the penetration of light by a factor of 3 (Stross and Hasler, 1960). This increased clarity of water enables photosynthesis of algae to occur in deeper water, and for zooplankton to flourish, hence permitting trout to forage for food in the cool but now oxygen-enriched thermocline, an area that previously had no oxygen. In the untreated twin reference lake (Paul), the oxygenless condition persisted, and trout were unable to feed in this zone, dying of lack of oxygen during the winter.

In an attempt to evaluate the effect of this treatment on trout, we (Stross *et al.*, 1961) studied the rate of turnover of the principal food of trout, planktonic *Daphnia*. The population of this crustacean waterflea renewed itself every 2.1 weeks in Peter Lake, but only every 4.6 weeks in Paul. In spite of this dramatic increase in rate of production of trout food, an increase in the growth rate of rainbow trout was not discernible (Johnson and Hasler, 1954), owing perhaps to the inadequacies of the field techniques then available, or perhaps of the difficulty in sampling a pelagic population of fish.

As simple as this ecological experiment appears to be, a multitude of events occur in the fish's habitat upon the addition of the simple compound $Ca(OH)_2$: (a) the pH is raised; (b) bog colloids are removed, hence increasing light penetration; (c) the HCO_3^- ion increases; (d) iron is precipitated (70% in the above treatment); (e) phosphorus is apparently made more available to algae; and (f) certain other unmeasured chemical interactions take place.

Obviously, a host of chemical and biological events are influenced, most of which lead to a more efficient utilization of the environment by the planktonic and litoral community. In addition, species that were not present before appear and their turnover rate increases.

More research needs to be done on the influence of the drainage basin of a lake, river, or estuary on the biota of the water because it is principally the nature of the runoff from the land which gives a water its chemical individuality and determines its productivity for fish.

A. Ecology of Rain and Cultural Additives

Rain carries sulfate into Swedish lakes from air systems wafting from Britain and central Europe. This affects softwater lakes by decreasing the pH from 5.6 to 4.6 and causes fish kills.

Dust-bearing mercury is also washed out of the atmosphere by rain. Fish absorb it from the water and it is passed on to fish-eating birds, affecting both adversely. Man gets a share of it too. Dead wild birds were found poisoned by mercury from seeds that had been impregnated with fungicides used by farmers. The study of this case revealed that, in addition, many freshwater fishes such as northern pike *Esox lucius* (Sjörstrand, 1967), contained much more mercury than the highest limit (0.05 mg Hg/kg fresh weight) permitted for food by the World Health Organization (WHO). As to fish and other aquatic organisms, the main mercury source did not prove to be agricultural, but industrial, for instance, paper pulp factories. They use in their processes phenylmercury, which predominates in the fish (Landell, 1968). Further studies disclosed that phenylmercury in nature is rapidly transformed into methylmercury, perhaps by microorganisms. In contrast to other mercury compounds, dimethylmercury is very volatile.

Some domestic effluents notably from food industries, pulp mills, and city sewage plants have an oxygen demand upon naturally available dissolved oxygen. These wastes are the greatest enemies of fish life, not only in rivers, but in estuaries and fjords; in fact, in the Oslo fjord, four species of fish previously common until 1897 are no longer able to exist because of the adverse action of urban and industrial wastes (Ruud, 1968). Moreover, the catches of the commercial fishery have also decreased.

Pesticides have rightfully received attention, and I hesitate to discuss them again. Importantly, it is believed that in Lake Michigan coho salmon eggs do not hatch because the insecticide DDT has accumulated to toxic levels from the environment (H. E. Johnson, 1968). The insecticide, when washed into lakes, is absorbed by the fish from food organisms, but Robert Reinert (1967) and Premdas and Anderson (1963) observed that fish absorb DDT and Dieldrin directly from the water and take it up faster than do either algae or *Daphnia*. Food chain transmission of chemicals in the habitat not only provides the ecological chemist with the opportunity to study basic ecology, but makes a contribution to environmental quality as well. D. W. Johnson (1968) and Kleinert *et al.* (1968) review the recent literature on this very grave problem as it relates to fish. There is a need to study the more subtle effects of prolonged exposure at sublethal levels of pesticides, for they doubtless interfere with biochemical systems, which may handicap the fish behaviorally and physiologically. Moreover, the toxicity varies widely with the solutes and suspensoids in water.

The introduction of soft or hard detergent into the water can seriously hamper the ability of certain fish (i.e., bullheads) to locate food sources (Bardach *et al.*, 1965). Detergents in concentrations as low as 0.5 ppm can cause histological damage of the taste buds on the barbels of these fish. This damage, in turn, leads to electrophysiological and behavioral impairments, the recovery from which is not realized even after the fish have been maintained for more than 6 weeks in water free of detergents.

These examples are but a few; they point up a host of problems in chemical ecology brought on by modern man as he changes the fish's environment.

B. Nitrogen from Rain

In light of an age-old practice of fertilizing fish ponds, it is astonishing to find that exhaust from combustion engines and industry contributes more nitrogen (in the form of NO_3^-) to Lake Mendota (Wisconsin) through rain to the drainage basin than does sewage and runoff from manured farms (Table I) (Lee *et al.*, 1966). This serves as a fertilizer to stimulate algae growth, thus making the lakes more nutrient rich (eutrophic). While such lakes produce more fish, the varieties change from noble fishes to less desirable species. Nevertheless, the chemical fertilization of ponds, lakes, and estuaries to increase protein production for food-scarce parts of the world provides chemical ecologists with new problems and opportunities to serve mankind.

VIII. ANALYSIS OF ECOSYSTEMS

Ideally, physical and chemical experiments conducted in the laboratory on cause and effect attempt to hold constant all variables except one. In ecological experiments out-of-doors, this refined specification is

TABLE I

Estimated Nitrogen and Phosphorus Reaching Wisconsin Surface Waters[a]

Source	Pounds/year		% of total	
	N	P	N	P
Municipal treatment facilities	20,000,000	7,000,000	24.5	55.7
Private sewage systems	4,800,000	280,000	5.9	2.2
Industrial wastes[b]	1,500,000	100,000	1.8	0.8
Rural sources				
Manured lands	8,110,000	2,700,000	9.9	21.5
Other cropland	576,000	384,000	0.7	3.1
Forest land	435,000	43,500	0.5	0.3
Pasture, woodlot, and other lands	540,000	360,000	0.7	2.9
Ground water	34,300,000	285,000	42.0[c]	2.3
Urban runoff	4,450,000	1,250,000	5.5	10.0
Precipitation on water areas	6,950,000	155,000	8.5	1.2
	81,661,000	12,557,500	100.0	100.0

[a] Schraufnagel *et al.* (1967).

[b] Excludes industrial wastes that discharge to municipal systems. Table does not include contributions from aquatic nitrogen fixation, waterfowl, chemical deicers, and wetland drainage.

[c] Source of nitrogen is principally from rain.

rarely met, for nature is exceedingly capricious and complex. A fish, for example, is affected by numerous biological factors: through its senses, by its associates, its school, predators, prey, algae and aquatic plants, in addition to the ever-changing physical and chemical characteristics of the water, which vary with depth and season.

While natural history and descriptive hydrobiology still leave many gaps, the modern ecology of fishes is superimposing on this tradition a need for a higher level of integrative research. A population of fish is part of an entire ecosystem, the analysis of which requires the simultaneous and continuous recording of measurements of the scores of interactions that take place in that system. For example, in order to evaluate the effect of a major environmental change upon the system or any of its parts, it is essential to plan and execute a large out-of-doors experiment in which a major environmental change is produced artificially. Such a grandiose program of evaluation is now being planned under the International Biological Program (IBP).

Ecologists, as a community of scholars, are now able to execute a study on a comprehensive scale which was never possible in the past because each ecologist was capable of coping only with his own limited study of a microcosm. We are now projecting toward a higher level of ecology — the analysis of ecosystems under the leadership of Frederick Smith, University of Michigan.

IBP ecologists are affiliating themselves with computer specialists and systems analysts in order to help them create models of an ecosystem and to plan an experiment that would make it possible to describe a normal and treated system, understand the operation of its diverse species and the flow of energy from the sun through the system to fish, and to identify the factors which influence the stability of the system. This seems to be possible; moreover, it is expected that if a wild drainage basin were needed in the future for an urban development, the landscape to be changed could be simulated by using facts from these intensively studied materials and artificially changed basins. Models would be constructed from which predictions could be made as to the detrimental effects of major treatments or other man-made changes. It would be possible to evaluate the effect it would have on the fish populations of a lake, river, or estuary. If, for example, it were planned to cover half of the drainage basin with concrete, it would be possible to predict the consequences. If an industry were located in a basin whose smoke or fluid effluent contained a contaminant such as mercury, this pollutant's possible impact on the basin in which the industry was located could be evaluated.

The new ecology lies ahead of us. It will be an intellectually exciting era. Moreover, ecologists can become more active in the decision-making process of managing our landscape and our waters more intelligently and less harmfully.

References

Backus, R. H., Craddock, J. E., Haedrich, R. L. Shores, D. L., Teal, J. M. Wing, A. S., Mead, G. W., and Clarke, W. D. (1968). *Ceratoscopelus maderersis*: Peculiar sound-scattering X layer identified with this Myctophid fish. *Science* **160**(3831), 991–993.

Bardach, J., Fujuja, M., and Holl, A. (1965). Detergents: effects on the chemical senses of the fish *Ictalurus natalis* (Le Sueur). *Science* **148**(3677), 1605–1607.

Bardach, J., Todd, J. H., and Crickmer, R. (1967). Orientation by taste in fish of the genus *Ictalurus*. *Science* **155**(3767), 1276–1278.

Beadle, L. C. (1968). Osmotic regulation and the adaptation of freshwater animals to inland saline waters. *Verhandl. Intern. Ver. Limnol.* **17.**

Brett, J. R. (1964). The respiratory metabolism and swimming performance of young sockeye salmon. *J. Fisheries Res. Board Can.* **21**(5), 1183–1226.

Carey, F. G., and Teal, J. M. (1966). Heat conservation in tuna fish muscle. *Proc. Natl. Acad. Sci. U.S.* **56**, 1464–1469.

Eipper, A. W., *et al.*, (1968). Thermal pollution of Cayuga Lake by a proposed power plant. Cornell University, Ithaca, New York.

Gordon, M. S., Amdur, B. H., and Scholander, P. F. (1959). Further observations on supercooling and osmoregulation in Arctic Fishes. *Intern. Oceanog. Congr., New York* pp. 234–356.

Hasler, A. D. (1966). "Underwater Guideposts," 155 pp. Univ. of Wisconsin Press, Madison, Wisconsin.

Hasler, A. D., and Wisby, W. J. (1950). Use of fish for the olfactory assay of pollutants (phenols) in water. *Trans. Am. Fisheries Soc.* **79**, 64–70.

Hasler, A. D., and Wisby, W. J. (1951). Discrimination of stream odors by fishes and relation to parent stream behavior. *Am. Naturalist* **85**, 223–238.

Hasler, A. D., Meyer, R. K., and Fields, H. M. (1939). Spawning induced prematurely in trout with the aid of pituitary glands of the carp. *Endocrinology* **25**(6), 978–983.

Hemmings, C. C. (1966). Olfaction and vision in fish schooling. *J. Exptl. Biol.* 45(3), 449–464.
Hüfner, G. (1892). Zur physikalischen Chemie der Schwimmblasengase. *Arch. Anat. Physiol., Physiol. Abt.* pp. 57–79.
Hunter, J. R., and Hasler, A. D. (1965). Spawning associations of the redfin shiner, *Notropis umbratilis*, and green sunfish, *Lepomis cyanellus. Copeia* No. 3, 265–281.
Idler, D. R., and Clemens, W. A. (1959). The energy expenditures of Fraser River sockeye salmon during the spawning migration to Chilko and Stuart Lakes. *Intern. Pacific Salmon Fisheries Comm., Progr. Rept.* No. 6, 80 pp. New Westminster, British Columbia.
Johnson, D. W. (1968). Pesticides in fishes – A review of selected literature. *Trans. Am. Fisheries Soc.* 97(4), 398–424.
Johnson, H. E. (1968). DDT infects coho in Lake Michigan. Wisconsin State Journal, Nov. 14.
Johnson, W. E., and Hasler, A. D. (1954). Rainbow Trout production in dystrophic lakes. *J. Wildlife Management* 18(1), 113–134.
Kleinert, S. J., Degurse, P. E., and Wirth, T. L. (1968). Occurrence and significance of DDT and Dieldrin residues in Wisconsin fish. Tech. Bull. No. 41. Wisconsin Dept. of Natural Resources, Madison, Wisconsin.
Kühme, W. (1963). Chemisch ausgelöste Brutflege-und Schwarm-reaktionen bei *Hemichromis bimaculatus* (Pisces). *Z. Tierpsychol.* 20, 688–704.
Landell, N. E. (1968). "Eageldöd fiskhot, kvicksilver," 133 pp. Aldus/Bonniers, Stockholm.
Lee, G. F., *et al.* (1966). Report on the nutrient sources of Lake Mendota. January 3, 37 pp. Nutrient Sources Subcommittee of the Technical Committee, Lake Mendota Problems Committee, Madison, Wisconsin. Available from University of Wisconsin Water Chemistry Laboratory, Madison, Wisc.
Marshall, N. B. (1966). "The Life of Fishes," 402 pp. World, Cleveland, Ohio.
Pickford, G. E., and Atz, J. W. (1957). "The Physiology of the Pituitary Gland of Fishes," 613 pp. New York Zool. Soc., New York.
Premdas, F. H., and Anderson, J. M. (1963). The uptake and detoxification of C^{14} labelled DDT in Atlantic Salmon, *Salmo salar. J. Fisheries Res. Board Can.* 20(3), 827–837.
Reinert, R. (1967). The accumulation of Dieldrin in the alga (*Scenedesmus obliquus*), daphnia (*Daphnia magna*), guppy (*Lebistes reticulatus*) food chain. Ph.D. Thesis, Univ. of Michigan, Ann Arbor, Michigan.
Ruud, J. T. (1968). Changes since the turn of the century in the fish fauna and fisheries of the Oslo-fjord. *Helgolaender Wiss. Meeresuntersuch.* 17, 510–517.
Schlaifer, A. (1939). An analysis of the effects of numbers upon the oxygen consumption of *Carrasius auratus. Physiol. Zool.* 12, 381–392.
Scholander, P. F., van Dam, L., and Enns, T. (1956). Nitrogen secretion in the swimbladder of whitefish. *Science* 123(3185), 59–60.
Scholander, P. F., Bradstreet, E., and Garey, W. F. (1962). Lactic acid response in the grunion. *Comp. Biochem. Physiol.* 6(3), 201–203.
Schraufnagel, F. H., Corey, R. B., Hasler, A. D., Lee, G. F. and Wirth, T. L. (1967). Excessive water fertilization. January 31, 58pp. (Mimeo). Report to the Water Subcommittee, Natural Resources Committee of State Agencies, Madison, Wisconsin.
Sjörstrand, B. (1967). Pike (*Esox lucius* L.) and some other aquatic organisms as indicators of mercury contamination in the environment. *Oikos* 18, 323–333.
Stross, R. G., and Hasler, A. D. (1960). Some lime induced changes in lake metabolism. *Limnol. Oceanog.* 5(3), 265–272.
Stross, R. G., Neess, J. C., and Hasler, A. D. (1961). Turnover time and production of the planktonic crustacea in limed and reference portions of a bog lake. *Ecology* 42(2), 237–245.
Teichmann, H. (1957). Das Riechvermögen des Aales (*Anguilla anguilla* L.). *Naturwissenschaften* 44, 242.
Uda, M. (1957). A consideration on the long years' trend of fisheries fluctuations in relation to sea conditions. *Nippon Suisan Gakkaishi* 23, 7–8.
von Frisch, K. (1941). Die Bedeutung des Geruchsinnes im Leben der Fische. *Naturwissenschaften* 29, 321–333.
Walker, T. J., and Hasler, A. D. (1949). Olfactory discrimination of aquatic plants by the bluntnose minnow (*Hyborhynchus notatus*). *Physiol. Zool.* 22(1), 45–63.

Wells, H. W., Wells, M. J., and Gray, I. E. (1961). Winter fish mortality in Pamlico Sound, North Carolina. *Ecology* **42**, 217–219.

Wittenberg, J. B. (1961). The secretion of oxygen into the swimbladder of fish. I. The transport of molecular oxygen. *J. Gen. Physiol.* **44**, 521–526.

Wohlschlag, D. E. (1963). An Antarctic fish with unusually low metabolism. *Ecology* **44**, 557–564.

Yashouv, A. (1968). Personal communication.

10

The Chemistry of Nonhormonal Interactions: Terpenoid Compounds in Ecology

RAYMOND B. CLAYTON

I. INTRODUCTION

There is very little of ecology that cannot be subsumed under the heading of "nonhormonal interactions." However, since the area of my own special interest has to do with terpene and steroid metabolism, I shall confine myself to topics that fall within this general field. It will already have been apparent from several preceding chapters of this book that this is justifiable by the fact that terpenoid and steroid compounds have a peculiarly important place in ecological interactions. It is in this area more than in most others that many organisms have exploited the biochemical labors of other species for their own ends, or have developed synthetic capacities of their own that lead to specifically ecologically active compounds.

Starting from this point of view then, I shall review in broad outline the present status of the biochemistry of terpenoid and steroid metabolism, with pauses at appropriate points to indicate the special relevance of these areas of biochemistry to some ecological questions. I shall necessarily be very selective in the emphasis of details which will often reflect my own past and present research interests. There have been a number of recent comprehensive reviews in which more detailed discussions of the general aspects of terpenoid and steroid metabolism can be found (1–7). A recent volume summarizes studies of terpenoids in plants (7a).

II. CHEMICAL ECOLOGY OF TERPENOIDS AND STEROIDS

A. Broad Outlines of Cholesterol Biosynthesis

The composite representation of the results of a number of studies (Fig. 1) carried out prior to the mid-1950's makes a suitable starting point for the discussion.

The fact that the pathway from acetate to cholesterol included squalene, a C_{30} triterpene hydrocarbon found in shark liver, and the C_{30} trimethyl sterol, lanosterol, was established largely by the work of Bloch and his colleagues (8, 9). His early studies of the distribution of labeled carbons of acetate in the side chain and angular methyls of cholesterol suggested the fundamentally terpenoid nature of cholesterol and prompted experimental tests (10) with labeling techniques which confirmed an older hypothesis (11, 12) that squalene was biogenetically related to cho-

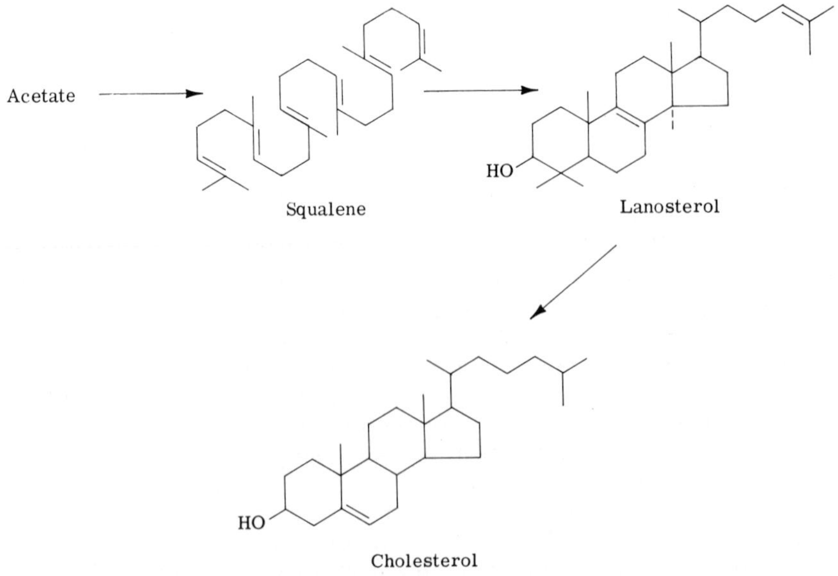

Acetate

Squalene

Lanosterol

Cholesterol

Fig. 1. Outline of cholesterol biosynthesis.

lesterol. Further experiments (13, 14) established that lanosterol, a C_{30} sterol of sheep's wool wax, whose structure was finally elucidated in 1952 (15), was an intermediate in the conversion of squalene to cholesterol. Bloch and co-workers restricted their degradative studies to certain key carbon atoms in cholesterol; the total degradation of squalene and cholesterol synthesized from labeled acetate that was carried out by Cornforth and Popjak and co-workers (16–18) fully substantiated the sequence indicated in Fig. 1.

B. General Scheme of Terpene Biosynthesis

The sequence of steps shown in Fig. 1 divides the biosynthesis of cholesterol into phases that embody fundamentally different mechanisms that recur repeatedly throughout the field of terpene and steroid metabolism.

The pathway by which acetate carbons are incorporated into squalene and cholesterol via β-hydroxy-β-methylglutaryl-CoA (19–21) and mevalonic acid (22–24) (the specific precursor of the isoprene unit) (25) is summarized in Fig. 2, which also indicates the absolute stereochemistry of mevalonic acid (26) as $3R(+)$ (27).

The importance of mevalonic acid as a specific terpenoid precursor lies in the fact that its formation from β-hydroxy-β-methylglutaryl-CoA is essentially irreversible (22), and its only significant pathway of utilization is in the direction of terpene synthesis, whereas β-hydroxy-β-methylglutaryl-CoA is open to attack by a cleavage enzyme (21, 28) which converts it back to acetyl-CoA and acetoacetate.

It became apparent from studies in Bloch's laboratory (29, 30) that the conversion of mevalonic acid to squalene required ATP and the biological equivalent of the five-carbon isoprene unit was subsequently identified (31–33) as isopentenyl pyrophosphate which was shown to arise by the sequence of phosphorylations (31, 34) and final dehydration-decarboxylation (32, 35–37), as shown in Fig. 3. The stereochemistry of this process, as elucidated by Cornforth, Popjak, and their co-workers (38) with the use of stereospecifically labeled mevalonic acid, is represented in the diagram and is interpreted as supporting a concerted trans elimination of CO_2 and inorganic phosphate from the hypothetical triphosphorylated intermediate.

Isopentenyl pyrophosphate, then, may be regarded as the immediate precursor of the isoprene unit, but its condensation into extended isoprenoid chains requires the presence of an allyl pyrophosphate as a receptor. Thus, in the formation of squalene, the condensation is initiated by the isomerization of isopentenyl pyrophosphate to dimethylallyl pyrophosphate (39, 40), which then condenses with a molecule of isopentenyl pyrophosphate to give geranyl pyrophosphate. Condensation with a further molecule of isopentenyl pyrophosphate yields farnesyl pyrophos-

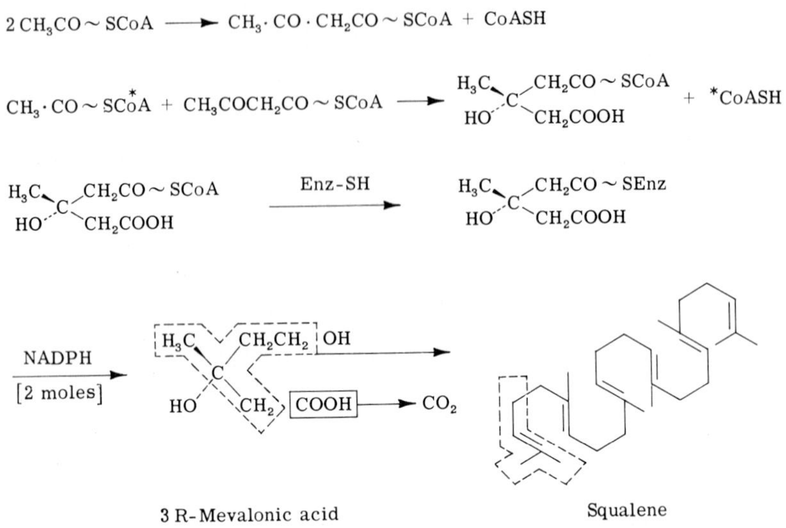

$$2 \, CH_3CO \sim SCoA \longrightarrow CH_3 \cdot CO \cdot CH_2CO \sim SCoA + CoASH$$

$$CH_3 \cdot CO \sim S\overset{*}{C}oA + CH_3COCH_2CO \sim SCoA \longrightarrow \quad \underset{HO}{\overset{H_3C}{>}} C \underset{CH_2COOH}{\overset{CH_2CO \sim SCoA}{<}} + {}^*CoASH$$

$$\underset{HO}{\overset{H_3C}{>}} C \underset{CH_2COOH}{\overset{CH_2CO \sim SCoA}{<}} \xrightarrow{\text{Enz-SH}} \underset{HO}{\overset{H_3C}{>}} C \underset{CH_2COOH}{\overset{CH_2CO \sim SEnz}{<}}$$

Fig. 2.   Biosynthesis of mevalonic acid and its pattern of incorporation into squalene.

phate (32, 41), whose carbon skeleton represents one-half of the squalene chain. These reactions, again with their stereochemical course, as elucidated by Cornforth, Popjak, and co-workers (38, 42), are depicted in Fig. 4.

Several points are established by these results and have been discussed by Cornforth and Popjak and co-workers.

1. In the isomerization of isopentenyl pyrophosphate to dimethylallyl pyrophosphate the stereochemistry of H⁺ elimination is the same as in

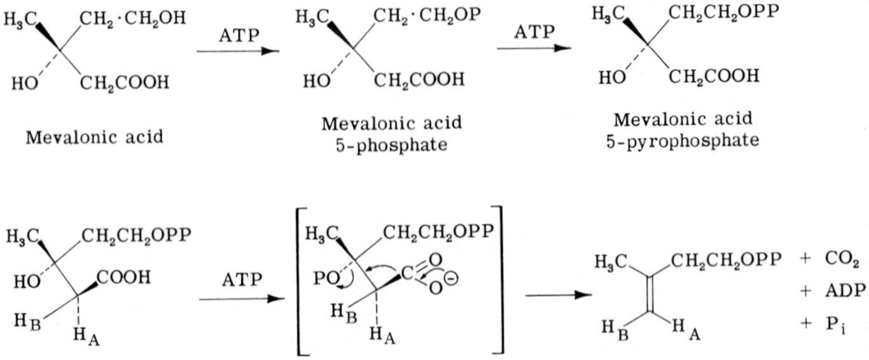

Fig. 3.   Conversion of mevalonic acid to isopentenyl pyrophosphate.

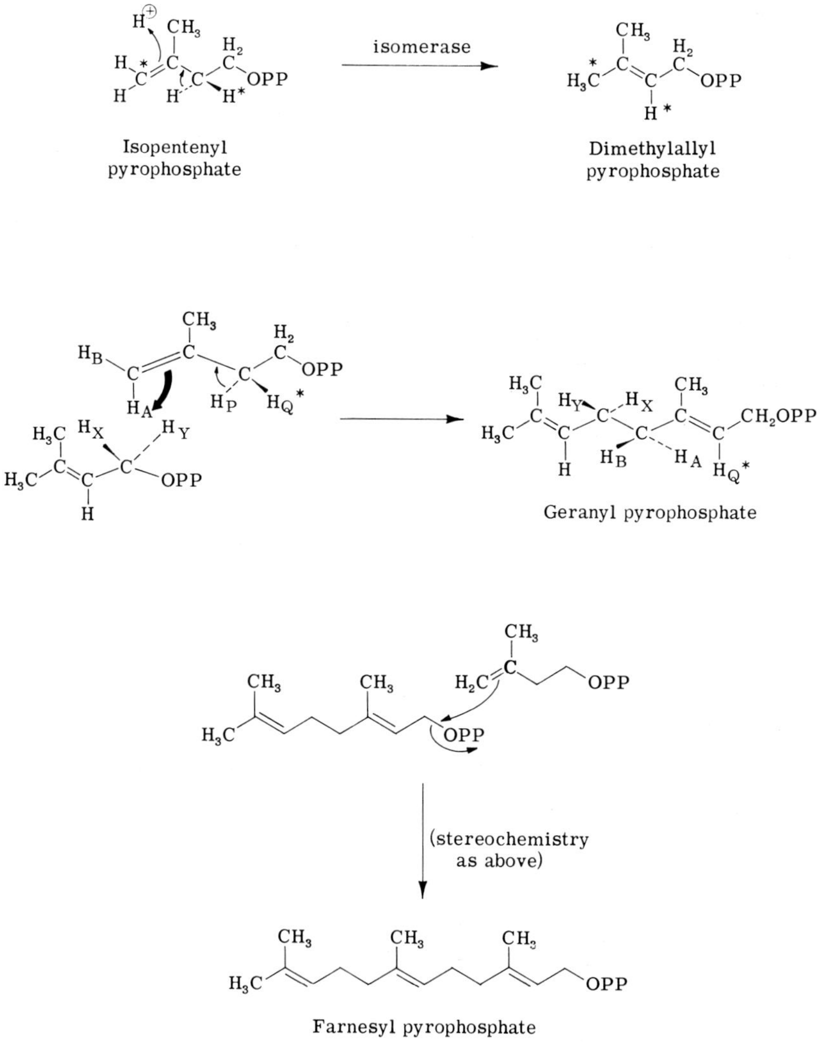

Fig. 4.   Incorporation of isopentenyl pyrophosphate into farnesyl pyrophosphate.

the condensation reaction. The result of this stereochemistry in the condensation process is to establish an all-trans structure in the product. In the formation of rubber, in which the corresponding bonds are all cis, the opposite stereochemistry of elimination in the condensation reaction prevails (43).

2. The hydrogens at C-1 of the allyl pyrophosphate are inverted. That is, the characteristics of the reaction are those of an $S_N 2$ displacement of OPP^- by the methylene of isopentenyl pyrophosphate. This is significant in that it argues against the existence of discrete resonance-stabilized carbonium ion forms derived from the allyl pyrophosphates.

3. The approach of the methylene group of isopentenyl pyrophosphate to C-1 of the allyl pyrophosphate occurs always from one side of the double bond. As shown in the diagram, this approach is from above.

These studies establish what is probably the general mechanism of coupling of isoprene units to form open chain terpenoid compounds. In the many cases in which the incorporation of labeled mevalonic acid and isopentenyl pyrophosphate into terpenoids has been studied (2, 5), the labeling pattern is in accord with the above scheme. The formation of cyclic terpenoids from acyclic precursors probably occurs by several cyclization mechanisms, hypothetical examples of which are shown in Fig. 5. In the formation of squalene, two farnesyl pyrophosphates react by an enzymatic mechanism (44), that appears to involve a C_{30} pyrophosphate intermediate, but which remains to be clarified (45, 45a).* The reaction is known to take place with inversion of configuration at C-1 of one farnesyl pyrophosphate, retention of configuration at the other, and incorporation of one atom of hydrogen from NADPH (44, 46, 47), as shown in Fig. 6.

Although the details are less clearly defined than in the case of squalene, there is evidence that symmetrical hydrocarbons of the carotene type (Fig. 9) (48) containing two C_{20}, geranylgeranyl units, are built up from two geranylgeranyl pyrophosphate molecules (49), which become united with similar stereochemical control (50) to that which operates in the formation of squalene from farnesyl pyrophosphate. The question of the involvement of reduced pyridine nucleotide in the synthesis of carotenes has been the subject of a number of conflicting reports (51). A number of long-chain, head-to-tail structures such as the side chains of ubiquinones and tocopherols are presumably elaborated in a similar fashion to the farnesyl and geranylgeranyl chains of these symmetrical terpenoids.

C. The Ecological Role of Some Simple Terpenoids

Before proceeding further with the discussion of how squalene becomes metabolized to sterols, we should briefly survey some of the structures which almost certainly arise by the routes we have discussed and which enter into environmental interactions. Among the most closely studied compounds of this type are the various lower terpenoids that are used by insects in communication and defense (52, 52a, 53). Examples

---

*More recent evidence (45b) indicates that this intermediate is a cyclic pyrophosphate of 10,11-dihydrosqualene 10,11-glycol.

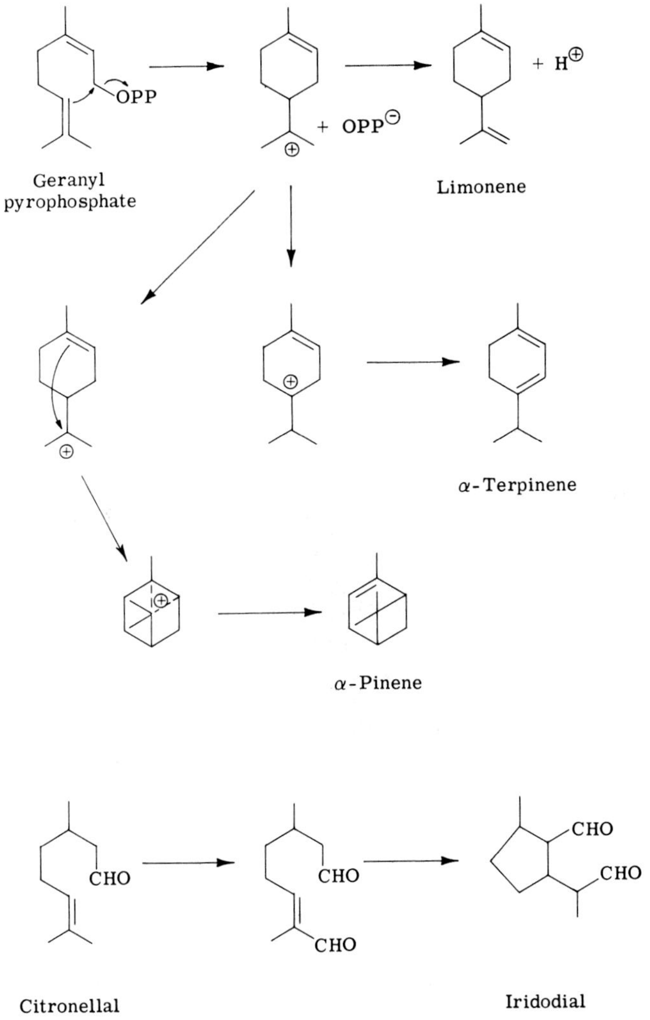
OPP
+ OPP⊖
+ H⊕
Geranyl pyrophosphate
Limonene
α-Terpinene
α-Pinene
CHO
CHO
CHO
CHO
CHO
Citronellal
Iridodial

Fig. 5. Some hypothetical mechanisms for the origin of cyclic monoterpenes from acyclic precursors.

are (Fig. 7) the terpenoid alarm substances produced by ants, the cantharidin of Meloidae beetles, dolichodial of the two-striped walkingstick, *Anisomorpha*, and isoamyl acetate which is released by a stinging bee and incites other bees to sting in the same area. The incorporation of mevalonic acid into citronellal and citral in *Acanthomyops* (54) and into dolichodial (anisomorphal) in *Anisomorpha* (55) has been reported.

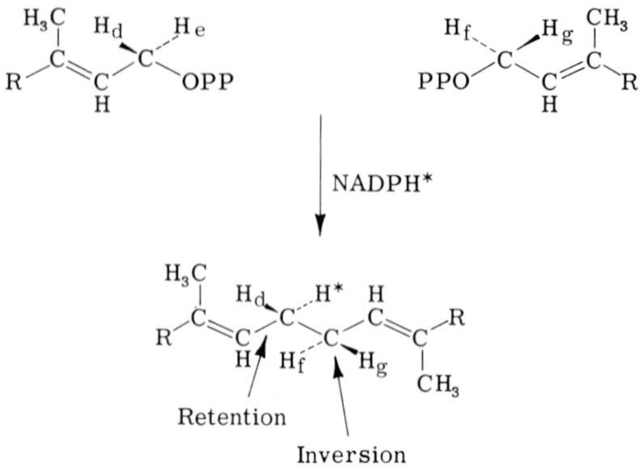

Fig. 6. Stereochemistry of the coupling of two farnesyl pyrophosphate groups in the biosynthesis of squalene.

In Fig. 8, some insect attractants are indicated. The sex attractants of the California five-spined ips, *Ips confusus* (Coleoptera: Scolytidae), which infests ponderosa pine, is a mixture (with synergistic activity) of at least three simple terpenes released in the feces and acting as an attractant and mating stimulant to males and females of the same species, as well as to some predators of *Ips confusus*. Figure 8 also shows a series of simple terpenes produced by the Nassanoff gland of bees and used to mark the way between rich food sources and the hive. Inasmuch as many simple terpenes of the types used by insects in communication and defense are also produced by plants, it seems reasonable to assume that such compounds also play an ecologically significant role in certain plant–insect interactions.

Among the higher terpenoids, β-carotene (Fig. 9) has long been recognized as an important precursor of vitamin A (retinol) (56), which is in metabolic equilibrium with retinal, an essential component of the visual system in mammals (57). This is also true of insects which, when reared aseptically on a carotene- (or retinol-) free diet were shown to be only 2.5% as sensitive to light as normal insects (58, 59). The function of retinol in the visual system requires its enzymic isomerization to 11-*cis*-retinol and its reconversion to the all trans form by light (57). While the visual function is the most clearly described role of vitamin A, it fulfills a variety of others, the detailed mechanisms of which are still far from being understood. These include effects on growth, maintenance of epithelial tissues, reproductive function and neural development (56, 60, 61).

α-Tocopherol (Fig. 10) is another higher terpene produced in vegetable material, which among other functions, plays a role in the reproduction of mammals (62, 63) and of several invertebrate species: a rotifer, *Asplanchna* (64), *Daphnia* (65), *Cyclops* (66), and the cricket, *Acheta* (67).

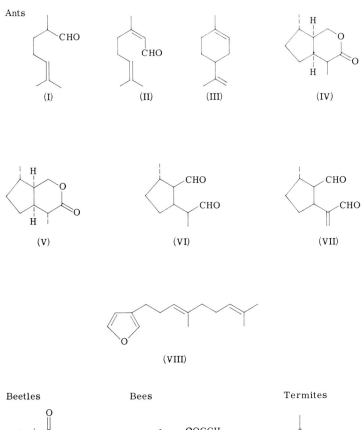

Fig. 7. Some terpenoid alarm and defense substances of insects.

Ants: *Acanthomyops claviger* (Formicinae) — Chandha, M. S., Eisner, T., Monro, A., and Meinwald, J. (1962). *J. Insect Physiol.* **8**, 175. I, citronellal; II, citral.   *Myrmicaria natalensis* (Myrmicinae) — Grünanger, P., Quilico, A., and Pavan, M. (1960). *Atti Accad. Nazl. Lincei, Rend. Classe Sci. Fis. Mat. Nat.* **28**, 293. III, Limonene.   Members of the Dolichoderinae — Cavill, G. W. K., and Hinterbarger, H. (1960). *Australian J. Chem.* **13**, 514. Cavill, G. W. K., Ford, D. L., and Locksley, H. D. (1956). *Australian J. Chem.* **9**, 288. IV, Iridomyrmecin; V, Isoiridomyrmecin; VI, Iridodial, VII, Dolichodial [or "anisomorphal" from *Anisomorpha buprestoides* (stick insect)]. Meinwald, J., Chadha, M. S., Hurst, J. J., and Eisner, T. (1962). *Tetrahedron Letters* p. 29.   *Lasius fuliginosus* (Formicinae) — Quilico, A., Piozzi, F., and Pavan, M. (1956). *Ric. Sci.* **26**, 177. Quilico, A., Piozzi, F., and Pavan, M. (1957). *Tetrahedron* **1**, 177. VIII Dendrolasin.

Beetles: Members of the Meloidae — Ude, W., and Heeger, E. F. (1941). *Pharm. Zentralhalle* **82**, 193. IX, Cantharidin.

Bees: *Apis mellifera* — Boch, R., Shearer, D. A., and Stone, B. C. (1962). *Nature* **195**, 1018. X, Isoamyl acetate.

Termites: *Nasutitermes* sp. — Moore, B. P. (1964). *J. Insect Physiol.* **10**, 371. XI, α pinene.

Beetle   *Ips confusus*

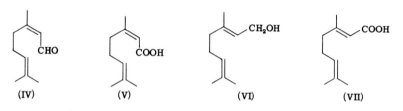

Bees   *Apis mellifera*

(IV)	(V)	(VI)	(VII)

Fig. 8.   Some terpenoid attractants of insects.

Beetle: *Ips confusus*. Attractants for both males and females and for some predators of *I. confusus*—Silverstein, R. M., Rodin, J. O., and Wood, D. L. (1966). *Science* **154**, 509. Wood, D. L., Browne, L. E., Bedard, W. D., Tilden, P. E., Silverstein, R. M., and Rodin, J. O. (1968). *Science* **159**, 1373. I, 2-methyl-6-methylene-7-octen-4-ol; II, cis-verbenol; III, 2-methyl-6-methylene-2,7-octadien-4-ol.

Bees: *Apis mellifera*. Food attractants—Boch, R., and Shearer, D. A. (1964). *Nature* **202**, 320. Shearer, D. A., and Boch, R. (1966). *J. Insect Physiol.* **12**, 1513. IV, citral; V, nerolic acid; VI, geraniol; VII, geranic acid.

Vitamins of the K group are required by mammals for prothrombin synthesis and hence for blood clotting. Their role in oxidative phosphorylation in mammals is questionable. There is evidence that intestinal microbial synthesis contributes significantly to the vitamin K intake of some animals but the dietary vitamin requirement and the ease of induction of hypovitaminosis vary considerably among different species (60, 68, 69).

From the foregoing paragraphs it is evident that the terpenoid biosynthetic pathway provides a wide variety of ecologically important substances that function as vitamins and as regulators of behavior among insects. The possibility of a further type of ecologically important interaction between animals and plants, which also involves terpenoid compounds, is suggested by recent work on the constitution of the juvenile hormone of insects and of plant products that mimic its action (70). The

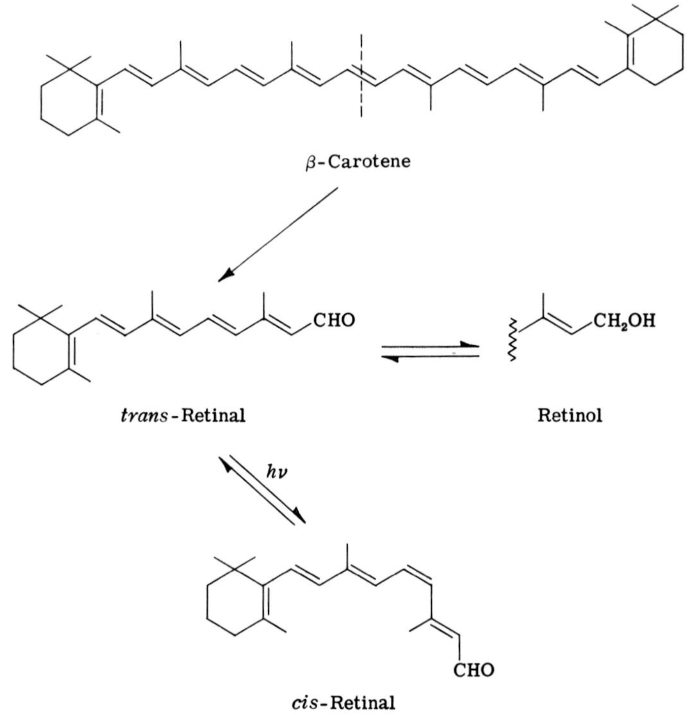

β-Carotene

trans-Retinal   Retinol

hv

cis-Retinal

Fig. 9. β-Carotene, a tetraterpene, and its relation to retinol in the visual process.

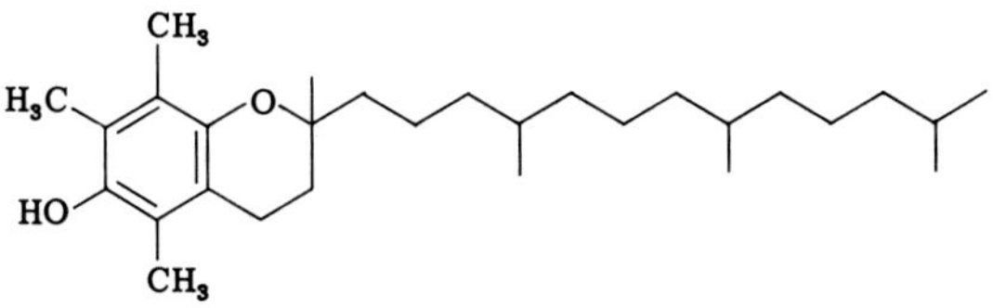

α-Tocopherol

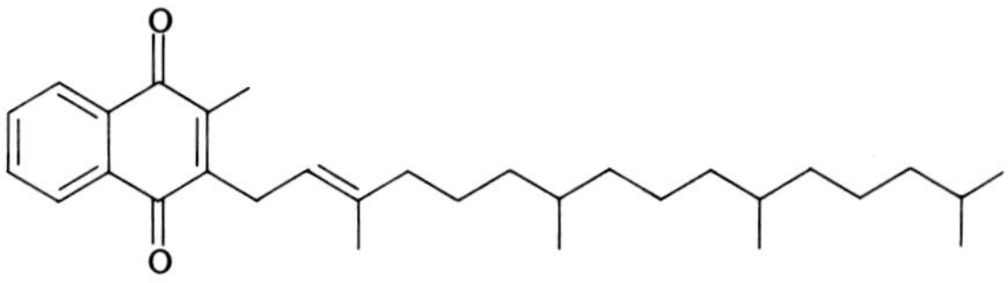

Vitamin K₁

Fig. 10. Isoprenoid vitamins.

structure of the juvenile hormone (Fig. 11) (71) suggests its terpenoid origin via a farnesol derivative that acquires an extra carbon atom at each of the second and third branching methyls which thereby become converted to ethyl groups. Mechanisms for methylations of branching methylene groups (that could arise in this case by double bond isomerizations) are known from studies of the synthesis of the phytosterol side chain, which bears a branching ethyl group at C-23 and is discussed below. The biological conversion of a trisubstituted ethylenic bond to an oxirane derivative such as is found in the juvenile hormone also has a precedent in the conversion of squalene to squalene-2,3-oxide, which is discussed below.

The mimics of the juvenile hormone, juvabione (72, 73) and dehydrojuvabione (74), that are extractable from the balsam fir have the structures as shown in Fig. 11. These compounds also may be synthesized via an intermediate that has a farnesyl skeleton and undergoes partial cyclization to the cyclohexenyl derivative. The remarkable species specificity of these terpenoid constituents of wood in inhibiting the larval and embryonic (75) development of *Pyrrhocoris apterus* suggests that protective mechanisms of plants against insect predators, based upon interference with the insect's normal endocrine controls, may be a significant ecological factor. This possibility is strengthened by the finding of insect molting hormone analogs in various plants (see below). Moreover, the

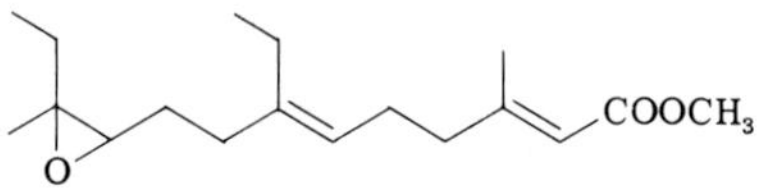

cis-   trans-   trans-
Juvenile hormone

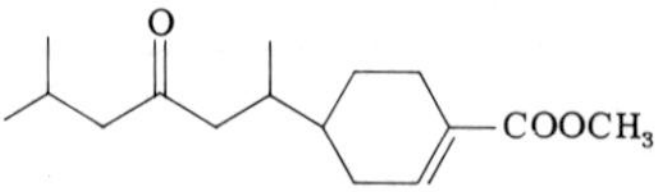

Juvabione

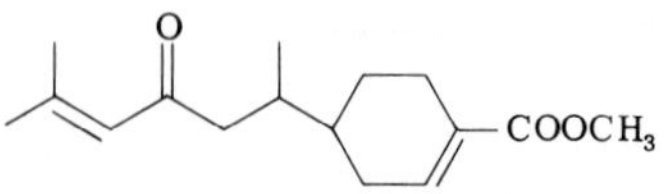

Dehydrojuvabione

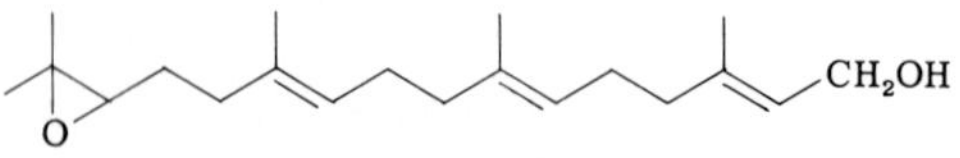

14,15-Epoxygeranylgeraniol

Fig. 11.   Juvenile hormone and some natural products with similar physiological activity.

recent report (76) of the potent activity of a diterpenoid epoxide (14,15-epoxygeranylgeraniol) in inhibiting the penetration and development of *Schistosoma mansoni* in mice, suggests that terminal epoxyterpenes may have a more general role than has been suspected hitherto in the control of development in invertebrates. The compound in question in this case was isolated from the oil of the leguminous plant *Pterodon pubescens.*

D. The Role of Squalene and Squalene-2,3-oxide

We may now proceed with tracing the biogenesis of the steroids which, themselves, have important roles in ecological interactions.

In 1953, Woodward and Bloch published their proposal (77) concerning the mechanism by which squalene cyclizes to give the sterol skeleton, via lanosterol. Immediately thereafter, the role of squalene as a precursor of numerous polycyclic triterpenes was proposed and extensively discussed by Ruzicka (78) and was later elaborated by Eschenmoser and co-workers (79) and by Stork and Burgstahler (80). An abbreviated form of these proposals is shown in Fig. 12. Lack of space makes it impossible to review their wide implications. They have been frequently reviewed (1–3, 5, 81) and it must suffice to say here that with one exception they have been supported by every experimental test to which they have been subjected so far. In particular, the proposed migrations of methyl groups and hydride ions involved in the formation of lanosterol have been fully established. The exception concerns the nature of the oxidative step and its relation to cyclization.

Eschenmoser *et al.* (79) and Stork and Burgstahler (80) had postulated the initiation of cyclization of squalene by attack of a hypothetical hydroxonium ion OH^+, as shown in Fig. 12. The possibility of generation of appropriately hydroxylated cyclic terpenoid products by proton attack on a terminal epoxide was discussed by Goldsmith (82) and at about the same time was established experimentally by van Tamelen and co-workers (83–87) (Fig. 13). The close analogy between the initial course of such nonenzymatic cyclizations and their enzymatic counterpart in the biosynthesis of lanosterol and other polycyclic 3β-hydroxylated terpenoids prompted a series of collaborative studies between our laboratories at Stanford (88–90) which has been closely paralleled by work carried out at Harvard by Corey and co-workers (91–93). From these studies it is now clear that the simple and elegant concept of a concerted oxidative cyclization of squalene to lanosterol is incorrect and that squalene is first converted to its 2,3-oxide as a stable intermediate in the formation of various polycyclic triterpenes (though not necessarily of all) (94, 95).

The biogenetic relationships between squalene, its 2,3-oxide and lanosterol, and two other polycyclic terpenoids that have been studied experimentally are illustrated in Fig. 14. Although the migrations of methyl groups and hydride ions in the formation of lanosterol have been

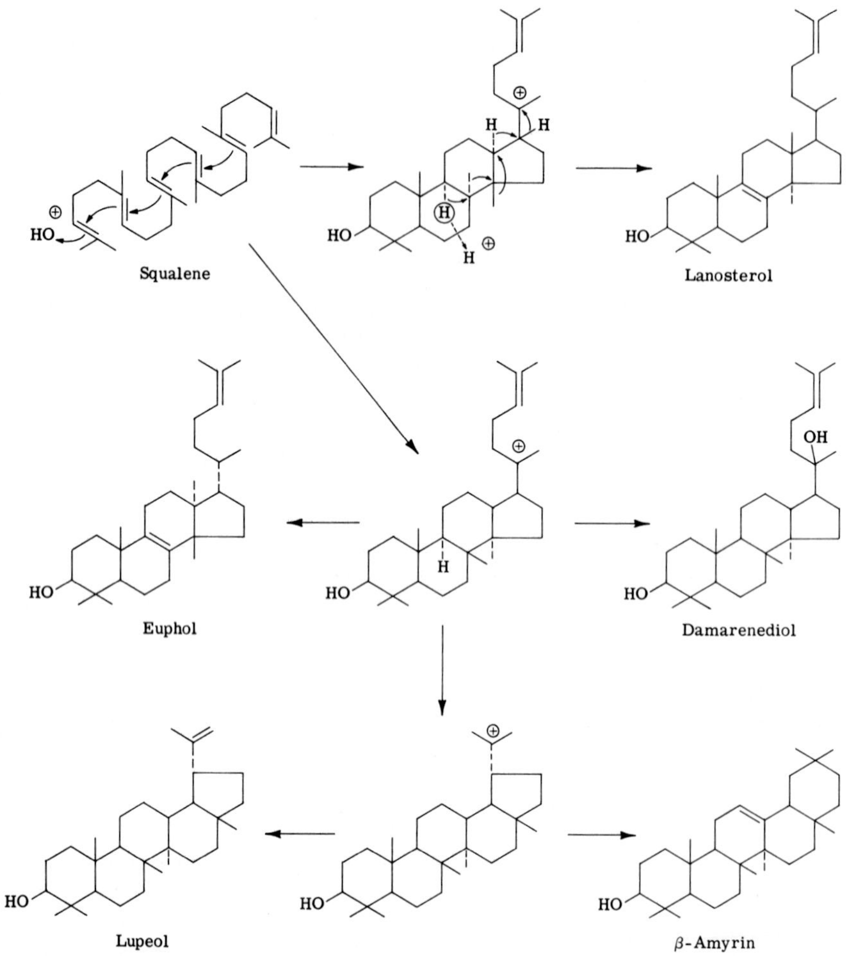

Fig. 12. The role of squalene as a precursor of several classes of polycyclic terpene (Eschenmoser et al.) (79).

shown by the work of Bloch and co-workers (96) and Cornforth and co-workers (97) (Fig. 12), the factors that control these rearrangements have remained obscure. In our laboratories and at Harvard, some characteristics of the cyclization have now been examined by studying the action of squalene-2,3-oxide lanosterol cyclase upon various structural analogs of squalene oxide.

E. The Mechanism of Action of Squalene-2,3-oxide Lanosterol Cyclase

Figure 15 shows the approximate relative efficiencies of enzymatic cyclization of squalene-2,3-oxide (arbitrarily = 100%) and three of its analogs in which structural modifications have been made in that part of the squalene chain that provides the side chain portion of the lanosterol molecule. It was found that under similar conditions, the 22, 23-dihydro- and 23,23′,24-trisnor derivatives (98), as well as the 21,22,23,23′,24-pentanor analog (99) were all converted to the corresponding tetracyclic analogs of lanosterol, but with decreasing efficiency, the further the structure of the substrate departed from that of squalene-2,3-oxide. Thus, the "side chain" terminus of squalene-2,3-oxide, as expected, is not critically involved in the mechanism of cyclization and rearrangement, but probably plays a significant role in binding the substrate to the active site.

The outcome of experiments in which substrates variously modified in the vicinity of the oxide ring were subjected to the action of the cyclase is shown in Fig. 16. In cases I–III, respectively, one methyl group, cis (I) or trans (II) to the squalene chain, or both methyls (III) of the gemdimethyl structure, were replaced by hydrogen atoms. Thus, in these analogs, C-2 no longer provided a tertiary center for the transient acceptance of the charge associated with proton-initiated opening of the oxide ring. Our finding (100) that two of these analogs (II and III) failed to cyclize and that the one example (I) which cyclized (to yield the 4α-methyl sterol, IV) did so with only 6% of the efficiency of squalene-2,3-oxide, no doubt reflects the overriding importance of the tertiary center of the normal substrate in initiating the cyclization. In the examples, V and VI, a methyl substituent at the tertiary center was replaced by an ethyl group, cis and trans to the chain, respectively (101). Only one product, the 4α-ethyl analog of lanosterol, VII, was obtained (102), a result that indicated that only the corresponding trans-ethyl oxide VI was cyclized, since other experiments in our laboratory have shown unequivocally that the cyclization of squalene-2,3-oxide occurs without transposition of the C-1 and C-2 methyl groups (103). It seems likely that ring closure of the cis-ethyl

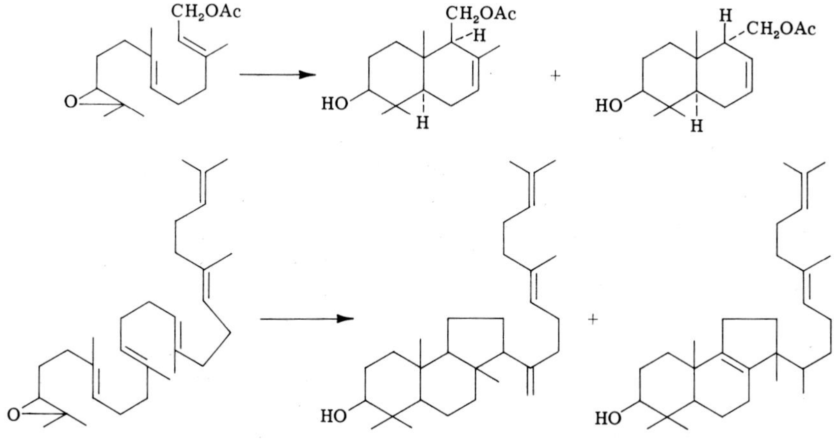

Fig. 13. The nonenzymatic cyclization of some terminal terpene oxides (van Tamelen and co-workers) (83–87).

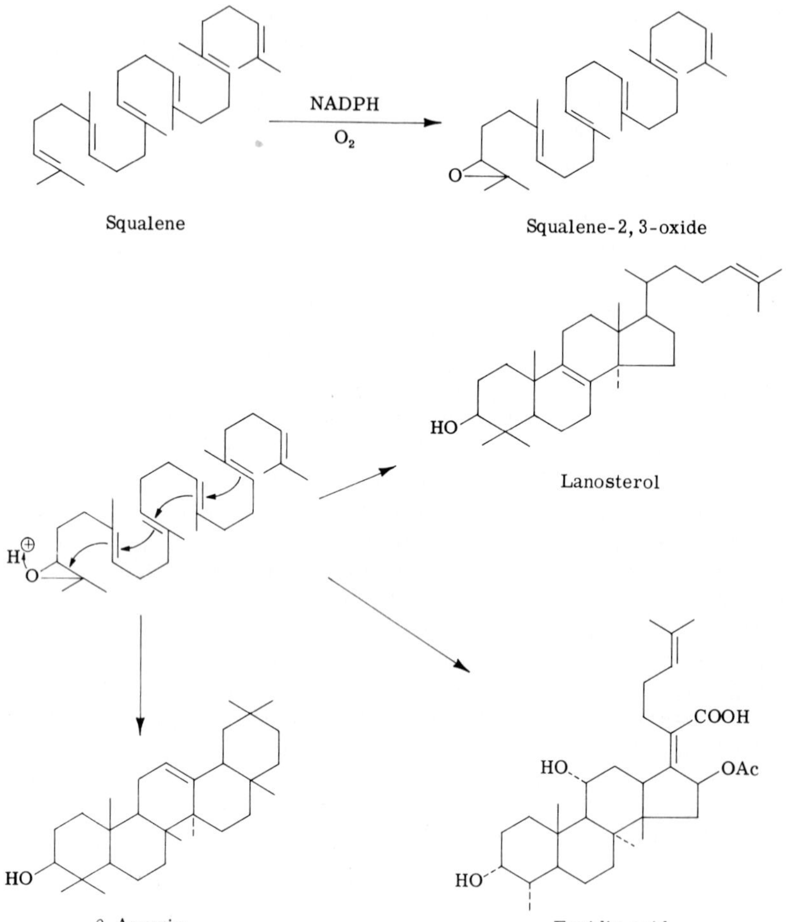

Fig. 14.   Some demonstrated biosynthetic relationships of squalene-2,3-oxide.

compound, V, to give the 4β-ethyl lanosterol analog is precluded by steric 1:3 interaction between the *cis*-ethyl group and the 6′-methyl group (potential C-19 of the sterol) or some similar unfavorable steric interaction between the *cis*-ethyl group and a region of the enzyme active site. Such steric factors may also be responsible for the different results obtained with the two 2′-desmethyl analogs, I and II.

An important question in reaching an understanding of the cyclization mechanism is whether the cyclization and rearrangement process, which can be written as a hypothetical sequence of events initiated by protonation of the oxide ring, is in fact separable into temporally discrete stages. The first evidence that the cyclization is indeed sequential came from experiments in our laboratories with the dihydro compound, I (Fig. 17) (104), which was shown to yield a tricyclic product, II, essentially similar to the nonenzymatic cyclization product, but differing from it in stereochemistry. More recently, Corey and co-workers (105) have obtained further evidence for the sequential nature of the cyclization rearrangement process in demonstrating the cyclization, presumably without rearrangement, of the bisdesmethyl squalene oxide analog, III, to a tetracyclic compound, IV. Presumably the formation of this compound reflects the stabilization of the carbonium ion bearing a charge on C-20 (sterol numbering) by deprotonation at C-22. In this connection it is interesting that we have found that the naturally occurring "protosterol", V, of the mold *Fusidium coccineum* (106), the stereochemistry of which is similar to that of compound IV, is a good inhibitor of squalene-2,3-oxide cyclization, though lanosterol is not (107).

The results obtained by Corey *et al.* with bisdesmethyl analog III illustrate the importance of methyl substitution at C-10 and C-15 for the promotion of the rearrangement, but do not give information as to whether one or both of these substituents is necessary. Further information on this point has come from experiments in our laboratory. The

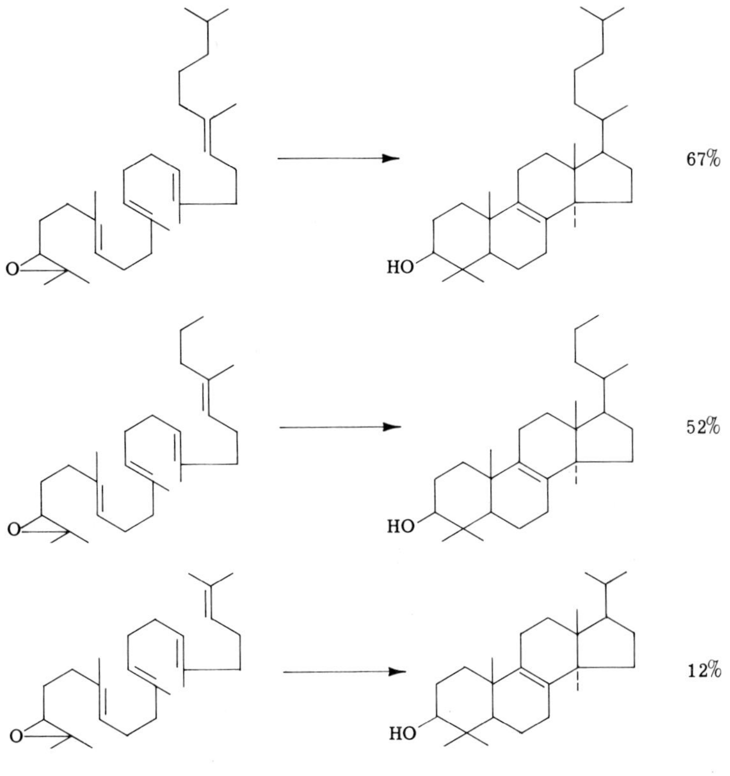

Fig. 15. Efficiency of cyclization of various terminally modified squalene-2,3-oxide analogs under standard conditions, related to squalene-2,3-oxide (arbitrarily = 100%).

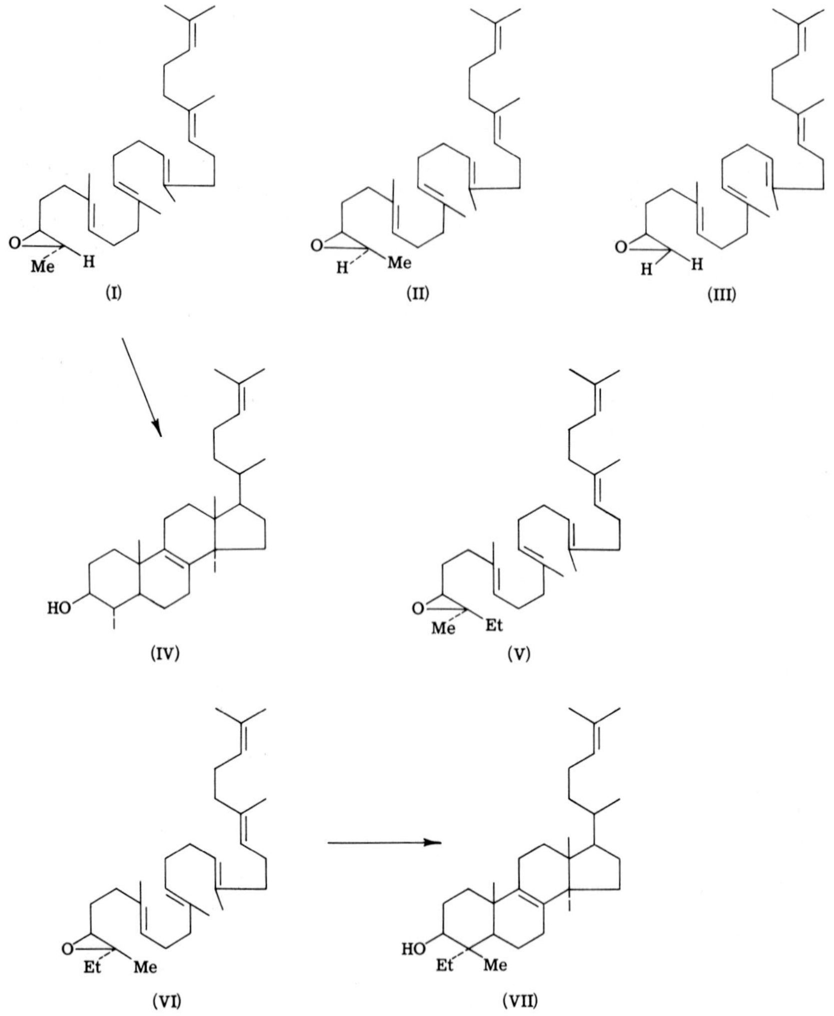

Fig. 16. Studies of squalene-2,3-oxide analogs modified in the vicinity of the oxide ring.

enzymatic cyclizations of the 15-desmethyl substrate VI to the rearranged, 18-norlanosterol VII (108) and of the 10-desmethyl substrate VIII to 14-norlanosterol IX (108a) have both been demonstrated. It must be concluded, then, that while the presence of one or other of these methyl groups is essential for rearrangement to occur, neither is indispensible. The evident interchangeability of either the C-10 or C-15 methyl with hydrogen, compatible with rearrangement, suggests that no special functional groups are present in the enzyme-active site for the translocation of these methyl groups. Rather, it seems probable that the rearrangement is promoted by steric interactions within the substrate

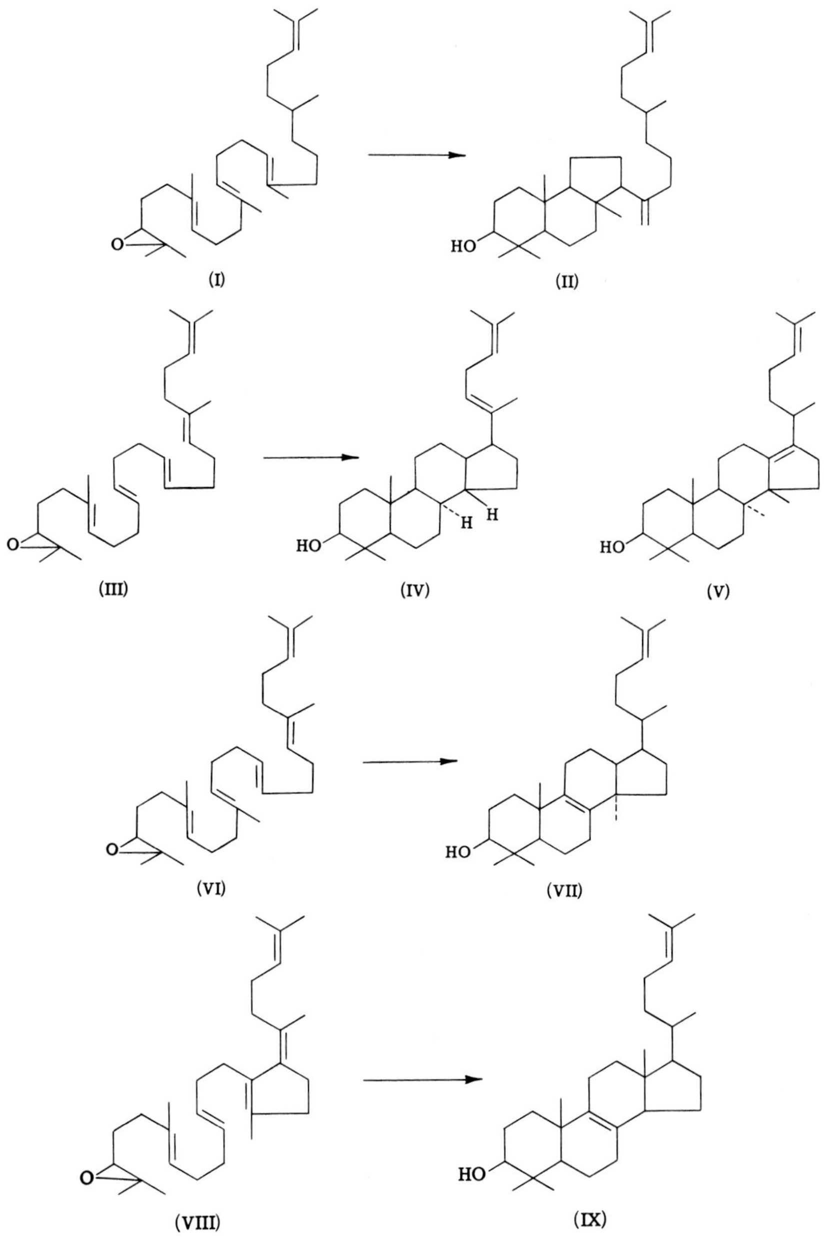

Fig. 17. Metabolic fate of some analogues of squalene-2,3-oxide modified in the interior portions of the chain and the structure (IV) of the unrearranged cyclization product of Corey, Ortiz de Montellano, and Yamamoto (105) compared with that of the naturally occurring protosterol (V).

molecule and between the substrate and the active site, to which both of the methyl groups contribute, but in such a manner that the presence of only one of them suffices to provide the necessary driving force.

In summary, these studies with squalene-2,3-oxide analogs suggest that the formation of lanosterol is in some important respects subject to stereoelectronic control mechanisms such as are operative in nonenzymatic systems. While the existence of stable intermediates between squalene-2,3-oxide and lanosterol cannot be ruled out on the basis of the available information, there is at present no clear evidence in favor of such intermediates. The data at hand are still most readily accommodated by a scheme of concerted cyclization and rearrangement such as that of Eschenmoser *et al.* (79) in which the outcome of the process is determined by the conformational constraints and steric pressures imposed upon the substrate by the active site. Whatever may be the final picture of these mechanisms to emerge from the present intensive series of investigations, it seems most likely that it will include most of the essential features of enzymatic cyclizations of terpenes in general.

F. The Conversion of Lanosterol to Cholesterol

The conversion of lanosterol to cholesterol takes place by a number of enzymatic steps, the principal features of which are shown in Fig. 18. These transformations have been studied by several research groups, notably in the laboratories of Bloch, Frantz, and Gaylor, and have recently been reviewed in some detail (4). Although the main outlines of the pathway as indicated in Fig. 18 are no doubt correct, many questions of detailed mechanism and exact sequence of events remain. Thus, the removal of the methyl groups in an enzyme system of rat liver was shown more than 10 years ago by workers in Bloch's laboratory to involve the release of their carbon atoms as CO_2 (109) and to be initiated by attack on the 14α-methyl group to yield, as an early intermediate, 4,4'-dimethyl-$\Delta^{8,24}$-cholestadien-3β-ol (110). The concept that each methyl group is oxidized in stepwise fashion to a carboxylic acid group and subsequently lost by decarboxylation which, in the case of the substituents at C-4, is facilitated by oxidation of the 3β-hydroxyl function to a 3-ketone (111), is supported by more recent studies (112). The exact mechanism of removal of the 4-methyl substituents, however, still remains to be elucidated. Figure 18 reflects the view (based upon studies carried out in our laboratories in collaboration with Dr. T. A. Spencer and coworkers) (113) that the 4α-methyl group is removed prior to attack on the 4β-methyl group. Although this view* conflicts with the conclusion reached earlier by Gaylor and Delwiche (114), it is strengthened by our demonstration of the existence of a hepatic isomerase which converts a

---

*Conclusion established for saturated sterol analogs now confirmed for lanosterol (115a). Goodwin *et al.* (115b) obtained similar results in study of demethylation of cycloartanol in plant tissues.

4β-methyl-3-keto substrate to a 4α-methyl derivative (115). Since 4α-monomethyl sterols are readily demethylated, whereas 4β-monomethyl sterols are not, a plausible metabolic pathway is indicated in which the oxidative attack involved in demethylation must always take place on a 4α-methyl group and the 4β-methyl that survives the first demethylation process must be isomerized to the 4α position before further attack can occur.

Greater uncertainty surrounds the exact mechanism of removal of the 14α-methyl carbon. Bloch (9) attributed an activating influence to the Δ^8-bond in the decarboxylation associated with this process, but it is only recently that new observations have been made which may bear on the mechanism. At some stage, the 15α-hydrogen atom is lost (116, 117) possibly with intermediary formation of the 8,14-diene (118) or participation of a $\Delta^{8(14)}$-compound (119, 120). That the Δ^8-bond is indeed essential for removal of the 14α-methyl group is confirmed by studies in our laboratory (113) which have also provided evidence for the inhibition by a 14α-methyl group of demethylations at C-4 in enzyme preparations from rat liver. This observation offers a possible explanation for the absence of 14α-methyl sterols lacking methyl substituents at C-4 in mammalian tissues and suggests a difference in this respect between C-4-demethylating systems of plants and animals.

The stereochemistry of the shift of the Δ^8-bond to Δ^7- has been studied and the process has been shown to involve loss of the 7β-proton without reincorporation elsewhere in the molecule (121, 122). In the subsequent reduction of the Δ^7-bond, the remaining 7α-hydrogen comes to occupy the 7β position. The stereochemistry of the desaturation of C-5 has also been established, but its mechanism, which requires oxygen, is still unresolved (123, 124). The saturation of the side chain in the biosynthesis of cholesterol is catalyzed by an enzyme system of the microsomes that requires NADPH and probably can act upon Δ^{24}-sterols of various nuclear structures (125), though this reduction probably normally occurs relatively late in the biosynthetic sequence (126), as indicated in Fig. 18.

G. The Biosynthesis of C_{28} and C_{29} Sterols of Fungi and Higher Plants

In fungi, the predominant sterol is ergosterol in which an extra methyl group (C-28) is incorporated into the side chain at C-24. Most of the evidence (4) obtained from studies with yeast indicates that the biosynthetic route to this sterol is similar in essentials to that in cholesterol in mammals, but that a Δ^{24} intermediate is methylated by reaction with S-adenosylmethionine (127) to yield first a 24-methylene derivative (128–131) with concomitant hydride ion migration (132) from C-24 to C-25 as shown in Fig. 19. Evidence has been published supporting both pathways, (a) (Ref. 133) and (b) (Refs. 131, 134), to the Δ^{22}-24-methyl side-chain structure of ergosterol. It should be noted, however, that the

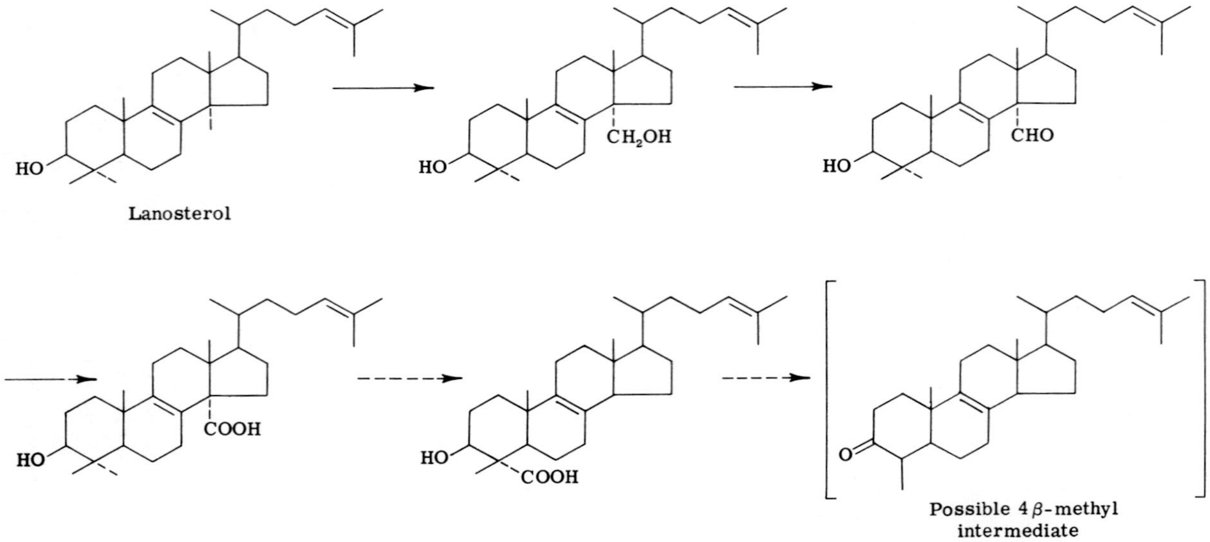

Fig. 18. Probable sequence of reactions in the conversion of lanosterol to cholesterol. (Broken arrows indicate the operation of more than one metabolic step.)

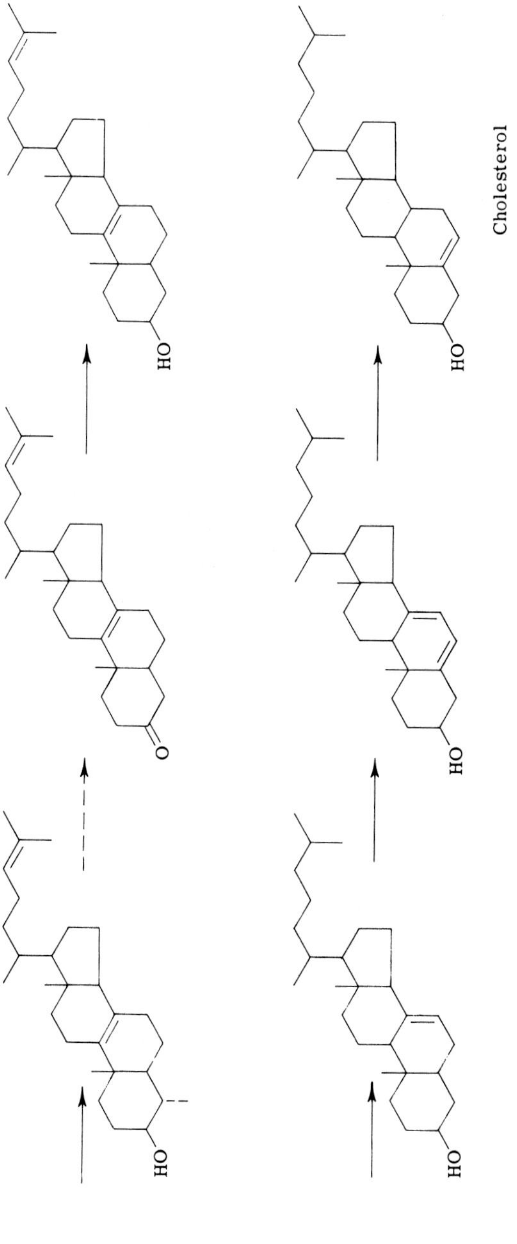

Fig. 18. (Continued)

identity of the Δ^{24} acceptor of the methyl group in the normally occurring biosynthetic pathway is still uncertain.

While many green plants also contain C_{28} sterols, e.g., campesterol (24α-methyl cholesterol) and brassicasterol (24β-methyl-22-dehydrocholesterol), the most abundant sterols of higher plants are C_{29} sterols such as sitosterol and its Δ^{22} analog, stigmasterol, in which a 24-methylene sterol apparently undergoes a further methylation by reaction with S-adenosylmethionine (135, 136). Again, the details of this further methylation process are in some doubt and possible alternatives indicated in Fig. 19 may operate in different species (137–139). Unlike the pathway to ergosterol in yeast, the route to the phytosterols of higher plants seems to include cycloartenol (Fig. 20), rather than lanosterol as the initial product of cyclization of squalene oxide (140–144).

H. Sterols as Dietary Requirements of Insects and Other Lower Organisms

While, so far as we know, all vertebrates (with the possible exception of elasmobranchs) (145) and higher plants can synthesize sterols, most bacteria appear not to be able to synthesize them, nor to require them (146, 147). The status of the blue-green algae in this respect is apparently questionable (147–149). There are, however, many lower organisms that lack the ability to make sterols and have a dietary requirement for them. Examples are shown in Table I.

The most intensively studied of these groups of lower organisms are the insects. No insect has so far been found to be independent of an exogenous sterol source (150). Those which are apparently self-sufficient in this respect have been shown to depend upon some form of symbiont for their supply of sterols. Examples of structural relationships of sterols as growth factors for several insect species are shown in Table II. It is evident that different species have different capacities to utilize sterols of different structures. Particularly noteworthy is the requirement of some strictly carnivorous species (exemplified by *Dermestes vulpinus*) for cholesterol and their failure to utilize plant sterols as a sole sterol supply. *Dermestes* can, nevertheless, utilize plant sterols quite well, provided that some percentage of the total dietary sterol is cholesterol (151, 152). Thus, in these insects, the normal requirement for cholesterol is "spared" by otherwise nonutilizable sterols. Under these conditions, both cholesterol and the sparing sterol will be incorporated into the tissues and the ratio of cholesterol to sparing sterol is increased in the tissues in comparison with that of the diet (153–155). Experiments carried out in our own laboratory (156) show that the sterols are utilized primarily by incorporation into the subcellular membrane structures of the insects' tissues, where they are retained with a remarkably long turnover time (154). The results strongly suggest that under the conditions just indicated, where the dietary cholesterol requirement is "spared" by other

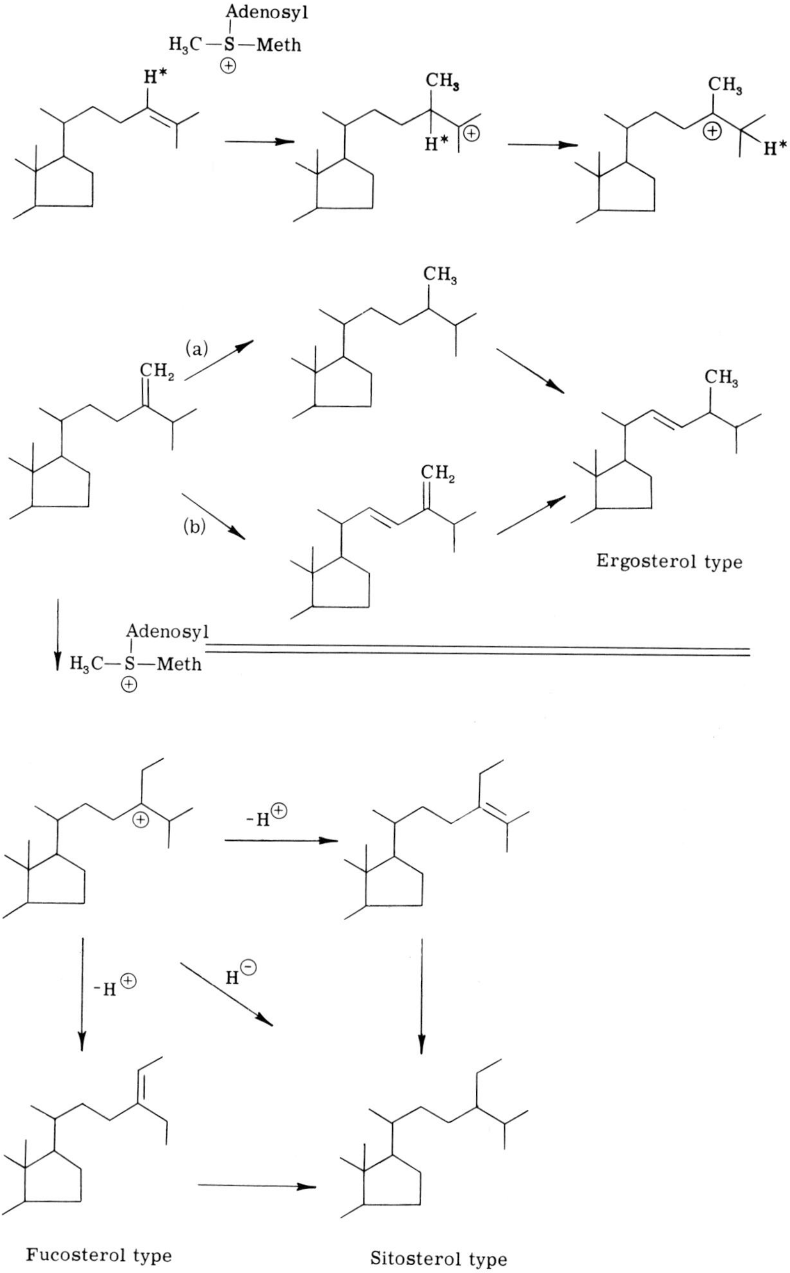

Fig. 19. Biogenesis of the side chains of C_{28} and C_{29} sterols.

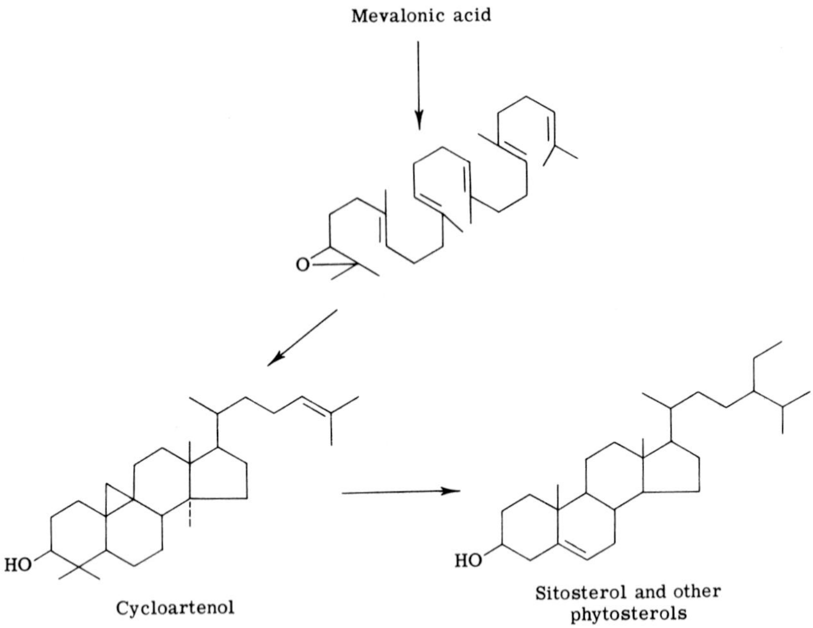

Fig. 20.   Cycloartenol as a precursor of C_{29} sterols in higher plants.

sterols, the two types of sterol serve different structural roles in the same subcellular membranes. Indeed, the apparently universal association of sterols and eucaryotic cytoplasmic membranes, which has been pointed out by Bloch (147), is probably the most general basis for sterol requirements among lower organisms. A structural role having broadly similar features in all the forms that have been studied is strongly suggested by the very similar structural prerequisites of virtually all organisms for physiologically functional sterols. In all well-authenticated cases the essential requirements are for (a) a planar ring system basically of the cholestane type, (b) a side chain at C-17 of the cholestane, ergostane, or stigmastane type, and (c) a 3β-hydroxyl group. It seems likely that these essentially similar requirements in all cases reflect limitations on the shape and dimensions of certain spaces in the structural matrices of subcellular membranes that also have much in common, but which vary in detail from species to species.

Among the insects there are many species (though not all) that can rapidly remove the C-24 alkyl substituents of the phytosterol side chain, thereby converting plant sterols to cholestane derivatives (150, 157, 158). Presumably the C_{27} products fit the intracellular membrane matrix in these cases more exactly than the phytosterols. It is very interesting

TABLE I
Sterol Requirements of Lower Organisms*

Protozoa
  Paramecium[a]
  Tetrahymena setifera and *paravorax*[b,c,d,e]
  T. pyriformis (partial requirement)[f]
Mycoplasmas (PPLO) (saprophytic but not nonsaprophytic types)[g]
Purple photosynthetic bacteria (*Rhodopseudomonas palustris*)[h]
Fungus—*Phytophthora cactorum*—29-isofucosterol or sitosterol required for sexual reproduction[i,j,k]
Yeast (anaerobically grown)[l]
Coelenterates—*Rhizostoma* sp.[m] and *Paracentrotus*[l]
Nematodes—*Turbatrix aceti* and *Caenorhabditis briggsae*, free-living forms utilizing *E. coli*, did well on phytosterols and ergosterol, but poorly on cholesterol[n,o,p]
Tapeworm (*Spirometra mansonoides*)[q]
Annelid—*Lumbricus terrestris*[r]
Mollusks (cuttlefish)—*Sepia officinalis*[s], *Ostrea*[t]
Crustaceans[u,v]
Insects[w,x]

---

*Key to superscript letters
[a] Conner, R. L., and Van Wagtendonk, W. J. (1955). *J. Gen. Microbiol.* **12**, 31.
[b] Wagner, B., and Erwin, J. A. (1961). *Comp. Biochem. Physiol.* **2**, 202.
[c] Holz, G. G., Erwin, J. A., and Wagner, B. (1961). *J. Protozool.* **8**, 297; Holz, G. G., Erwin, J. A., and Wagner, B. (1962). *J. Protozool.* **9**, 359.
[d] Holz, G. G., Wagner, B., Erwin, J. A., Britt, J. J., and Bloch, K. (1961). *Comp. Biochem. Physiol.* **2**, 202.
[e] Hutner, S. H., and Holz, G. G. (1962). *Ann. Rev. Microbiol.* **16**, 189.
[f] Conner, R. L., and Ungar, F. (1964). *Exptl. Cell Res.* **36**, 134.
[g] Smith, P. F., and Rothblat, G. H. (1960). *J. Bacteriol.* **80**, 842.
[h] Aaronson, S. (1964). *J. Gen. Microbiol.* **37**, 225.
[i] Elliott, C. G., Hendrie, M. R., Knights, B. A., and Parker, W. (1964). *Nature* **203**, 427.
[j] Hendrix, J. W. (1964). *Science* **144**, 1028.
[k] Elliott, C. G., Hendrie, M. R., and Knights, B. A. (1966). *J. Gen. Microbiol.* **42**, 425.
[l] Andreasen, A. A., and Stier, T. J. B. (1953). *J. Cellular Comp. Physiol.* **41**, 23.
[m] Van Aarem, H. E., Vonk, H. J., and Zandee, D. I. (1964). *Arch. Intern. Physiol. Biochim.* **72**, 606.
[n] Hieb, W. F., and Rothstein, M. (1968). *Science* **160**, 778.
[o] Rothstein, M. (1968). *Comp. Biochem. Physiol.* **27**, 309.
[p] Cole, R. J., and Krusberg, L. R. (1968). *Life Sci.* **7**, 713.
[q] Meyer, F., Kimura, S., and Mueller, J. F. (1966). *J. Biol. Chem.* **241**, 4224.
[r] Wooton, J. A. M., and Wright, L. D. (1962). *Comp. Biochem. Physiol.* **5**, 253.
[s] Zandee, D. I. (1967). *Arch. Intern. Physiol. Biochim.* **75**, 487.
[t] Salaque, A., Barbier, M., and Lederer, E. (1966). *Comp. Biochem. Physiol.* **19**, 45.
[u] Zandee, D. I. (1966). *Arch. Intern. Physiol. Biochim.* **74**, 435.
[v] Van der Oord, A. (1964). *Comp. Biochem. Physiol.* **13**, 461.
[w] Clark, A. J., and Bloch, K. (1959). *J. Biol. Chem.* **234**, 2578.
[x] Clayton, R. B. (1964). *J. Lipid Res.* **5**, 3.

that such evidence as is available concerning the mechanism of these dealkylations suggests that desmosterol is formed as an intermediate (159, 160). Moreover, in collaborative studies with Dr. F. J. Ritter some years ago, we obtained evidence for the formation of $\Delta^{24(28)}$-desaturated sterols as intermediates in the dealkylation of ergostane derivatives

TABLE II

Sterol Requirements of Some Insects*

	Cholesterol	7-Dehydrocholesterol	Cholestanol	Sitosterol	Stigmasterol	Ergosterol
Lucilia sericata[a]	+++	– –	– –	++	– –	+
Dermestes vulpinus[b,c]	+++	0	0	0	0	0
Attagenus piceus[d,e]	+++	++	0	– –	– –	0
Lasioderma serricorne†[f]	+++	+++	++	+++	– –	+++
Stegobium paniceum†[f]	+++	+++	+	+++	– –	+++
Blatella germanica[g,h]	+++	0	0	+++	+++	+
Drosophila[i,j]	+++	– –	++	+++	– –	+++
Bombyx mori[k]	+	– –	– –	+++	++	0

*Key to superscript letters and symbols. Efficiency of individual sterols in supporting development in different species is represented as: +++ good, ++ moderate, + poor, or 0 ineffective. The sign – – indicates that the sterol was not studied for this species.

†Can depend entirely on gut symbionts, cf. *Aphids: Myzus persicae,*[l] *Neomyzus circumflexus.*[m]

[a] Hobson, R. P. (1935). *Biochem. J.* **29**, 1292.
[b] Clark, A. J., and Bloch, K. (1959). *J. Biol. Chem.* **234**, 2583.
[c] Clayton, R. B., and Bloch, K. (1963). *J. Biol. Chem.* **238**, 586.
[d] McKennis, H., Jr. (1947). *J. Biol. Chem.* **167**, 645.
[e] McKennis, H., Jr. (1954). *Proc. Soc. Exptl. Biol. Med.* **87**, 289.
[f] Fraenkel, G., and Blewett, M. (1943). *Biochem. J.* **37**, 692.
[g] Gordon, H. T. (1959). *Ann. N. Y. Acad. Sci.* **77**, 290.
[h] Clayton, R. B. (1964). *J. Lipid Res.* **5**, 3.
[i] Van't Hoog, E. G. (1935). *Z. Vitaminforsch.* **4**, 300.
[j] Van't Hoog, E. G. (1936). *Z. Vitaminforsch.* **5**, 118.
[k] Ito, T. (1961). *Nature* **191**, 882.
[l] Dadd, R. H., and Mittler, T. E. (1966). *Experientia* **22**, 832.
[m] Ehrhardt, P. (1968). *Experientia* **24**, 82.

(161). The possible relationship between the dealkylation pathway and the route of biosynthesis of C-24 alkylated sterols in plants which is suggested by these results remains to be studied.

Certain insects have the necessary enzymatic machinery for the saturation of double bonds or their introduction into the sterol nucleus (150). If the concept of the "best fit" of sterol molecules for incorporation into cellular structures is correct, the types of sterol available in insect food sources, and the versatility or otherwise of the insect's enzymatic mechanisms for effecting modifications in their structure may have important ecological consequences. The strict requirement (151) of a carnivorous beetle, *D. vulpinus*, for cholesterol (Table II) is a case in point. An interesting (though so far unconfirmed) example is *Drosophila pachea* which lives in a highly restricted environment on the stems of senita cactus (*Lophocereus schottii*) and has been reported (162) to require a Δ^7-sterol for growth. If this observation is correct, this is the first insect species for which cholesterol has been shown to be entirely ineffective in maintaining development. The influence of plant steroids on the feeding behavior and survival of the two-striped grasshopper, *Malanoplus bivattatus*, has been studied with results that support the view that these substances are important components in insect–plant associations (163).

The structural role of sterols in insects is quantitatively their most important function, but the sterols also provide the starting material for the ecdysones, the steroid hormones that promote molting and adult development in insects and crustaceans, and which are also found in many plants. Since these compounds have been discussed fully elsewhere (164), the structures of only four examples are given (Fig. 21) to illustrate the point that they may be either C_{27} or C_{29} compounds. The possibility of the influence of phytoecdysones upon the development of predatory insects, a matter of obvious ecological significance, has been studied with conflicting results (165, 166).

While there is so far no evidence to implicate steroids of the vertebrate hormone type in insect endocrinology, the well-known degradative route from cholesterol to C_{21} steroids that operates in mammals (Fig. 22) may be functional in some insects in the context of defense. A number of pregnane derivatives (Fig. 23) are secreted by special glands of certain species of water beetles and are released into the water as deterrents to predatory fish (167). Schildknecht *et al.* (167) have examined the possibility that the effectiveness of these compounds is related to the anaesthetic action (168) which many pregnane derivatives are known to have in mammals, and their findings strikingly recall the "fish assay" used by Selye and Heard (169) in some of the original studies of steroid anaesthesia. Though it seems uncertain whether, in fact, the defensive effect should be properly attributed to anaesthetic action, there is direct electrophysiological evidence that some steroid hormones effect the spontaneous and evoked neural discharges from the olfactory system of fish (170, 171). There is, therefore, a possibility that the olfactory effects of such compounds are interpreted as repellent by the fish, quite independently of their potency as anaesthetic agents.

Although evidence for the presence of steroid estrogens and progesterone in a number of primitive forms has been reported (see references cited in 172), it is only recently that studies have been described that suggest the involvement of such compounds in reproductive function in an invertebrate. The hermaphrodite gland of the slug, *Arion ater rufus*, is reported (172, 173) to produce estrogenic steroids during the female phase and androgenic steroids during the male phase of the reproductive cycle. Though the unequivocal chemical identification of these and other steroid products from the reproductive tissues of this organism remains to be described and their physiological function is uncertain, this work suggests that in certain mollusks, cholesterol may be metabolized to hormonal products by routes essentially the same as those found in vertebrates. It is not clear whether *A. ater rufus* depends upon a dietary sterol supply. It has been reported that sterols are synthesized in this species (174) and in some terrestrial and aquatic snails (175, 176), but the relevant experiments do not exclude contributions to synthesis by symbionts, a factor which, as indicated earlier in this discussion, is of major importance in some insects.

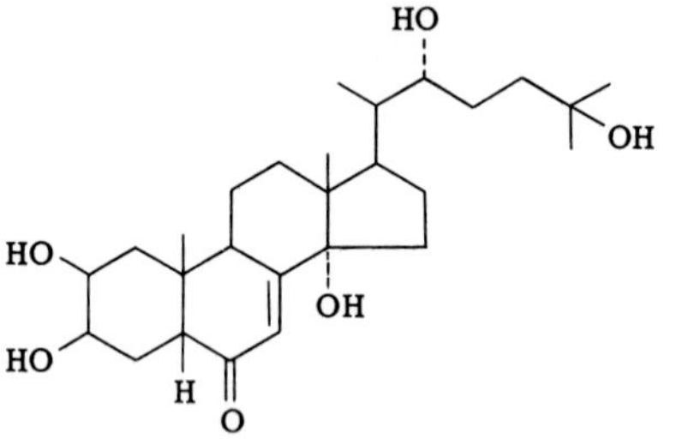

α-Ecdysone[1]

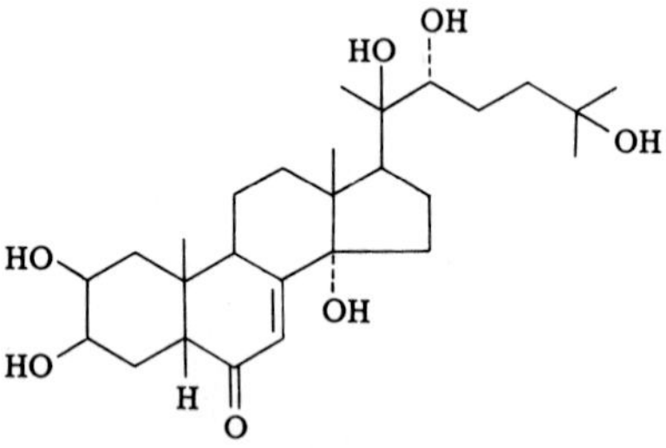

Ecdysterone
(crustecdysone)[2]

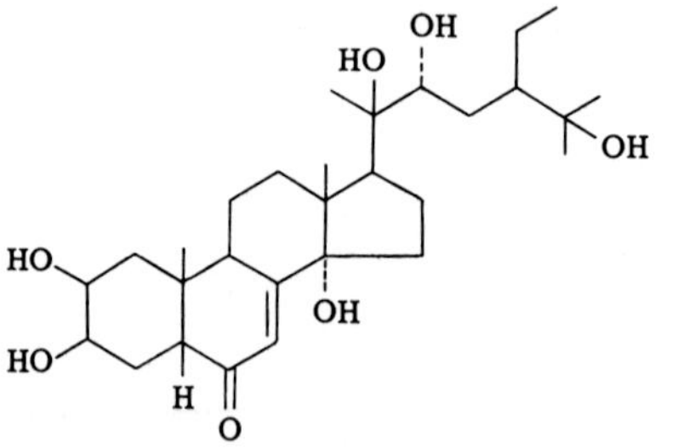

Podecdysone[3]

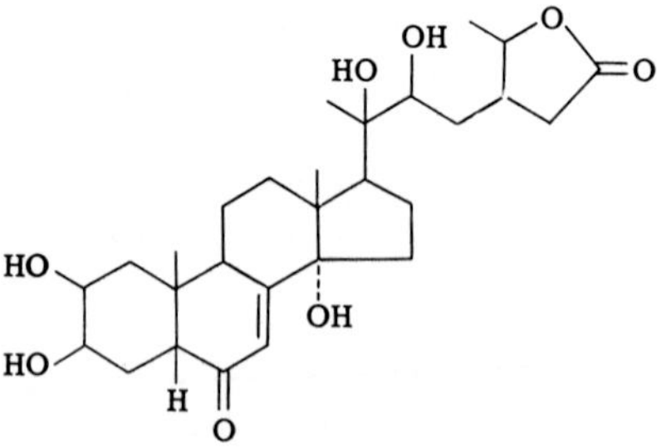

Cyasterone[4]

Fig. 21.   Some examples of ecdysones and their sources.

1. From *Bombyx mori* — Butenandt, A., and Karlson, P. (1964). *Z. Naturforsch.* **9b**, 389. Huber, R., and Hoppe, W. (1965). *Chem. Ber.* **98**, 2403.   From bracken fern, *Pteridium aquilinum* — Kaplanis, J. N., Thompson, M. J., Robbins, W. E., and Bryce, B. M. (1967). *Science* **157**, 1436.

2. From *B. mori* — Hocks, P., and Wiechert, R. (1966). *Tetrahedron Letters* p. 2989.   From crayfish, *Jasus lalandii* — Hampshire, F., and Horn, D. H. S. (1966). *Chem. Commun.* p. 37.   From *P. aquilinum* — Kaplanis, J. N., Thompson, M. J., Robbins, W. E., and Bryce, B. M. (1967). *Science* **157**, 1436.   From *Podocarpus elatus* — Galbraith, M. N., and Horn, D. H. S. (1966). *Chem. Commun.* p. 905.

3. From *Podocarpus elatus* (Taxaceae) — Galbraith, M. N., Horn, D. H. S., Porter, Q. N., and Hackney, R. J. (1968). *Chem. Commun.* p. 971.

4. From *Cyathula capitata* (Amaranthaceae) — Takemoto, T., Hikino, Y., Nomoto, K., and Hikino, H. (1967). *Tetrahedron Letters* p. 3191.

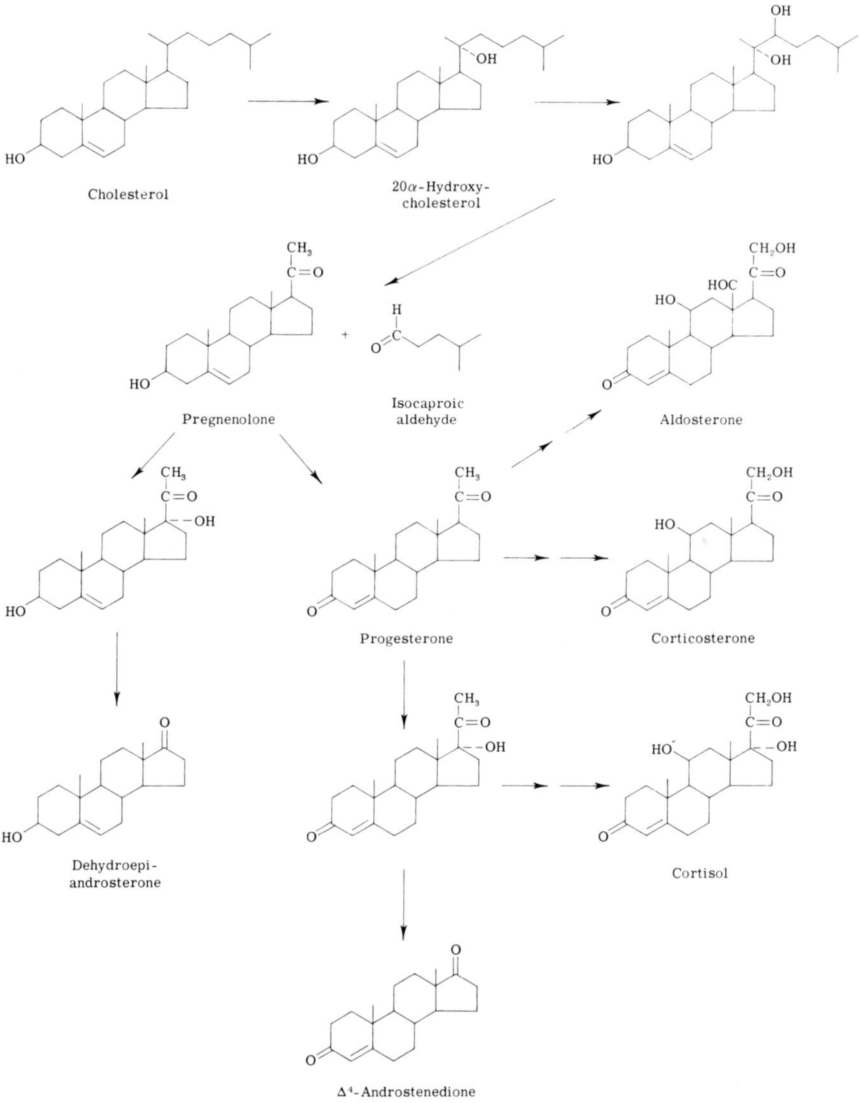

Fig. 22. Some major biogenetic interrelationships in mammalian steroid metabolism.

Recent work suggests that sterols function as precursors of hormonal metabolites in certain fungi. The structure of antheridiol (177) (Fig. 24), a product of the female hyphae of the fungus, *Achlya bisexualis*, which promotes antheridium formation in the male (178), has been shown to be a lactone with the carbon skeleton of 29-isofucosterol. The particular

Steroids of water beetles

Acilius sulcatus (a)

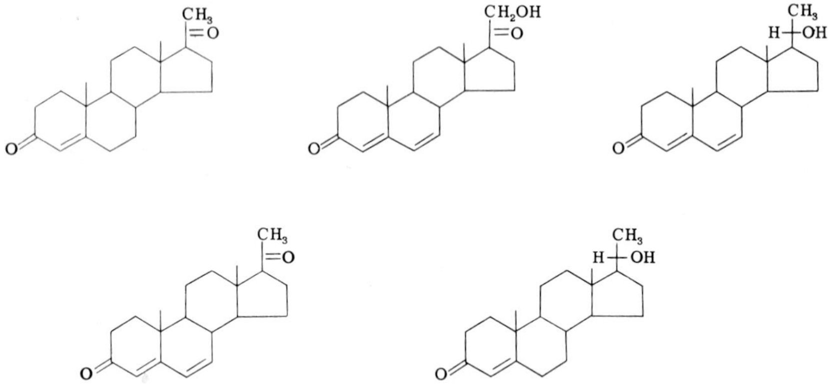

Cybister sp. (b, c)

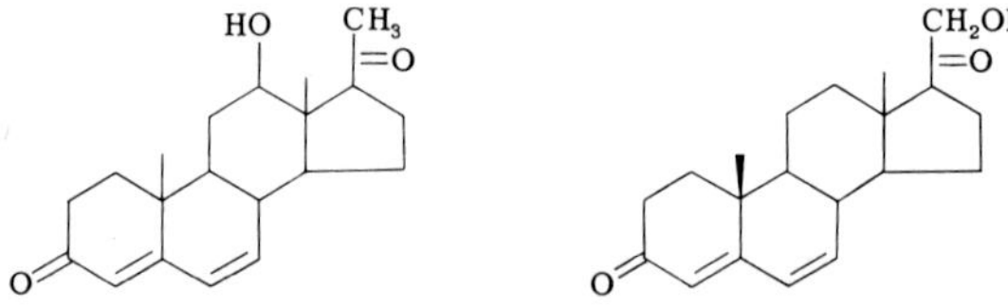

Dytiscus marginalis (d, e)

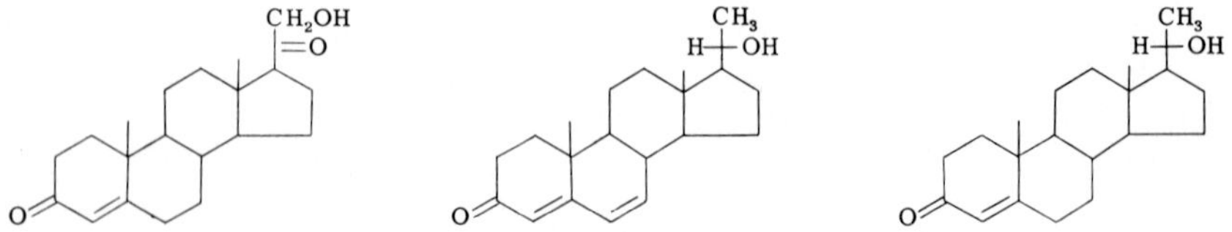

Fig. 23. Steroid defensive secretions of water beetles. (a) Schildknecht, H., Hotz, D., and Maschwitz, U. (1967). *Z. Naturforsch.* **22b**, 938. (b) Schildknecht, H., Siewerdt, R., and Maschwitz, U. (1967). *Ann. Chem.* **703**, 182. (c) Schildknecht, H., and Kornig, W. (1968). *Angew. Chem.* **80**, 45. (d) Schildknecht, H., Siewerdt, R., and Maschwitz, U. (1966). *Angew. Chem.* **78**, 392. (e) Schildknecht, H., and Hotz, D. (1967). *Angew. Chem.* **79**, 902.

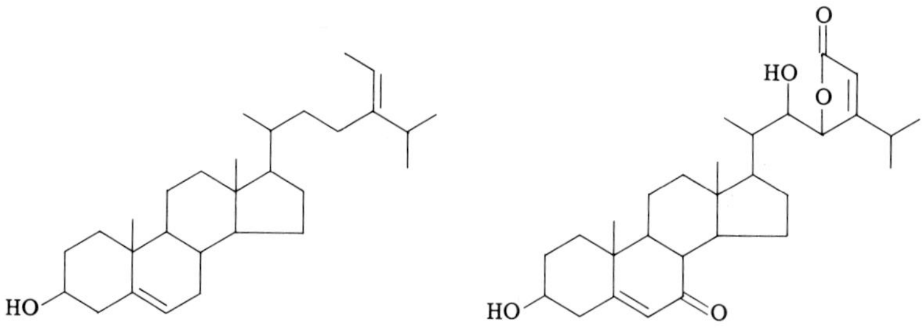

29-Isofucosterol
(Pythium)

Antheridiol
(Achlya)

Fig. 24. Steroids with reproductive function in molds.

potency of 29-isofucosterol (but not fucosterol) in stimulating sexual reproduction in another fungus, the plant pathogen, *Phytophthora cactorum* (179) is suggestive of a precursor role of this sterol for antheridiol or some similar metabolite in this case also. It is interesting that sterol-induced sexual reproduction in another parasitic fungus, *Pythium periplocum*, is suppressed by the mammalian female sex hormone, estradiol (180), but it must be kept in mind in evaluating the possible physiological significance of this observation, that many quite varied influences of steroid hormones on the growth and development of microorganisms have been reported (146).

That terpenoids of types other than steroids are involved in the reproductive processes of certain fungi is indicated by the recent structural identification of the sex pheromone, sirenin (146a,b) (Fig. 25). This compound, released from the oogonia of the aquatic fungus, *Allomyces*, strongly attracts the male gametes.

I. Steroid Metabolism in Higher Plants

The foregoing references to the possible occurrence of the vertebrate steroid hormone synthesizing pathway in invertebrates and microorga-

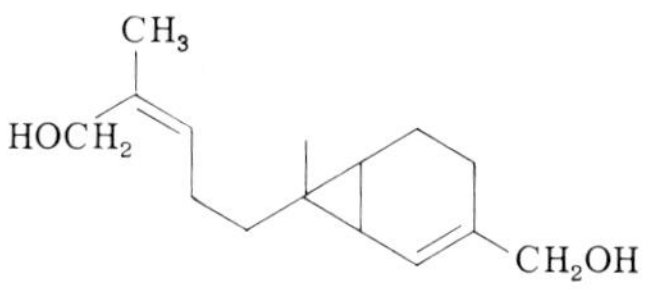

Sirenin (allomyces)

Fig. 25. Sirenin, a sesquiterpene reproductive pheromone of an aquatic fungus.

nisms all indicate the very tentative state of knowledge in these areas. This is not true, however, in the case of higher plants which, only a few years ago, were considered to be devoid of cholesterol, the "characteristic" sterol of animal tissues, and were scarcely suspected of containing enzyme systems capable of its degradation to vertebrate steroids. Following the discovery of cholesterol in plant tissues (181, 182), it has been shown that a number of transformations of cholesterol that take place in vertebrate endocrine tissues also occur in higher plants as stages in the biosynthesis of plant steroids, many of which have powerful physiological effects upon vertebrates. A detailed review of this field, which has developed dramatically in recent years (183–185, 185a) is not possible here, but some salient findings are summarized in Fig. 26.

The route, mevalonic acid → cholesterol → pregnenolone → progesterone, has been established in *Digitalis* by the work of Jacobsohn and Frey (186) and Caspi *et al.* (187, 188), and conversions of cholesterol into solanidin, in the potato (189) and into neotigogenin and Δ^{16}-5α-pregnenolone in tomato plants (190) have also been described. Studies of the distribution of labeling from 2-^{14}C-mevalonate into digitoxigenin synthesized in *Digitalis* plants led Leete and co-workers (191) to the conclusion that the butenolide ring in this compound must arise by condensation of a C_{21} steroid intermediate with a C_2 unit. The details of the sequence of reactions involving reduction of ring A and hydroxylation at C-12,C-14, are still undetermined. There is, however, experimental evidence to suggest that prior to its conversion to the cardenolides, the C_{21} precursor is hydroxylated at C-14 (192) and possibly also at C-12, in the case of digoxigenin synthesis (193). It should perhaps be pointed out that the origin of the 14β-hydroxyl group in these compounds requires a mechanism that must differ from the usual direct hydroxylation processes encountered in steroid metabolism and elsewhere, in which a hydrogen atom is replaced by a hydroxyl group that has the same steric orientation (194, 195), since in this case the product would then be a 14α-hydroxysteroid.

This rapidly developing field of steroid biochemistry in plants has raised many questions concerning the significance of steroids in plant metabolism. Variations in metabolic patterns are evident in different parts of the plant (e.g., as between leaves and roots) (193), and the relationship of glycoside intermediates to the pathways under discussion appears to be an important area for future clarification (196, 197). These and other problems in this area are critically discussed by Heftmann (185).

J. Defensive Role of Toxic Steroids

The various cardiac-active steroids of plants occur as glycosides that render them more readily soluble and lead to more rapid physiological

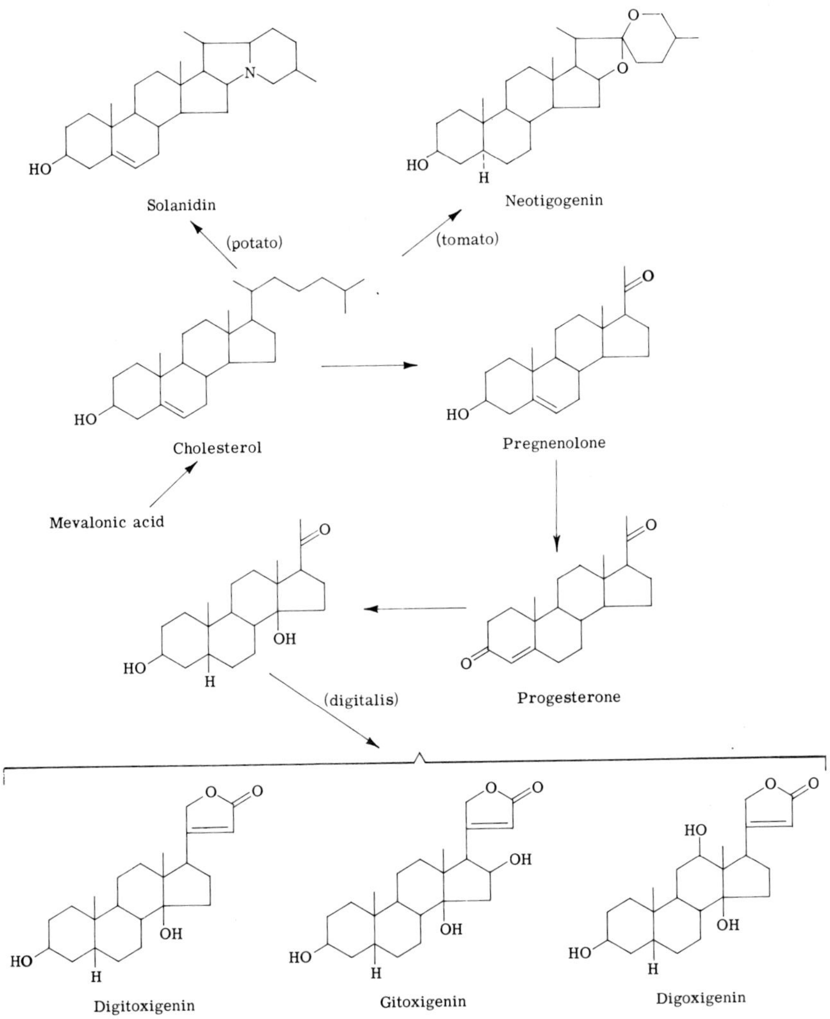

Fig. 26. Some biogenetic interrelationships among steroids in higher plants.

actions, which, besides their well-known and medicinally important effects in stimulating the heart rate, include vomiting, diuresis, CNS effects, and muscle spasms. It seems clear that these highly unpleasant effects are protective against grazing animals. A major site of physiological action of these compounds appears to be the sodium–potassium-activated ATPase of mammalian cells (198), which is integral to the mechanism of active transport of sodium and potassium through the cell membrane (199). In all cases, however, these compounds have more gross membrane-destructive effects, as evidenced by hemolysis, that are no doubt attributable to their detergent properties. An intriguing aspect of the physiological action of cardiac-active steroids is the difference in susceptibility of different vertebrate species toward them (200), which appears to reside in the properties of the receptors rather than in differences in the metabolic fate of the steroid. In general, the toxicity is greater for higher vertebrates. It is of particular interest that the toad is relatively immune to the effects of digitalis while its own defense mechanism employs highly toxic heart-active steroids, the bufogenins, such as bufotalin, Fig. 27, (201) which, as its 14β-suberylarginine conjugate, is the major genin of bufotoxin from the common European toad, *Bufo vulgaris*.

The deadly alkaloid neurotoxin of the skin of salamanders, samandarin (Fig. 28) (202), which is also biosynthesized from cholesterol (202a) is also increasingly toxic in progressively higher vertebrate forms. Other steroid alkaloids, the batrachotoxins of the frog, *Phyllobates aurotoenia*, are reputed to be the most powerful known naturally occurring neurotoxins ($LD_{50} = 1.15-2.7\,\mu g/kg$ in mice) (203). As well as serving a protective function in the frog, these compounds, as arrow poisons, have played a role in human ecology. Their structures have yet to be fully elucidated,* but that of a closely related material, batrachotoxinin A, which apparently arises on decomposition of one of the native toxins, pseudobatrachotoxin, is shown in Fig. 29 (204). This compound, though only 1/500 as toxic as batrachotoxin, is still about as toxic as strychnine.

The neurotoxic agents produced by certain members of the Holothurioideae (sea cucumbers) have interesting structures based, presumably, upon lanosterol as the biogenetic precursor. 22,25-Oxidoholothurinogenin (205), a major representative of the group, is shown in Fig. 30.†

*While this manuscript was in the proof stage the complete structure of the major naturally occurring component, batrachotoxin, was published (203a). The compound is a pregnane derivative of the same, unusual nuclear structure as batrachotoxinin A:

The reported finding that the compound is much less abundant in terrarium-reared frogs than in animals taken from the wild state, and failure in preliminary studies, to demonstrate its biosynthesis from cholesterol, suggests that its derivation from a dietary precursor, available in the natural habitat, should not be ruled out.

†For other, more recently characterized members of this group, see Reference 205a.

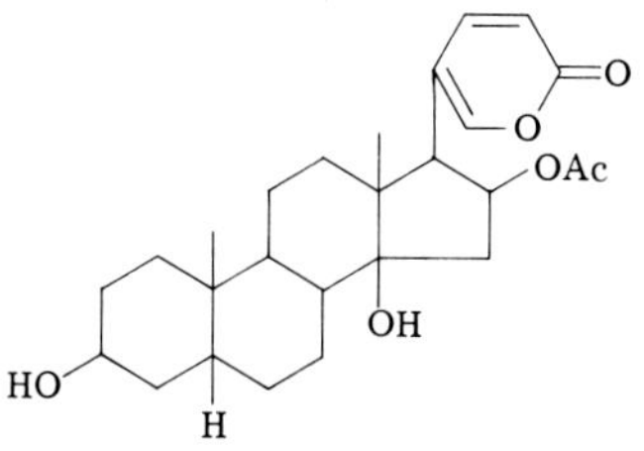

Bufotalin

Fig. 27.   Bufotalin, the bufogenin of the common toad.

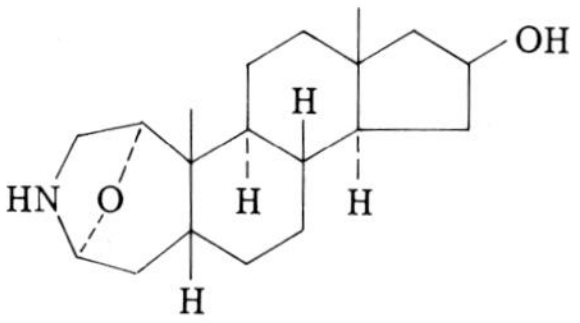

Samandarin

Fig. 28.   Samandarin, a defensive secretion of salamanders.

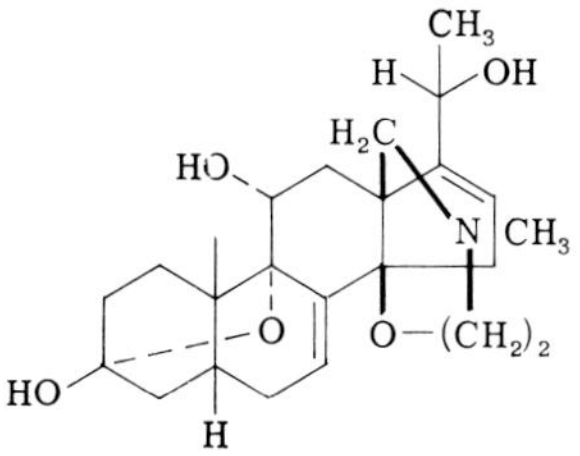

Batrachotoxinin A

Fig. 29.   Batrachotoxinin A, a derivative of a toxin of a Colombian frog.

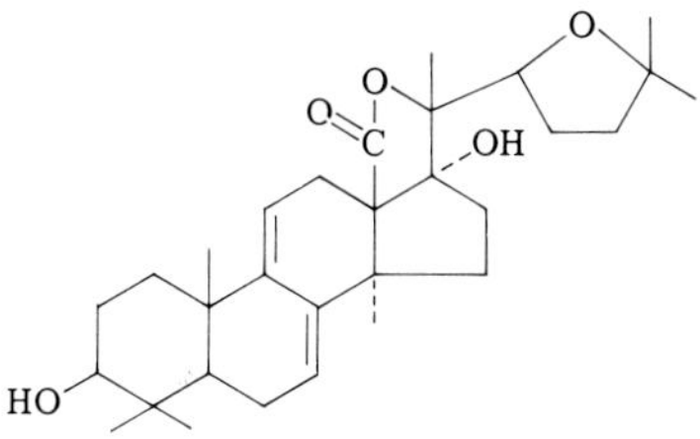

22, 25-Oxidoholo-
thurinogenin

Fig. 30.   22,25-Oxidoholothurinogenin, a major aglycone of holothurin from sea cucumbers.

The aglycone occurs naturally in special glands as a sulfated tetraglycoside, holothurin, which acts both as a powerful and irreversible neural blocking agent and as a hemolytic agent five times as powerful as digitonin (206). Holothurin is, in consequence, a potent toxin for vertebrates, including fish. There have so far been no reported studies of the biosynthesis of this material, and its relationship to the overall scheme of sterol biosynthesis in these echinoderms presents an interesting problem.

Although the synthesis of such highly toxic steroid derivatives by insects for use in defense against vertebrate predators has not so far been reported, insects are known to make defensive use of such compounds that have been ingested from plant food sources and accumulated in their tissues. The grasshopper, *Poekilocerus bufonius*, apparently accumulates cardenolides from milkweed, upon which it feeds and is adapted to eject the material as a noxious foam or jet, from special dorsally situated poison glands. When ingested by birds, these insects and their toxic contents promptly induce vomiting. The monarch butterfly, *Danaus plexippus*, is similarly protected (207, 207a). The cinnabar moth, *Callimorpha jacobaeae*, is similarly unacceptable to most vertebrate predators because of its accumulation of poisonous alkaloids of the senecio group that are ingested in feeding upon groundsel and ragwort (*Senecio vulgaris* and *S. jacobaeae*) (208). The compounds accumulated in this case are the partially terpenoid senecionine (Fig. 31) and several congeners of closely related structure. In considering the mechanism of protection afforded to insects by the ingestion of such toxic food constituents, it may be difficult to distinguish between the effect of deterrence per se, due to the predator's distaste for the insect, and the protection afforded by the foliage which is avoided by most grazing animals. These and other aspects of the problem are discussed in references (207) and (208).

III. CONCLUSION

Although much of the foregoing discussion has implications for ecological interactions that are suggestive rather than definitively demonstrated, there can be no doubt of the wide-ranging importance of the many classes of terpenoid compounds in communication and defense mechanisms, as agents for the control of growth and development, and as structural components of biological membranes. In conclusion, it seems appropriate to summarize very briefly, and speculatively, the reasons for this preeminent position of the terpenoids in the interaction and organization of living systems.

The most noteworthy characteristic of the terpenoid biosynthetic pathway is the extraordinary structural diversity of products attainable through the operation of a remarkably small number of enzymatic steps, reflecting a corresponding economy of enzymatic machinery and its

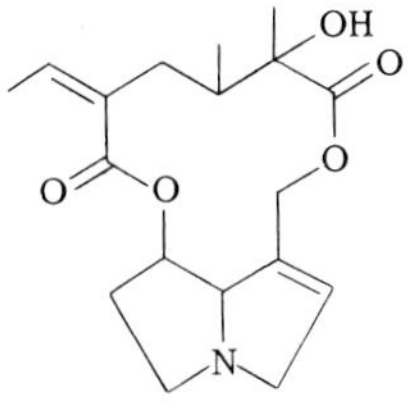

Senecionine

Fig. 31. Senecionine, a toxic alkaloid of groundsel and ragwort.

genetic determinants. The construction of the fundamental "isoprene unit" of five carbon atoms, in the form of isopentenyl pyrophosphate, from acetyl-CoA requires only 7 reactions. The synthesis of geranyl pyrophosphate entails 16, and of farnesyl pyrophosphate, 24, and of squalene, probably 50 separate steps, in all cases with the same isoprene-forming sequence operating in repetitive fashion. The product yielded by each consecutive condensation of an isopentenyl unit has a number of alternative fates open to it, depending of course, upon the enzymes available. It may be dephosphorylated to yield the parent prenol, or may serve as the substrate for a further isoprene addition. A third, and most important alternative, is enzymatic cyclization, for which the terpenoid structure by virtue of its inherent chemical reactivity is uniquely fitted. In fact, so closely do the biological cyclization reactions in general conform to patterns that can be rationalized in terms of nonenzymatic organic chemistry (209, 210), that it is reasonable to view the role of the terpenoid cyclases as being relatively simple and involving two essential functions: initiation of the reaction (usually, presumably, by protonation), and exclusion of all possible end products but the one that is biologically desirable. The exclusive aspect of the cyclization may be achieved by the provision of a hydrophobic site of appropriate topography to specify the required conformation of the terpenoid chain and to prevent the intervention of unwanted anions (such as OH⁻) in the reaction. Since the same acyclic precursor (squalene-2,3-oxide) can yield both rearranged products such as lanosterol and unrearranged products such as the protosterol (V, Fig. 17), presumably from essentially the same conformation, one may speculate that the detailed profile of the active site also contributes to the subtle steric pressures that conduce to rearrangements in those cases where they occur.

Thus, the terpenoid scheme of synthesis makes available to the organism a potential array of metabolic products that increases in geometric fashion with the addition of each isoprene unit. This extensive range of structural possibilities encompasses a correspondingly wide range of volatility, sufficient to provide for the diverse requirements of different chemical communication systems (52, 211). It is no doubt of particular importance for purposes of communication between organisms that, at all levels of complexity, the terpenoid compounds have a high degree of structural specificity which lends itself to the transfer of information. For example, in a monoterpene such as geraniol the total of ten carbon atoms comprises two groups of five within which all interatomic distances are rigidly fixed. This ordered structure is, of course, achieved at the expense of metabolic energy, largely in the form of the ATP which is used in the generation and condensation of isopentenyl pyrophosphate groups from mevalonic acid. It seems likely that this element of structural specificity, possibly coupled with a contribution by the double bonds of such simple terpenes to their attachment in receptor organelles, accounts for their selection as communication agents in preference to fatty acid derivatives which may have comparable volatility. It is indeed interesting that where long chain fatty acids and their corresponding primary alcohols are used by insects as attractants, they are most often desaturated and sometimes also substituted, almost as if in an attempt to simulate some of these essential features of terpene structure (212–216).*

The structural specificity of terpenes is increased still further by cyclization. Further transformations of the cyclic products, as for example, by hydroxylation, may also take place in a wide variety of structurally and stereochemically specific ways which may increase the informational value of the molecule. This aspect of terpene function is most clearly exemplified by the hormonal action of steroids in which the steroid nucleus may be regarded as a "blank" upon which information relevant to biological control is encoded in the form of substituent groups in specific positions and orientations. It seems clear that the size and rigidity of the steroid nucleus make it particularly suitable for this general role, subserving a multiplicity of regulatory functions. It is likely, moreover, that certain structural similarities to the hormonally active steroids play some role in determining the toxicity of the various steroidal defensive agents that have been discussed.

Finally, it should be noted that the terpenoids are of immense biological age (218–220). Their presence may, in fact, be considered synonymous with "life" (cf. 218 and other references therein). It is entirely reasonable that the extremely versatile and relatively simple pathway by which they are synthesized should have been exploited throughout the biological continuum. It seems clear that it has played a profoundly significant role in the evolution of those life forms whose interactions are the concern of modern chemical ecology.

---

*A saturated fatty acid, valeric acid, has recently been identified by Jacobson and co-workers (217) as the sex attractant of the sugar beet wireworm (Coleoptera: Elateridae) *Limonius californicus*. Among the known compounds with this type of biological activity it is also unusual, however, in having only five carbon atoms.

Acknowledgments

The development of my interest in the several different areas represented in this review has been stimulated by discussions with many past and present colleagues, in particular, Dr. Konrad Bloch, Dr. Eugene van Tamelen, Dr. John Law, and Dr. Barry Sharpless.

Work carried out in these laboratories and discussed here has been generously supported by Grants in Aid from the American Heart Association and by USPHS grants GM 12493 and GM 10421.

References

1. Popjak, G., and Cornforth, J. W. (1960). *Advan. Enzymol.* **22**, 281.
2. Clayton, R. B. (1965). *Quart. Rev. (London)* **19**, 168.
3. Olson, J. A. (1965). *Ergeb. Physiol. Biol. Chem. Exptl. Pharmakol.* **56**, 173.
4. Frantz, I. D., Jr., and Schroepfer, G. J., Jr. (1967). *Ann. Rev. Biochem.* **36**(II), 691.
5. Richards, J. H., and Hendrickson, J. B. (1964). "The Biosynthesis of Steroids, Terpenes and Acetogenins." Benjamin, New York.
6. Tchen, T. T., and Danielsson, H. (1968). In "Metabolic Pathways" (D. M. Greenberg, ed.), 3rd Ed., Vol. II, p. 117. Academic Press, New York.
7. Samuels, L. T., and Eik-Nes, K. B. (1968). In "Metabolic Pathways" (D. M. Greenberg, ed.), 3rd Ed., Vol. II, p. 169. Academic Press, New York.
7a. Pridham, J. B., ed. (1967). "Terpenoids in Plants." Academic, New York.
8. Bloch, K. (1952). *Harvey Lectures Ser.* **48**, 68.
9. Bloch, K. (1957). *Vitamins Hormones* **15**, 119.
10. Langdon, R. G., and Bloch, K. (1953). *J. Biol. Chem.* **200**, 129.
11. Heilbron, I. M., Kamm, E. D., and Owens, W. M. (1926). *J. Chem. Soc.* p. 1630.
12. Robinson, R. (1934). *J. Soc. Chem. Ind. (London)* **53**, 1062.
13. Clayton, R. B., and Bloch, K. (1956). *J. Biol. Chem.* **218**, 305, 319.
14. Tchen, T. T., and Bloch, K. (1957). *J. Biol. Chem.* **226**, 921, 931.
15. Voser, W., Mijovic, M. V., Heusser, H., Jeger, O., and Ruzicka, L. (1952). *Helv. Chim. Acta* **35**, 2414.
16. Cornforth, J. W., and Popjak, G. (1954). *Biochem. J.* **58**, 403.
17. Cornforth, J. W., Hunter, G. D., and Popjak, G. (1953). *Biochem. J.* **54**, 590, 597.
18. Cornforth, J. W., Gore, I. Y., and Popjak, G. (1957). *Biochem. J.* **65**, 94.
19. Ferguson, J. J., Jr., and Rudney, J. (1959). *J. Biol. Chem.* **234**, 1072.
20. Rudney, H., and Ferguson, J. J., Jr. (1959). *J. Biol. Chem.* **234**, 1076.
21. Lynen, F., Henning, U., Bublitz, C., Sorbö, B., and Kroplin-Rueff, L. (1958). *Biochem. Z.* **330**, 269.
22. Knappe, J., Ringelmann, E., and Lynen, F. (1959). *Biochem. Z.* **332**, 195.
23. Durr, I. F., and Rudney, H. (1960). *J. Biol. Chem.* **235**, 2572.
24. Linn, T. C. (1967). *J. Biol. Chem.* **242**, 984.
25. Tavormina, P. A., Gibbs, M. H., and Huff, J. W. (1956). *J. Am. Chem. Soc.* **78**, 4498.
26. Eberle, M., and Arigoni, D. (1960). *Helv. Chim. Acta* **43**, 1508.
27. Cahn, R. S., Ingold, C. K., and Prelog, V. (1956). *Experientia* **12**, 81.
28. Bachhawat, B. K., Robinson, W. G., and Coon, M. J. (1955). *J. Biol. Chem.* **216**, 727.
29. Amdur, B. H., Rilling, H. C., and Bloch, K. (1957). *J. Am. Chem. Soc.* **79**, 2646.
30. Tchen, T. T. (1958). *J. Biol. Chem.* **233**, 1100.
31. Chaykin, S., Law, J., Phillips, A. H., Tchen, T. T., and Bloch, K. (1958). *Proc. Natl. Acad. Sci. U.S.* **44**, 998.
32. Lynen, F., Eggerer, H., Henning, U., and Kessel, I. (1958). *Angew. Chem.* **70**, 739.
33. de Waard, A., Phillips, A. H., and Bloch, K. (1959). *J. Am. Chem. Soc.* **81**, 2913.
34. Levy, H. R., and Popjak, G. (1960). *Biochem. J.* **75**, 417.
35. Bloch, K., Chaykin, S., Phillips, A. H., and de Waard, A. (1959). *J. Biol. Chem.* **234**, 2595.
36. Hellig, H., and Popjak, G. (1961). *Biochem. J.* **80**, 47P.
37. Lindberg, M., Yuan, C., de Waard, A., and Bloch, K. (1962). *Biochemistry* **1**, 182.

38. Cornforth, J. W., Cornforth, R. H., Popjak, G., and Yengoyan, L. (1966). *J. Biol. Chem.* **241**, 3970.
39. Agranoff, B. W., Eggerer, H., Henning, U., and Lynen, F. (1960). *J. Biol. Chem.* **235**, 326.
40. Holloway, P. W., and Popjak, G. (1968). *Biochem. J.* **106**, 835.
41. Lynen, F., Agranoff, B. W., Eggerer, H., Henning, U., and Moslein, E. M. (1959). *Angew. Chem.* **71**, 657.
42. Cornforth, J. W., Cornforth, R. H., Donninger, C., and Popjak, G. (1966). *Proc. Roy. Soc. (London)* **B163**, 492.
43. Archer, B. L., Barnard, D., Cockbain, E. G., Cornforth, J. W., Cornforth, R. H., and Popjak, G. (1966). *Proc. Roy. Soc. (London)* **B163**, 519.
44. Popjak, G., Goodman, D. S., Cornforth, J. W., Cornforth, R. H., and Ryhage, R. (1961). *J. Biol. Chem.* **236**, 1934.
45. Rilling, H. C. (1966). *J. Biol. Chem.* **241**, 3233.
45a. Corey, E. J., and Ortiz de Montellano, P. R. (1968). *Tetrahedron Letters* p. 5113.
45b. Popjak, G., Edmond, J., Clifford, K., and Williams, V. (1969). *J. Biol. Chem.* **244**, 1897.
46. Cornforth, J. W., Cornforth, R. H., Donninger, C., Popjak, G., Ryback, G., and Schroepfer, G. J., Jr. (1966). *Proc. Roy. Soc. (London)* **B163**, 436.
47. Donninger, C., and Popjak, G. (1966). *Proc. Roy. Soc. (London)* **B163**, 465.
48. Porter, J. W., and Anderson, D. G. (1967). *Ann. Rev. Plant Physiol.* **18**, 197.
49. Kandutsch, A. A., Paulus, H., Levin, E., and Bloch, K. (1964). *J. Biol. Chem.* **239**, 2507.
50. Williams, R. J. H., Britton, G., Charlton, J. M., and Goodwin, T. W. (1967). *Biochem. J.* **104**, 767.
51. Jungalwala, F. B., and Porter, J. W. (1967). *Arch. Biochem. Biophys.* **119**, 209.
52. Wilson, E. O., and Bossert, W. H. (1963). *Recent Progr. Hormone Res.* **19**, 673.
52a. Weatherston, J. (1967). *Quart. Rev. (London)* **21**, 287.
53. Regnier, F. E., and Law, J. H. (1968). *J. Lipid Res.* **9**, 541.
54. Happ, G. M., and Meinwald, J. (1965). *J. Am. Chem. Soc.* **87**, 2059.
55. Meinwald, J., Happ, G. M., Labows, J., and Eisner, T. (1966). *Science* **151**, 79.
56. Olson, J. A. (1967). *Pharmacol. Rev.* **19**, 559.
57. Wald, G. (1968). *Nature* **219**, 800.
58. Goldsmith, T. H., and Warner, L. T. (1964). *J. Gen. Physiol.* **47**, 433.
59. Goldsmith, T. H., Barker, R. J., and Cohen, C. F. (1964). *Science* **146**, 65.
60. Bro-Rasmussen, F., and Hjarde, W. (1961). *Ann. Rev. Biochem.* **30**, 447.
61. Ames, S. R. (1958). *Ann. Rev. Biochem.* **27**, 371.
62. Roels, O. A. (1967). *Nutr. Rev.* **25**, 33.
63. *Vitamins Hormones* (1962). **20** (a series of papers by several authors on various aspects of vitamin E).
64. Gilbert, J. J., and Thompson, G. A. (1968). *Science* **159**, 734.
65. Viehover, A., and Cohen, I. (1938). *Am. J. Pharm.* **110**, 1.
66. Jakobi, H. (1957). *Tribuna Farm. (Brazil)* **25**, 73.
67. Meikle, J. E. S., and McFarlane, J. E. (1965). *Can. J. Zool.* **43**, 87.
68. Pitt, G. A. J., and Morton, R. A. (1962). *Ann. Rev. Biochem.* **31**, 491.
69. Gloor, U., and Wiss, O. (1964). *Ann. Rev. Biochem.* **33**, 313.
70. Williams, C. M., see chapter 6.
71. Dahm, K. H., Röller, H., and Trost, B. M. (1968). *Life Sci.* **7**, 129.
72. Sláma, K., and Williams, C. M. (1965). *Proc. Natl. Acad. Sci. U.S.* **54**, 411.
73. Bowers, W. S., Fales, H. M., Thompson, M. J., and Uebel, E. C. (1966). *Science* **154**, 1020.
74. Černy, V., Dolejs, L., Lábler, L., Sorm, F., and Sláma, K. (1967). *Tetrahedron Letters* p. 1053.
75. Sláma, K., and Williams, C. M. (1966). *Nature* **210**, 329.
76. Mors, W. B., dos Santos fo, M. F., Monteiro, H. J., Gilbert, B., and Pellegrino, J. (1967). *Science* **157**, 950.
77. Woodward, R. B., and Bloch, K. (1953). *J. Am. Chem. Soc.* **75**, 2023.
78. Ruzicka, L. (1953). *Experientia* **9**, 357, 362.

79. Eschenmoser, A., Ruzicka, L., Jeger, O., and Arigoni, D. (1955). *Helv. Chim. Acta* **38**, 1890.

80. Stork, G., and Burgstahler, A. (1955). *J. Am. Chem. Soc.* **77**, 5068.

81. Wright, L. D. (1961). *Ann. Rev. Biochem.* **30**, 525.

82. Goldsmith, D. J. (1962). *J. Am. Chem. Soc.* **84**, 3913.

83. van Tamelen, E. E., and Curphey, T. J. (1962). *Tetrahedron Letters* **3**, 121.

84. van Tamelen, E. E., Storni, A., Hessler, E. J., and Schwartz, M. (1963). *J. Am. Chem. Soc.* **85**, 3295.

85. van Tamelen, E. E., Schwartz, M. A., Hessler, E. J., and Storni, A. (1966). *Chem. Commun.* **13**, 409.

86. van Tamelen, E. E., and Coates, R. M. (1966). *Chem. Commun.* **13**, 413.

87. van Tamelen, E. E., Willett, J. D., Schwartz, M., and Nadeau, R. (1966). *J. Am. Chem. Soc.* **88**, 5937.

88. van Tamelen, E. E., Willett, J. D., Clayton, R. B., and Lord, K. E. (1966). *J. Am. Chem. Soc.* **88**, 4752.

89. Willett, J. D., Sharpless, K. B., Lord, K. E., van Tamelen, E. E., and Clayton, R. B. (1967). *J. Biol. Chem.* **242**, 4182.

90. Godtfredsen, W. O., Lorck, H., van Tamelen, E. E., Willett, J. D., and Clayton, R. B. (1968). *J. Am. Chem. Soc.* **90**, 208.

91. Corey, E. J., and Russey, W. E. (1966). *J. Am. Chem. Soc.* **88**, 4750.

92. Dean, P. D. G., Ortiz de Montellano, P. R., Bloch, K., and Corey, E. J. (1967). *J. Biol. Chem.* **242**, 3014.

93. Corey, E. J., and Ortiz de Montellano, P. R. (1967). *J. Am. Chem. Soc.* **89**, 3362.

94. Caspi, E., Zander, J. M., Greig, J. B., Mallory, F. M., Conner, R. L., and Landrey, J. R. (1968). *J. Am. Chem. Soc.* **90**, 3563.

95. Caspi, E., Grieg, J. B., and Zander, J. M. (1968). *Biochem. J.* **109**, 931.

96. Maudgal, R. K., Tchen, T. T., and Bloch, K. (1958). *J. Am. Chem. Soc.* **80**, 2589.

97. Cornforth, J. W., Cornforth, R. H., Pelter, A., Horning, M. G., and Popjak, G. (1959). *Tetrahedron* **5**, 311.

97a. Yamamoto, S., Lin, K., and Bloch, K. (1969). *Proc. Natl. Acad. Sci. U.S.* **63**, 110.

98. van Tamelen, E. E., Sharpless, K. B., Willett, J. D., Clayton, R. B., and Burlingame, A. L. (1967). *J. Am. Chem. Soc.* **89**, 3920.

99. Anderson, R. J., Hanzlik, R. P., Sharpless, K. B., van Tamelen, E. E., and Clayton, R. B. (1969). *Chem. Commun.* p. 53.

100. Clayton, R. B., van Tamelen, E. E., and Nadeau, R. G. (1968). *J. Am. Chem. Soc.* **90**, 820.

101. Corey, E. J., Lin, K., and Jautelat, M. (1968). *J. Am. Chem. Soc.* **88**, 2727.

102. Crosby, L. O., van Tamelen, E. E., and Clayton, R. B. (1969). *Chem. Commun.* p. 532.

103. Stone, K. J., Roeske, W. R., Clayton, R. B., and van Tamelen, E. E. (1969). *Chem. Commun.* p. 530.

104. van Tamelen, E. E., Sharpless, K. B., Hanzlik, R., Clayton, R. B., Burlingame, A. L., and Wszolek, P. C. (1967). *J. Am. Chem. Soc.* **89**, 7150.

105. Corey, E. J., Ortiz de Montellano, P. R., and Yamamoto, H. (1968). *J. Am. Chem. Soc.* **90**, 6254.

106. Arigoni, D. (1968). IUPAC Conference on Natural Products, London, July.

107. Stone, K. J., Wientjes, C., Clayton, R. B., and van Tamelen, E. E. (1969). (in preparation).

108. van Tamelen, E. E., Hanzlik, R. P., Sharpless, K. B., Clayton, R. B., Richter, W. J., and Burlingame, A. L. (1968). *J. Am. Chem. Soc.* **90**, 3284.

108a. van Tamelen, E. E., Hanzlik, R. P., and Clayton, R. B. (1969) unpublished observations.

109. Olson, J. A., Lindberg, M., and Bloch, K. (1957). *J. Biol. Chem.* **226**, 941.

110. Gautschi, F., and Bloch, K. (1958). *J. Biol. Chem.* **233**, 1343.

111. Lindberg, M., Gautschi, F., and Bloch, K. (1963). *J. Biol. Chem.* **238**, 1661.

112. Swindell, A. C., and Gaylor, J. L. (1968). *J. Biol. Chem.* **243**, 5546.

113. Sharpless, K. B., Snyder, T. E., Spencer, T. A., Maheshwari, K. K., Guhn, G., and Clayton, R. B. (1968). *J. Am. Chem. Soc.* **90**, 6874.

114. Gaylor, J. L., and Delwiche, C. V. (1964). *Steroids* **4**, 207.

278 RAYMOND B. CLAYTON

115. Sharpless, K. B., Snyder, T. E., Spencer, T. A., Maheshwari, K. K., Nelson, J. A., and Clayton, R. B. (1969). *J. Am. Chem. Soc.* **91**, 3394.

115a. Rahman, R., Sharpless, K. B., Spencer, T. A., and Clayton, R. B. (1970). *J. Biol. Chem.*, submitted for publication.

115b. Ghisalberti, E. L., de Souza, N. J., Rees, H. H., Goad, L. J., and Goodwin, T. W. (1969). *Chem. Commun.* p. 1403.

116. Gibbons, G. F., Goad, L. J., and Goodwin, T. W. (1968). *Chem. Commun.* p. 1458.

117. Canonica, L., Fiechi, A., Galli Kienle, M., Scala, A., Galli, G., Grossi Paoletti, E., and Paoletti, R. (1968). *Steroids* **12**, 445.

118. Akhtar, M., Watkinson, I. A., Rahimtula, A. D., Wilton, D. C., and Munday, K. A. (1968). *Chem. Commun.* p. 1406.

119. Fried, J., Dudowitz, A., and Brown, J. W. (1968). *Biochem. Biophys. Res. Commun.* **32**, 568.

120. Lee, W.-H., and Schroepfer, G. J., Jr. (1968). *Biochem. Biophys. Res. Commun.* **32**, 635.

121. Gibbons, G. F., Goad, L. J., and Goodwin, T. W. (1968). *Chem. Commun.* p. 1212.

122. Caspi, E., Greig, J. B., Ramm, P. J., and Varma, K. R. (1968). *Tetrahedron Letters* p. 3829.

123. Akhtar, M., and Marsh, S. (1967). *Biochem. J.* **102**, 462.

124. Paliokas, A. M., and Schroepfer, G. J., Jr. (1968). *J. Biol. Chem.* **243**, 453.

125. Avigan, J., Goodman, D. S., and Steinberg, D. (1963). *J. Biol. Chem.* **238**, 1283.

126. Goodman, D. S., Avigan, J., and Steinberg, D. (1963). *J. Biol. Chem.* **238**, 1287.

127. Parks, L. W. (1958). *J. Am. Chem. Soc.* **80**, 2023.

128. Lederer, E. (1964). *Biochem. J.* **93**, 449.

129. Akhtar, M., Parvez, M. A., and Hunt, P. F. (1966). *Biochem. J.* **100**, 38c.

130. Barton, D. H. R., Harrison, D. M., and Moss, G. P. (1966). *Chem. Commun.* p. 595.

131. Katsuki, H., and Bloch, K. (1967). *J. Biol. Chem.* **242**, 222.

132. Akhtar, M., Hunt, P. F., and Parvez, M. A. (1967). *Biochem. J.* **103**, 616.

133. Akhtar, M., Parvez, M. A., and Hunt, P. F. (1968). *Biochem. J.* **106**, 623.

134. Petzoldt, K., Kühne, M., Blanke, E., Kieslich, K., and Kaspar, E. (1967). *Ann. Chem.* **709**, 203.

135. Castle, M., Blondin, G., and Nes, W. R. (1963). *J. Am. Chem. Soc.* **85**, 3306.

136. Bader, S., Guglielmetti, L., and Arigoni, D. (1964). *Proc. Chem. Soc.* p. 16.

137. Lenfant, M., Zissmann, E., and Lederer, E. (1967). *Tetrahedron Letters* p. 1049.

138. Smith, A. R. H., Goad, L. J., and Goodwin, T. W. (1967). *Biochem. J.* **104**, 56c.

139. van Aller, R. T., Chikamatsu, H., de Souza, N. J., John, J. P., and Nes, W. R. (1968). *Biochem. Biophys. Res. Commun.* **31**, 842.

140. von Ardenne, M., Osske, G., Schreiber, K., Steinfelder, K., and Tummler, R. (1965). *Kulturpflanze* **13**, 101.

141. Benveniste, P., Hirth, L., and Ourisson, G. (1966). *Phytochemistry* **5**, 31, 45.

142. Ehrhardt, J. D., Hirth, L., and Ourisson, G. (1967). *Phytochemistry* **6**, 815.

143. Baisted, D. J., Gardner, R. L., and McReynolds, L. A. (1968). *Phytochemistry* **7**, 945.

144. Rees, H. H., Goad, L. J., and Goodwin, T. W. (1968). *Biochem. J.* **107**, 417.

145. Diplock, A. T., and Haslewood, G. A. D. (1965). *Biochem. J.* **97**, 36P.

146. Buetow, D. E., and Levendahl, B. H. (1964). *Ann. Rev. Microbiol.* **18**, 167.

146a. Machlis, L., Nutting, W. H., Williams, M. W., and Rapoport, H. (1966). *Biochemistry* **5**, 2147.

146b. Machlis, L., Nutting, W. H., and Rapoport, H. (1968). *J. Am. Chem. Soc.* **90**, 1674.

147. Bloch, K. (1965). *In* "Evolving Genes and Proteins" (V. Bryson and H. J. Vogel, eds.), p. 53. Academic Press, New York.

148. Holm-Hansen, O. (1968). *Ann. Rev. Microbiol.* **22**, 47.

149. de Souza, N. J., and Nes, W. R. (1968). *Science* **162**, 363.

150. Clayton, R. B. (1964). *J. Lipid Res.* **5**, 3.

151. Clark, A. J., and Bloch, K. (1959). *J. Biol. Chem.* **234**, 2583.

152. Clayton, R. B., and Bloch, K. (1963). *J. Biol. Chem.* **238**, 586.

153. Clayton, R. B., and Edwards, A. M. (1961). *Biochem. Biophys. Res. Commun.* **6**, 281.

154. Lasser, N. L., Edwards, A. M., and Clayton, R. B. (1966). *J. Lipid Res.* **7**, 403.

155. Thompson, M. J., Louloudes, S. J., Robbins, W. E., Waters, J. A., Steele, J. A., and Mosettig, E. (1962). *Biochem. Biophys. Res. Commun.* **9**, 113.

156. Lasser, N. L., and Clayton, R. B. (1966). *J. Lipid Res.* 7, 413.
157. Schaefer, C. H., Kaplanis, J. N., and Robbins, W. E. (1965). *J. Insect Physiol.* 11, 1013.
158. Ikekawa, N., Saito-Suzuki, M., Kobayashi, M., and Tsuda, K. (1966). *Chem. Pharm. Bull. (Tokyo)* 14, 834.
159. Svoboda, J. A., Thompson, M. J., and Robbins, W. E. (1967). *Life Sci.* 6, 395.
160. Svoboda, J. A. and Robbins, W. E. (1967). *Science* 156, 1637.
161. Ritter, F. J., and Wientjens, W. H. T. J. M. (1967). *TNO Nieuws* 22, 381.
162. Heed, W. B., and Kircher, H. W. (1965). *Science* 149, 758.
163. Harley, K. L. S., and Thosteinson, A. J. (1967). *Can. J. Zool.* 45, 305.
164. Siddall, J., see chapter 11.
165. Carlisle, D. B., and Ellis, P. E. (1968). *Science* 159, 1472.
166. Robbins, W. E., Kaplanis, J. N., Thompson, M. J., Shortino, T. J., Cohen, C. F., and Joyner, S. C. (1968). *Science* 161, 1158.
167. Schildknecht, H., Hotz, D., and Maschwitz, U. (1967). *Z. Naturforsch.* 22b, 938.
168. Kappas, A., and Palmer, R. H. (1963). *Pharmacol. Rev.* 15, 123.
169. Selye, H., and Heard, R. D. H. (1943). *Anesthesiology* 4, 36.
170. Hara, T. J. (1967). *Comp. Biochem. Physiol.* 22, 209.
171. Oshima, K., and Gorbman, A. (1968). *J. Endocrinol.* 40, 409.
172. Gottfried, H., and Lusis, O. (1966). *Nature* 212, 1488.
173. Gottfried, H., Dorfman, R. I., and Wall, P. E. (1967). *Nature* 215, 409.
174. Voogt, P. A. (1967). *Arch. Intern. Physiol. Biochim.* 75, 492.
175. Addink, A. D. F., and Ververgaert, P. H. J. T. (1963). *Arch. Intern. Physiol. Biochim.* 71, 797.
176. Voogt, P. A. (1968). *Comp. Biochem. Physiol.* 25, 943.
177. Arsenault, G. P., Biemann, K., Barksdale, A. W., and McMorris, T. C. (1968). *J. Am. Chem. Soc.* 90, 5635.
178. McMorris, T. C., and Barksdale, A. W. (1967). *Nature* 215, 320.
179. Elliott, C. G., Hendrie, M. R., and Knights, B. A. (1966). *J. Gen. Microbiol.* 42, 425.
180. Hendrix, J. W., and Guttman, S.-M. (1968). *Science* 161, 1252.
181. Tsuda, K., Agaki, S., Kishida, Y., Hayatsu, R., and Sakai, K. (1958). *Chem. Pharm. Bull. (Tokyo)* 6, 724.
182. Johnson, D. F., Bennett, R. D., and Heftmann, E. (1963). *Science* 140, 198.
183. Heftmann, E. (1963). *Ann. Rev. Plant Physiol.* 14, 225.
184. Tschesche, R. (1965). *Bull. Soc. Chim. France* p. 1219.
185. Heftmann, E. (1967). *Lloydia* 30, 209.
185a. Heftmann, E. (1968). *Lloydia* 31, 293.
186. Jacobsohn, G. M., and Frey, M. J. (1967). *J. Am. Chem. Soc.* 89, 3338.
187. Caspi, E., Lewis, D. O., Piatak, D. M., Thimann, K. V., and Winter, A. (1966). *Experientia* 22, 506.
188. Caspi, E., and Lewis, D. O. (1967). *Science* 156, 519.
189. Tschesche, R., and Hulpke, H. (1967). *Z. Naturforsch.* 22b, 791.
190. Bennett, R. D., Lieber, E. R., and Heftmann, E. (1967). *Phytochemistry* 6, 837.
191. Leete, E., Gregory, H., and Gros, E. G. (1965). *J. Am. Chem. Soc.* 87, 3475.
192. Tschesche, R., Hulpke, H., and Scholten, H. (1967). *Z. Naturforsch.* 22b, 677.
193. Tschesche, R., and Brassat, B. (1967). *Z. Naturforsch.* 22b, 679.
194. Hayano, M., Gut, M., Dorfman, R. I., Sebek, O. K., and Peterson, D. H. (1958). *J. Am. Chem. Soc.* 80, 2336.
195. Corey, E. J., Gregoriou, G. A., and Peterson, D. H. (1958). *J. Am. Chem. Soc.* 80, 2338.
196. Sauer, H. H., Bennett, R. D., and Heftmann, E. (1967). *Naturwissenschaften* 54, 226.
197. von Euw, J., and Reichstein, T. (1966). *Helv. Chim. Acta* 49, 1468.
198. Repke, K., and Portius, H. J. (1963). *Experientia* 19, 452.
199. Skou, J. C. (1964). *Progr. Biophys. Mol. Biol.* 14, 133.
200. Repke, K., Est, M., and Portius, H. J. (1965). *Biochem. Pharmacol.* 14, 1785.
201. Meyer, K. (1952). *Helv. Chim. Acta* 35, 2444.
202. Habermehl, G. (1966). *Naturwissenschaften* 53, 123.
202a. Habermehl, G., and Haaf, A. (1968). *Berichte* 101, 198.
203. Daly, J. W., Witkop, B., Bommer, P., and Biemann, K. (1965). *J. Am. Chem. Soc.* 87, 124.

203a. Tokuyama, T., Daly, J., and Witkop, B. (1969). *J. Am. Chem. Soc.* **91**, 3931.
204. Tokuyama, T., Daly, J. Witkop, B., Karle, I. L., and Karle, J. (1968). *J. Am. Chem. Soc.* **90**, 1918.
205. Chanley, J. D., Mezzetti, R., and Sobotka, H. (1966). *Tetrahedron* **22**, 1857.
205a. Roller, P., Djerassi, C., Cloetens, R., and Tursch, B. (1969). *J. Am. Chem. Soc.* **91**, 4918.
206. Sobotka, H., Friess, S. L., and Chanley, J. D. (1964). *In* "Comparative Neurochemistry" (D. Richter, ed.), p. 471. Macmillan (Pergamon), New York.
207. von Euw, J., Fishelson, L., Parsons, J. A., Reichstein, T., and Rothschild, M. (1967). *Nature* **214**, 35.
207a. Reichstein, T. (1967). *Naturwiss. Rundsch.* **20**, 499.
208. Aplin, R. T., Benn, M. H., and Rothschild, M. (1968). *Nature* **219**, 747.
209. van Tamelen, E. E. (1968). *Accounts Chem. Res.* **1**, 111.
210. Johnson, W. S. (1967). *Trans. N.Y. Acad. Sci.* **29**, 1001.
211. Wilson, E. O. (1965). *Science* **149**, 1064.
212. Jacobson, M., Beroza, M., and Jones, W. A. (1960). *Science* **132**, 1011.
213. Butenandt, A., and Hecker, E. (1961). *Angew. Chem.* **73**, 349.
214. Butler, C. G., Callow, R. K., and Chapman, J. R. (1964). *Nature* **201**, 733.
215. Matsumura, F., Coppel, H. C., and Tai, A. (1968). *Nature* **219**, 963.
216. Silverstein, R. M., Rodin, J. O., Burkholder, W. E., and Gorman, J. E. (1967). *Science* **157**, 85.
217. Jacobson, M., Lilly, C. E., and Harding, C. (1968). *Science* **159**, 208.
218. Eglinton, G., Scott, P. M., Belsky, T., Burlingame, A. L., Richter, W., and Calvin, M. (1966). *Advan. Org. Geochem., Proc. Intern. Meeting*, in Rueil-Malmaison, *1964* p. 41.
219. MacLean, I., Eglinton, G., Douraghi-Zadeh, K., Ackman, R. G., and Hooper, S. N. (1968). *Nature* **218**, 1019.
220. Henderson, W., Wollrab, V., and Eglinton, G. (1968). *Chem. Commun.* p. 710.

11

Chemical Aspects of Hormonal Interactions

JOHN B. SIDDALL

I. INTRODUCTION

Ecology may be viewed as a study of the interactions of organisms and their environments and the interrelationships of different organisms. In the special connection of two most important organisms, man and the insect, it may be noted that vast efforts have been expended in attempts to control the various contributions made by insects to the environment (1). The control of insects themselves is now a major economic and ecological problem complicated by the short-sighted approaches born of expediency.

It is remarkable that such a vast effort on the part of chemical industry should continue while relatively little attention is paid to the most elegant control that the insects exert over their own development. An understanding of this endocrine control would be a major contribution not only to developmental biology but to chemical ecology in general. Endocrine control in insects has been reviewed recently by Engelmann (2). By means of a deceptively simple endocrine system consisting basically of the brain, the prothoracic glands, and the corpora allata, insects are able to complete postembryonic development with periodic molting followed by metamorphosis and adult emergence. Some aspects of reproduction in female adults are demonstrably influenced by hormones of the same endocrine system (3). Interference with the endocrine system, by extirpation or implantation of glands and now by injection or even topical application of the known hormones of the system, has remarkable consequences which have recently diverted attention from the basic aspects of hormonal interactions.

The extrinsic factors, hormones, presumably produced by the endocrine organs (4) are inconsequential in the absence of a better understanding of the intrinsic properties of cells to divide and to differentiate. This basic study of cell biology and molecular biology of the cell nucleus is making rapid strides, but the complexity of the simplest cell is astounding. We may profitably survey some of the current views of insect hormone action before turning to the more chemical aspects of such studies.

Early experiments by Wigglesworth (5) demonstrated the presence of a juvenile hormone which appeared to inhibit adult development, since removal of corpora allata from immature insects resulted in earlier appearance of adult or more advanced characters. Implantation of active corpora allata at pupal or last larval stages resulted in partial or complete suppression of more mature characters, often with earlier molting, indicating a possible prothoracotropic effect of the corpus allatum hormone. Removal of prothoracic glands led to inhibition of molting, suggesting their production of hormones (now known as ecdysones) which induce molting with or without metamorphosis.

II. JUVENILE HORMONES

A. Mode of Action

Recent experimentation (6) with epidermal cells of a lepidopterous insect *Galleria mellonella* has led to the conclusion that the epidermal cells of last instar larvae, which are programmed to secrete pupal cuticle after ecdysis or separation of the old larval cuticle, can be induced to secrete larval cuticle if subjected to excess of Cecropia moth juvenile hormone at the correct time in the last larval instar, but not at all points in time in this instar. The epidermal cells appear to undergo determination in the first (one) third of the last larval instar at a stage which may coincide temporally with synthesis of new DNA, just before a first determinative cell division. The consequences of this determination are manifested in the nature of the cuticle which is subsequently secreted following stimulation by ecdysone which induces molting. The point of sensitivity to juvenile hormone clearly precedes the phased sensitivity to ecdysone. After the molt these cells again become sensitive to juvenile hormone. Apparently, juvenile hormone dictates (at determination) which information may be released later by a call from ecdysone. Earlier in larval life, the period of sensitivity to ecdysone is less clear but always begins in the latter stage of a larval instar. Further work is necessary to correlate the point in time at which larval cells are susceptible to juvenile hormone with discrete biochemical processes which are at that time directly related to the action of the hormone.

The differentiation of hemipteran larval cells into adult cells without intervening cell division (5, 7) (and probably without replication of DNA) poses a special problem for theoreticians, but may provide a simpler system for study of the shifts in capabilities of differentiated and undifferentiated cells for protein synthesis. Preliminary work with an *in vitro* protein synthesizing system from *Tenebrio* (8) indicates that messenger RNA for adult cuticular protein is present on ribosomes isolated from first day pupae, but the nature of the mechanism of suppression of this information in juvenile hormone treated insects is not understood.

Speculatively, juvenile hormone may indirectly or directly exert influence on the replication of DNA, which determines form after ecdysis. Many other possibilities exist. Perhaps juvenile hormone influences, by derepression, the modulation sequence for transcription of selected larval regions of the genome or activates synthesis of specific pupal and adult region repressors in relation to hormone titer. In some presently unknown way this hormone leads to suppression of undesirable information and a later absence of the hormone leads to suppression of information for larval syntheses perhaps by successful competition of repressors.

B. Isolation and Structure

Following the early discovery by Wigglesworth of the existence of a juvenile hormone (5), no progress toward isolation and chemical structure elucidation was made until Williams (9) found that adult male abdomens of the silk moth *Hyalophora cecropia* could be ether extracted to provide a relatively rich source of hormone, now estimated at $1.5\mu g$/abdomen. Although data on hormone titers are sparse, it still appears that male adults of these moths are unique in their peculiar ability to store this hormone.

Events leading to the isolation of Cecropia moth juvenile hormone by Röller and co-workers (10) have been adequately reviewed (11). Considerable surprise attended the announcement of the unusual structure I (Fig. 1) for at least two reasons. First, the hormone bears a remarkable resemblance to the sesquiterpene terminal epoxide (Fig. 2) of methyl farnesoate synthesized much earlier by Bowers and co-workers (12) who

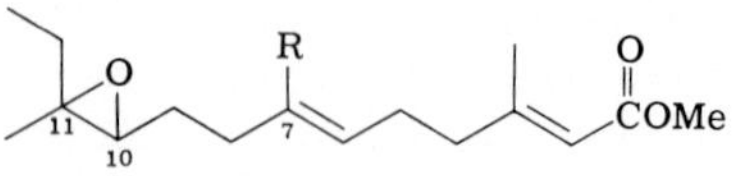

(I) R = Et   Röller *et al*. 1967
(II) R = Me   Meyer *et al*. 1968

Fig. 1.   The natural juvenile hormones.

predicted such a resemblance on the basis of high morphogenetic and gonadotropic activity shown by their compound. Second, the presence of two ethyl groups attached to carbons 7 and 11 (Fig. 1) instead of the normal sesquiterpene methyl groups raised doubts about the correctness of structure I but these were dispelled when the first synthesis by Dahm and co-workers (13) confirmed the structure and stereochemistry of I.

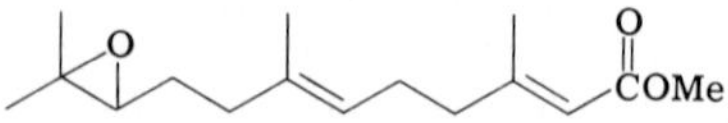

Fig. 2.   Methyl farnesoate epoxide.

When the isolation and elucidation of a second Cecropia moth juvenile hormone II (Fig. 1) was announced by Meyer *et al.* (14), any remaining doubts were removed, since this hormone II also shows ethyl branching at C-11 but is otherwise a regular isoprenoid.

C. Synthetic Methods

Considerable attention has been given to synthesis of the juvenile hormones, since these molecules provide diverse challenges. The difficulty of isolation dictated total synthesis before anything more than preliminary endocrinological and biological study of this hormone could be carried out. Clearly the promises of most interesting discoveries still to be made in understanding how metamorphosis is controlled and in how such hormones might be turned to practical use in insect control have added impetus to chemical synthetic research. From the purely chemical point of view, the presence of two trisubstituted olefinic linkages, which must have trans geometry and an epoxide function derivable from a third trisubstituted double bond, presents particularly challenging problems for the synthetic chemist.

The first phase of synthesis (13) required preparation of several geometrical isomers to provide confirmation of the suspected trans, trans, cis geometry of I (Fig. 1) and was accomplished by introducing double bonds at C-10 and later C-6 and C-2 in the form of α,β-unsaturated esters whose cis and trans isomers are completely separable and identifiable with relative ease. After generation by modified Wittig reaction the required isomer was separated and its geometry fixed by conversion to an allylic halide for chain extension (Fig. 3). In this way, eight geometrical isomers of the hormone I were prepared and their biological activities compared (11). Interestingly, the naturally occurring isomer is most potent biologically but simple conversion to an ethyl ester analog increased activity in mealworms eightfold. Other nonstereoselective syntheses have appeared (15, 16) but suffer greatly from the difficulty of separation of isomers of isolated olefinic bonds.

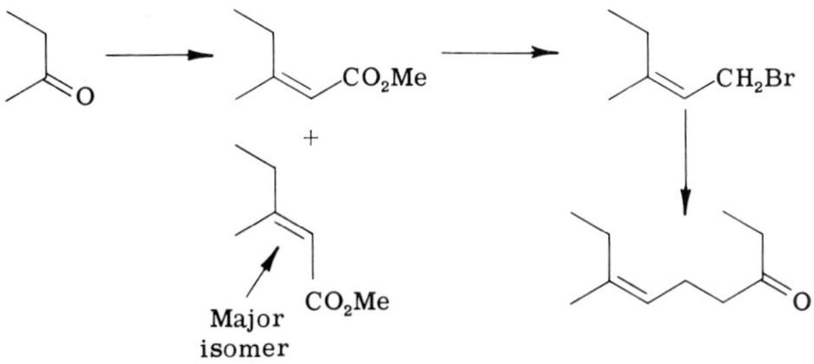

Fig. 3. Nonstereoselective olefin synthesis.

A major objective of stereospecific synthesis of the hormone I (Fig. 1) was achieved independently in the laboratories of Harvard (17), Stanford (18), and Syntex (19) by widely different methods. Though the details cannot be included in this brief review, the new synthetic methodology which is of general utility is well worthy of mention. In the synthesis of Corey and co-workers (17), the stereospecific conversion of propargyl alcohols to trisubstituted olefins via organo aluminum and organo copper reagents and a novel two-step homologation of primary halides (Fig. 4) are notable among the five new methods that were used. An elegant combination of a modified Julia method for stereospecific synthesis of homoallylic halides (20) with the stereoselective Cornforth method for epoxides (Fig. 5) was employed in the Stanford synthesis (18) which develops the hormone chain from the methoxycarbonyl terminus. The scarcity of convenient methods for stereospecific synthesis of trisubstituted olefins led the author's group (19) to use sequential fragmentation (Fig. 6) of a bicyclic precursor which incorporates two diol systems in 1,3- relationships. These permit opening of the rings to provide double bonds whose geometry is determined entirely by relative stereochemistry in the ring system which is more easily controlled.

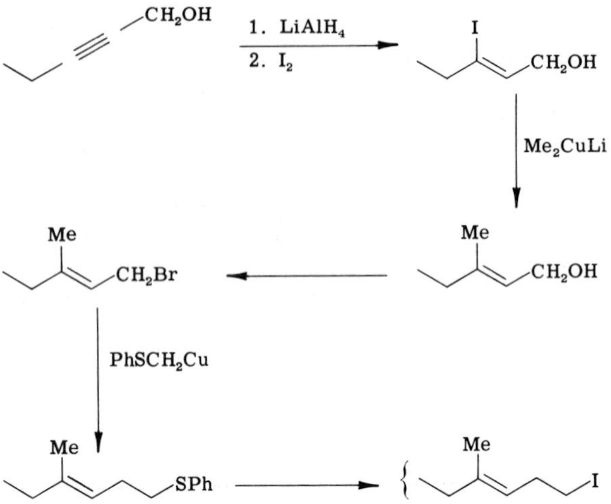

Fig. 4. Stereospecific olefin synthesis and homologation.

On-going research in several laboratories will undoubtedly provide shorter syntheses of the C_{17} juvenile hormone I (Fig. 1). Absolute configuration of the oxirane rings in natural juvenile hormones remains to be determined.

Synthesis of the C_{16} juvenile hormone II (Fig. 1) presents fewer problems since naturally occurring sesquiterpenes provide starting materials embodying both double bonds of II. In the author's laboratory (21) the

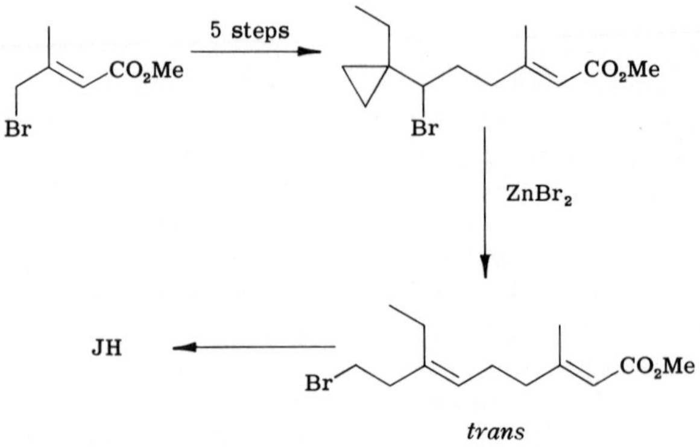

Fig. 5. Stereospecific olefin synthesis.

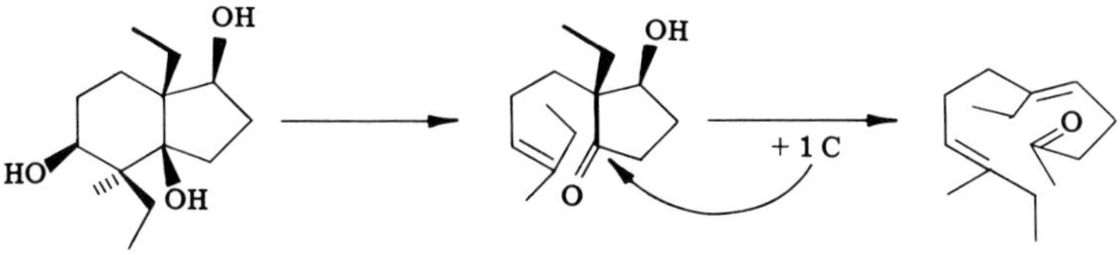

Fig. 6. Internal fragmentation for stereospecific olefin synthesis.

terminal epoxide of methyl farnesoate (12) available from all-*trans*-farnesol was converted stereospecifically in five steps (Fig. 7) to the C_{16} juvenile hormone. Biological assay by injection in olive oil into fresh pupae of the yellow mealworm *Tenebrio molitor* showed (22a) the C_{16} hormone II (Fig. 1) to have three-fifths of the morphogenetic activity of the C_{17} hormone I (Fig. 1). Topical application (22b) in acetone to yellow mealworm pupae similarly showed II to have one-half of the activity of C_{17} hormone I. In topical assays (22c) on last larvae of the hemipteran *Rhodnius prolixus*, 0.6 μg of the C_{16} hormone II gave the same median score, stage ten, as 0.4 μg of the C_{17} hormone I. Further considerations of synthetic chemical aspects of the juvenile hormones have led to a new synthetic method for trisubstituted olefinic esters (23, 24) by direct alkylation of acetylenic esters with the versatile organo copper reagents.

D. Modified Hormones

Chemical modification of hormone structure coupled with a quantitative bioassay which measures some conveniently observable end point affords a datum for each modification which is really a summation of interdependent factors. Considering only the influence of hormone analogs on epidermal cells where the nature of newly secreted cuticle is a biological end point of prior topical application, several factors must be considered before interpretation of "biological activity" is possible: (a) solubility in cuticular waxes; (b) partition from lipid into aqueous solution; (c) ability to bind (hypothetical) lipoprotein carriers for transport; (d) stability to metabolic inactivation; (e) molecular conformation in solution; and (f) (hypothetical) receptor fit, binding constants and kinetics.

Physicochemical parameters can be measured and related to factors (a) and (b) but the solution conformation (e) is not amenable to simple numerical description. The presently shady areas of carrier binding (c) for transport and binding to receptor sites (f) are becoming major lines of investigation in hormone biology and require a multidisciplinary attack. Metabolism of insect hormones is still only poorly understood in chemical terms and will be discussed later.

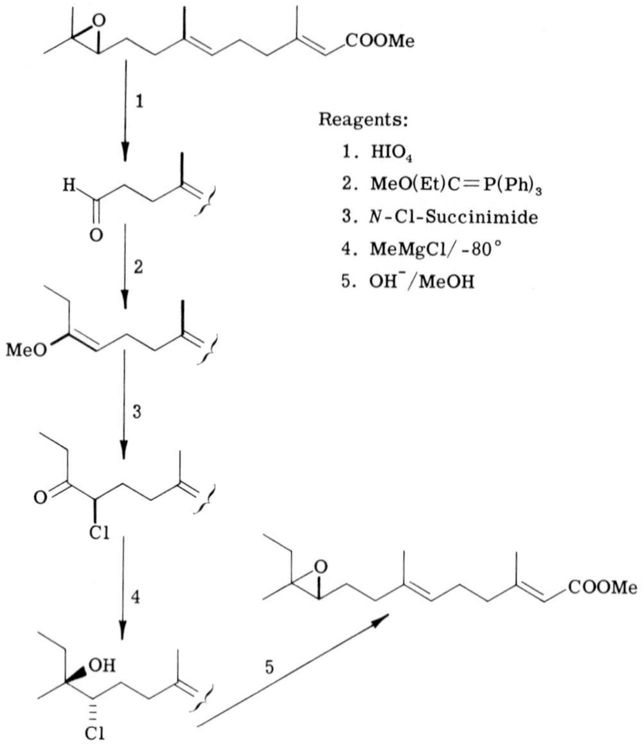
COOMe
1
Reagents:
1. HIO_4
2. $MeO(Et)C{=}P(Ph)_3$
3. N-Cl-Succinimide
4. MeMgCl/-80°
5. OH^-/MeOH
H
O
2
MeO
3
O
Cl
4
O
COOMe
OH
5
Cl

Fig. 7. Stereospecific synthesis of C_{16} juvenile hormone.

In modifications of the natural Cecropia moth juvenile hormone I (Fig. 1), the importance of trans geometry at C-2 and C-6 is clearly evident (4, 15). In physiological aqueous solution the all-trans form may adopt a minimum-volume helical conformation that exposes only the terminal ester and epoxide functions as possible binding sites. The relative unimportance of stereochemistry in the epoxide system is not inconsistent with this view. Chemical support is present in the observation (25) of pronounced solvent effects on relative reactivity of terminal versus in-chain double bonds in polyolefin systems (26).

Since squalene terminal epoxide is probably the last acyclic intermediate in biogenesis of sterols and triterpenes (27), the possibility that the acyclic natural juvenile hormones are merely precursors for enzymatic cyclization to authentic hormones was not overlooked. However, removal of the central double bond (which would be essential for cyclization) of the juvenile hormones leads to only a small decrease in

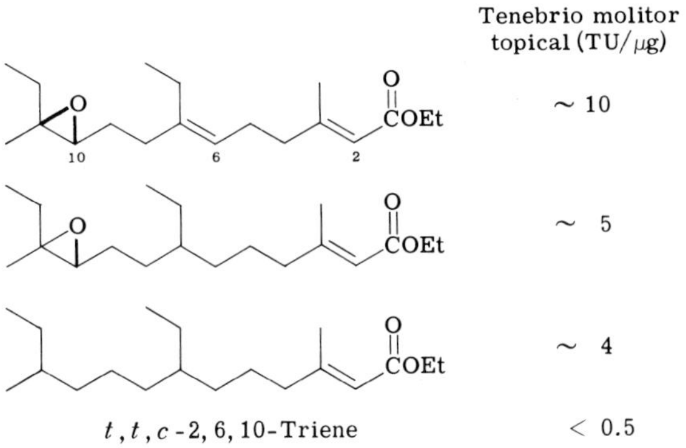

Fig. 8. Biological activities of juvenile hormone analogs.

biological activity (Fig. 8) and in the sesquiterpene series actually increases activity in *Rhodnius* (28). Such findings imply that cyclization is unimportant except as a possible pathway for metabolic inactivation of the natural hormones. The retention of considerable activity after removal of both the epoxide function and the central double bond (Fig. 8) is noteworthy and may reflect resistance to metabolic inactivation resulting in a more persistent effect.

It is unlikely that a unifying concept of structure–activity relationships can be evolved unless an inordinately large number of parameters is included in regression analysis of biological activities in many orders of insects. Nevertheless, progress in devising highly specific modified hormones can continue as a partly intuitive process.

E. Biosynthesis

Speculation that S-adenosyl methionine is responsible for methylation of a sesquiterpene at the branching methyl groups (probably by alkylation of methylene groups prior to double bond migration) has analogy in the biosynthesis of C_{29} sterols (27). Such a pathway from a sesquiterpene could account for biosynthesis of both the C_{17} hormone I (Fig. 1) and the C_{16} hormone II (Fig. 1). The present absence of published evidence for such a pathway to juvenile hormones probably attests to the difficulty of isolation of sufficient hormone for chemical degradation and location of radio label. Advances in organ culture technique may later simplify such work and presently provide an avenue for fruitful research.

F. Mimics

The discovery of a multitude of apparently unrelated compounds and mixtures which possess some degree of juvenile hormone activity has added much confusion to the extensive literature of insect hormone chemistry, but this is far outweighed by the considerable stimulation of interest in insect endocrinology and mechanism of hormone action. An excellent appraisal of the situation in 1963 and the implications of such discoveries is due to Schneiderman and Gilbert (29).

During the isolation and elucidation of structure of the natural juvenile hormones (10, 14), several findings of considerable significance were reported from many laboratories. Events leading to the discovery of paper factor by Sláma (30) have been described by Williams (31). The structure of the major active principle of the balsam fir *Abies balsamea* was elucidated by Bowers and co-workers (32) in 1966. Surprisingly this

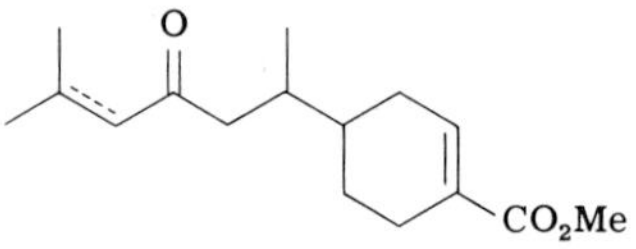

Fig. 9. Juvabione [paper factor(s)], the major active principle of the balsam fir *Abies balsamea* (L), is a sesquiterpenoid α,β-unsaturated ester.

compound, named "juvabione" (Fig. 9), is a sesquiterpenoid α,β-unsaturated ester in common with the derivative (Fig. 2) of methyl farnesoate prepared earlier by Bowers *et al.* (12), but juvabione differs greatly in possessing a cyclohexene ring, which may be a major reason for its high morphogenetic activity in pyrrhocoris bugs. However, description of juvabione as a specific hormone mimic for Pyrrhocoridae is inaccurate (12). In 1967 a related sesquiterpene dehydrojuvabione (indicated in Fig. 9 by a dotted line) was isolated from a Czechoslovakian balsam by Černy and co-workers (33) and shows similar biological activity although considerable increase in polarity is conferred by the α,β-unsaturated ketone system. Several syntheses of juvabione have appeared recently (42–45). Notable for ingenuity are those of Birch and co-workers (45) and Pawson's group (44). At the Czechoslovak National Academy of Sciences in Prague, considerable effort has been expended in modification (34, 35) of both juvabione and farnesoic acid (36) with notable increases in biological activity (Fig. 10) for selected hemipteran insects. Species specificity within one family is not seen (34) in the genus *Dysdercus* but selectivity of action on the family Pyrrhocoridae of hemipterans is apparent.

Chemical structure elucidation (37) of a dihydrochloride derivative (Fig. 11) of methyl farnesoate by the Czech workers clarified some of the mysteries surrounding a mixture prepared earlier by Law, Yuan, and Williams (38). The remarkably high morphogenetic and ovicidal activities of this dichloride (Fig. 11) in certain hemipterans (28, 36, 37, 39, 40) appear to be related to its ability to withstand metabolic inactivation (41). While pure samples of this dichloride (Fig. 11) show remarkable

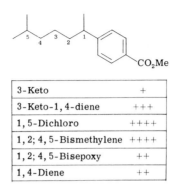

3-Keto	+
3-Keto-1,4-diene	+++
1,5-Dichloro	++++
1,2;4,5-Bismethylene	++++
1,2;4,5-Bisepoxy	++
1,4-Diene	++

Fig. 10.   Morphogenetic activities of aromatic juvabione analogs in *Pyrrhocoris apterus*.

selectivity for hemipterans, the complex mixtures of Law, Yuan, and Williams (38) exhibit morphogenetic and ovicidal activity in a variety of insect orders (3, 31, 38, 46–53), but the structures of individual components responsible for the many biological effects remain to be elucidated.

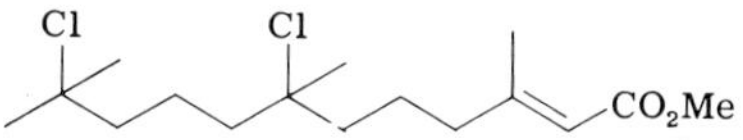

Fig. 11.   Methyl farnesoate dihydrochloride.

A recent development, which has two aspects of interest in the chemistry of hormonal interactions, was reported by Bowers (54), who observed that certain pyrethrin and carbamate pesticide synergists, notably the 3,4-methylenedioxyphenol derivative sesoxane (Fig. 12), appear to mimic the effects of juvenile hormones. Structural resemblances between sesoxane (Fig. 12) and the natural hormones are difficult to see, but in a more recent paper (55), Bowers examined 3,4-methylenedioxyphenyl ethers of terminally epoxidized geraniol and homologs. These compounds, drawn in Fig. 13 alongside a natural hormone, show a striking structural resemblance to natural Cecropia moth hormone and exhibit high activity in yellow mealworms. At the present, one still cannot distinguish between the possibilities that such mimics (a) duplicate all the functions of a natural hormone or (b) merely suppress oxidative metabolic inactivation of traces of endogenous juvenile hormone (which are present, even in allatectomized insects). The former possibility is not only more likely but considerably more interesting.

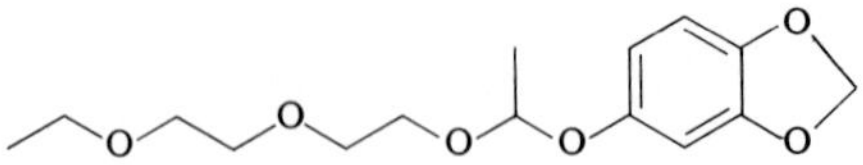

Fig. 12.   Sesoxane.

III. STEROIDAL MOLTING HORMONES

A. Occurrence

Since Butenandt and Karlson (56) first isolated α-ecdysone (Fig. 14) from silkworm pupae, other workers have isolated and structurally eluci- dated five more zooecdysones, Fig. 15 — crustecdysone (57, 68) or β-ecdy- sone, II, (58–64), 26-hydroxy-β-ecdysone, IV, (65, 66), 2-deoxy-β-ecdysone, III, (67), callinecdysone-A, V, (68), and callinecdysone-B, VI, (68) from various insects and crustaceans. The probable hormonal function of β- ecdysone II (20-hydroxy-α-ecdysone) in crustaceans and other arthro- pods is indicated by work of Lowe et al. (69), Krishnakumaran and Schneiderman (70), and Kurata (71).

B. Structures

Although some 10 years elapsed between the first isolation (56) and the x-ray crystallographic elucidation of the structure of α-ecdysone (Fig. 14) by Huber and Hoppe (72) in 1965, the proper use of physical methods for structure determination has led since 1965 to a knowledge of more than twenty ecdysones found in animals (Section III, A) and plants (Section III, C). The advent of highly sophisticated mass spectroscopy and nuclear magnetic resonance (NMR) spectroscopy is primarily re- sponsible for this knowledge.

At present, all zooecdysones with the sole exception of 2-deoxy-β-ecdy- sone (67) possess a tetracyclic nucleus bearing hydroxyl groups in posi- tions 2β, 3β, and 14α, an unsaturated (Δ^7)-6-ketone system and cis fusion between rings A and B (Fig. 15). This nucleus is also common to almost all phytoecdysones and the great diversification of structures lies almost entirely in the oxidation state of the cholestane side chain (Fig. 15). Sur- prisingly, the 5α-isomers of zooecdysones have not been found in vivo,

Fig. 13.   Planar structural similarities between a hormone and a potent analog.

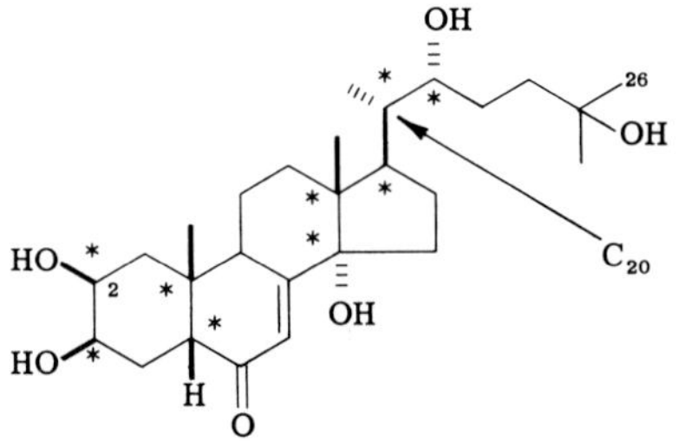

Fig. 14.   Structure of α-ecdysone. * = asymmetry.

although isomerization of the A/B-cis ring junction to A/B-trans is chemically easy and would provide a simple inactivation pathway.

Although the nuclear stereochemistry of most ecdysones is firmly established by correlations with α-ecdysone or synthetic compounds (73, 74), little is known with certainty of the absolute configurations in the oxidized cholestane side chains of zooecdysones (other than α-ecdysone) and phytoecdysones.

C. Phytoecdysones

Even before the first zooecdysone had been synthesized (section III, D), a remarkable paper from Nakanishi's group (75) reported the isolation from *Podocarpus nakaii* plants of a steroid, ponasterone-A (Fig. 16), having a structure almost identical with that of α-ecdysone and possessing full molting hormone activity.

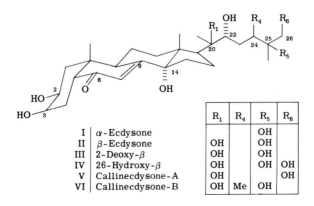

		R_1	R_4	R_5	R_6
I	α-Ecdysone			OH	
II	β-Ecdysone	OH		OH	
III	2-Deoxy-β	OH		OH	
IV	26-Hydroxy-β	OH		OH	OH
V	Callinecdysone-A	OH			OH
VI	Callinecdysone-B	OH	Me	OH	

Fig. 15.   Naturally occurring molting hormones.

Structures for other ponasterones have now been elucidated (76) and present a significant divergence from the usual $2\beta,3\beta$-glycol system of the majority of ecdysones. Ponasterone-B has hydroxyl groups in the 2α and 3α positions but nevertheless shows activity as molting hormone. Nakanishi (41) has rationalized this observation by suggesting that this isomer adopts a boat conformation (Fig. 16) in ring A which places the hydroxyls at C-2α and 3α in virtually the same exposed positions as the normal chair form β-glycols (Fig. 15).

Almost concurrent with the elucidation of ponasterone-A, the group of Takemoto (77) isolated inokosterone (Fig. 17) and β-ecdysone itself from *Achyranthis radix*. Horn's group meanwhile found β-ecdysone in Australian *Podocarpus elatus* leaves (78) and shortly afterward even α-ecdysone was isolated by the USDA group (79) from bracken fern, *Pteridium aquilinum*. In 1967, β-ecdysone was found in *Polypodium vulgare* by the Czech group (80) and in *Vitex megapotamica* by Rimpler and Schulz (81). Since these early discoveries, a veritable battery of phytoecdysones has been assembled from numerous plants, largely through the efforts of Takeda Industries, Japan and Takemoto's group in Sendai (41).

From the chemical viewpoint, the phytoecdysones fall into three categories: (a) variously hydroxylated or C-24 alkylated coprostanes (Fig. 17, 75, 77, 84–93) having the α-ecdysone nucleus; (b) lactonized stigmastanes (Fig. 18); and (c) nuclear modified versions of group (a), such as polypodine-B (82), ponasterone B (76), and sengosterone (93a).

The most striking deviant is rubrosterone (83) which has a simple keto group at C-17 of the 5β-androstane skeleton, and may be a distant metabolite of ecdysone.

Phytoecdysones of category (a) are tabulated for convenience in Fig. 17. All are hydroxylated at C-20 and C-22 probably in the 20R, 22R configuration with the exception of shidasterone (90). This compound appears to be 20-iso-β-ecdysone having 22R configuration in view of its high biological activity and nonidentity with a synthetic isomer, 20R,22S-β-ecdysone (94). Inkosterone appears to be racemic at C-25 (95) contrary to the usual outcome of biological hydroxylation. Ajugasterone (Fig. 17) represents the first phytoecdysone to possess unsaturation additional to the normal 7,8-double bond but a recent structure elucidation of podecdysone-B by Horn's group (92) shows it to be a nuclear diene possibly arising from dehydration of β-ecdysone in the plant. The lactonized stigmastane derivatives cyasterone (96) and capitasterone (97) of category (b) are illustrated in Fig. 18 and present complex problems for the synthetic chemist in view of their large number of asymmetric centers.

A fundamental relationship between the zooecdysones and phytoecdysones was established when the author's group (74) showed that β-ecdysone and ponasterone-A are identical in absolute configuration at all ten asymmetric centers, by synthesis of both compounds from a common immediate precursor (Fig. 19).

Ponasterone-A

Ponasterone-B

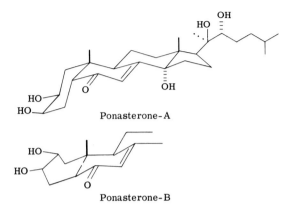

Fig. 16. Structures and probable conformations of ponasterones A and B.

Phytoecdysones (Category a)

	Substituent at					
	24	25	26	28	29	Refs.
Ponasterone-A						75
Inokosterone			OH			77
Pterosterone, viticosterone-D	OH					84, 85
Amarasterone-A	Et		OH			86
Amarasterone-B	Et				OH	86
Lemmasterone, makisterone C, podecdysone-A	Et	OH				87–89
Shidasterone (iso-β-ecdysone)		OH				90
Makisterone-A, podecdysone-D	Me	OH				91, 92
Makisterone-B	Me		OH			89
Viticosterone-E		OAc				85
Makisterone-D	Et			OH		89
Ajugasterone-B	Et	$\Delta^{25(27)}$	OH			93

Fig. 17. Table of phytoecdysones having the same steroid nucleus as α-ecdysone. All are hydroxylated at C-20 and C-22.

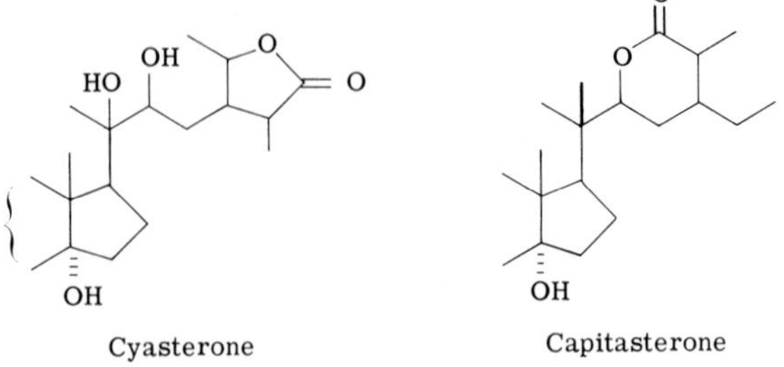

Cyasterone   Capitasterone

Fig. 18.   Lactonic phytoecdysones having the α-ecdysone nucleus.

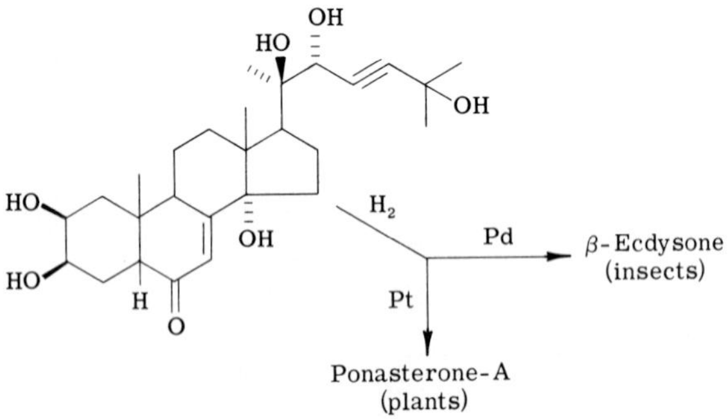

β-Ecdysone
(insects)

Ponasterone-A
(plants)

Fig. 19.   Correlation of absolute configurations of plant and insect molting hormones.

D. Synthetic Methods

Shortly after the announcement (72) of structure elucidation of α-ecdysone (Fig. 14), two independent syntheses were reported simultaneously from this laboratory (98, 99) and from a combined Schering A.G.–Hofmann La Roche group (100). The extreme scarcity of natural ecdysone and its potential interest as a *bona fide* invertebrate steroidal hormone were factors additional to the unusual challenge presented by this complex structure to the synthetic chemist. Planning for the synthesis of ecdysones involved recognition of: (1) instability in the A/B ring junction; (2) polyfunctionality, demanding protection of functions introduced at early stages; (3) a lack of suitable side chain functionalized cholestane starting materials; (4) stereochemistry, requiring stereoselective operations particularly at C-2, C-3, and C-14; and (5) choice of timing for addition of a side chain fragment to an evolving nucleus.

Surprisingly similar methods were employed by the two groups (98–100) in the synthesis of the nucleus and these have been reviewed in detail by Horn (101). In the short space available here, only certain aspects of methodology will be mentioned. The Syntex group recognized (98) that the A/B-cis fusion (characteristic of all molting hormones) would be available from the A/B-trans systems by equilibration via an enol of the 6-keto group provided that a bulky substituent is present at C-2. However, if such equilibration occurs prematurely, isomer mixtures could result at several stages with resulting inefficiency. For this reason a stereochemical holding group in the form of a 5α-hydroxyl was employed (98) and removed selectively (after acetylation) at a later stage by chromous chloride reduction.

Choice of a bisnorcholenic ester as starting material hinged on a requirement for a very stable function, able to survive some drastic changes in the nucleus but available for later alkylation when side chain synthesis was performed. This was carried out directly (99) by use of an α-sulfinyl carbanion reagent and indirectly (100) by prior transformation of the C-24 carbomethoxyl to a formyl group for Grignard alkylation.

Hydroxylation at C-14 is easily effected by selenium dioxide after a Δ^7-6-ketone system has been placed. Other methods employing epoxidation of dienol acetates (102) and photochemical oxygenation of unconjugated $\Delta^{8(14)}$-6-ketone systems (103) are available.

During a synthesis of α-ecdysone by the Teikoku group (101), a lactone precursor of the dihydroxylated α-ecdysone side chain was employed. Several sequences (99, 100, 104, 105) led to 22-isoecdysone (Fig. 20) as a major by-product which surprisingly is devoid of biological activity. Following early syntheses of α-ecdysone, the more complex β-ecdysone was first synthesized (106) from a 20-hydroxy-22,23-bisnorcholan-24-aldehyde derivative and later (107) from a 20-keto pregnane precursor. Ponasterone-A has been synthesized (74) and its stereochemical identity with β-ecdysone established (Section III,C), but the absolute configurations remain to be determined in the many phytoecdysones now known.

E. Modified Hormones

A feature of all ecdysones presently known is the presence of oxygen at C-22 of the side chain, usually as a hydroxyl group with the sole exception of capitasterone (Fig. 18) where lactonization has occurred to C-26. It is likely that this oxygen has always the β (Fischer) configuration as in α-ecdysone, since modification of α-ecdysone to its 22-isomer (Fig. 20) leads to a complete loss of biological activity. However, removal of this hydroxyl, by reduction of an acetylenic carbinol to an allene (Fig. 21) and thence to 22-deoxyecdysone, does not remove molting hormone activity (Fig. 22) in housefly assays (108). In fact, considerable biological activity (Fig. 23) is associated with 22,25-dideoxyecdysone (109) and with 25-deoxyecdysone (108) synthesized as shown in Fig. 24. These compounds might serve as biosynthetic precursors for β-ecdysone during the per-

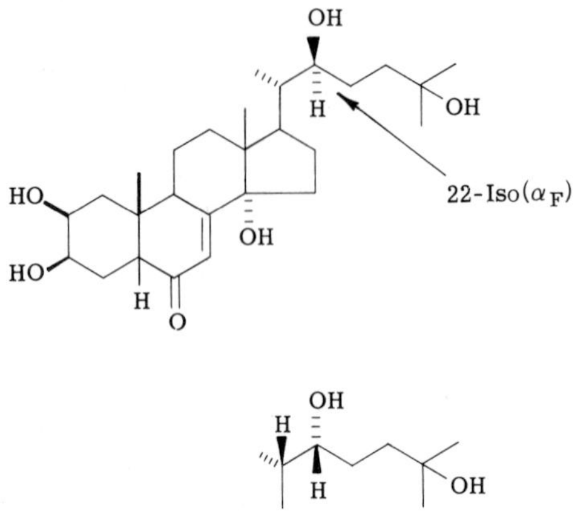

Fig. 20.   Biologically inactive 22-isoecdysone. Broken lines represent bonds to atoms lying below the plane of the paper.

formance of dipteran bioassays. By catalytic tritiation of the unsaturated hormone derivative (Fig. 24), a labeled modified hormone was prepared in the Syntex laboratory for such biosynthetic studies (110).

The specialized function of ecdysones is clearly connected with the total molecule since removal of the cholestane side chain leads to complete loss of insect molting hormone activity (Fig. 25) even when a (primary) hydroxyl group is present at the C-22 terminus of bisnorcholane derivatives. Rubrosterone (83) is apparently no exception to this generalization and presently stands somewhere between the steroidal invertebrate and the vertebrate hormones.

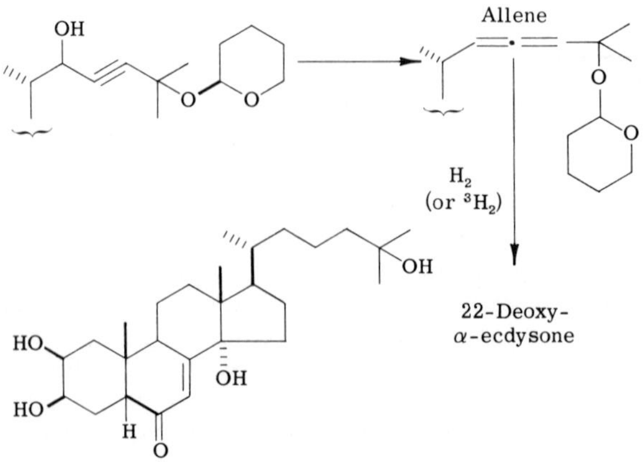

Fig. 21.   Synthesis of 22-deoxyecdysone.

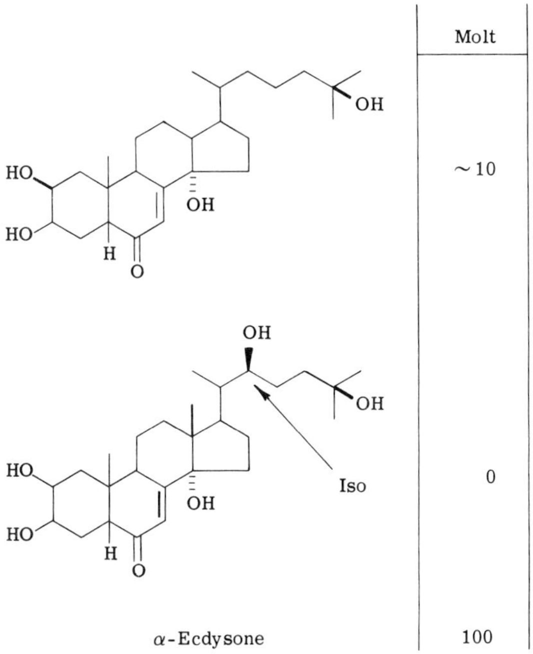

Fig. 22.   Relative potencies in the *Calliphora* assay for molting hormone activity.

F. Biosynthesis of Ecdysones

While the ecdysones have long been known as prothoracic gland hormone, the site or sites of their biosynthesis in invertebrates remain to be defined. Clearly the ecdysial glands (insect prothoracic glands or crustacean Y-organs) are involved but the extent of their contribution to ecdysone biosynthesis is unknown (111). Since the ecdysial glands could, in principle, control molting by carrying out only a single crucial step or a very small number of key transformations in the pathway from simple sterols to the complex ecdysones, a detailed knowledge of these biosynthetic steps and of the sites of their *in vivo* occurrence is most desirable.

Present knowledge of the biosynthesis of molting hormones is somewhat limited to three studies. Karlson and Hoffmeister (112) showed that cholesterol is a precursor of α-ecdysone in mature *Calliphora* larvae and suggested (113) that cholesterol, which is an essential dietary sterol in many insects, may be the starting material for molting hormone biosynthesis, possibly by dehydrogenation (114) to 7-dehydrocholesterol. Prob-

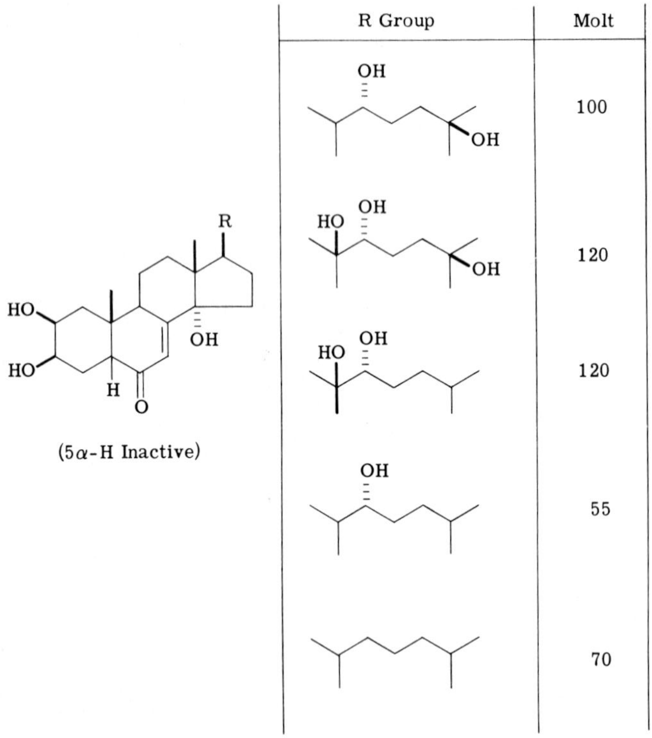

R Group	Molt
	100
	120
	120
	55
	70

(5α-H Inactive)

Fig. 23.   Relative potencies in the *Calliphora* assay for molting hormone activity.

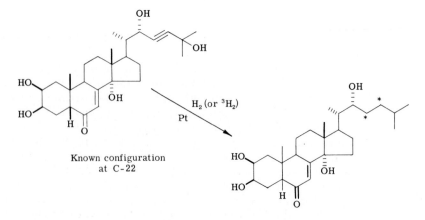

Known configuration
at C-22

H₂ (or ³H₂)
Pt

Fig. 24.   Synthesis of tritiated 25-deoxyecdysone.

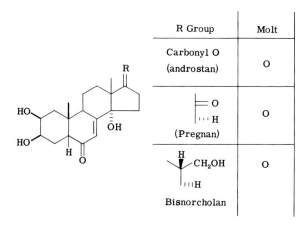

R Group	Molt
Carbonyl O (androstan)	O
⊨O ⋯H (Pregnan)	O
H CH₂OH ⋯H Bisnorcholan	O

Fig. 25. Biologically inactive analogs of ecdysone.

lems of excessive dilution of the radio label of injected cholesterol might be avoided by subtle use of a (presently unknown) slightly more advanced precursor to allow isolation of later intermediates. A similar demonstration of the precursor role of cholesterol in biosynthesis of ecdysterone in *Podocorpus elata* seedlings was accomplished by Sauer and his co-workers (115).

The last step in ecdysterone biosynthesis in crustaceans and insects has recently been found by Siddall and King (116) and by Thomson *et al.* (117) to be hydroxylation at C-20 of α-ecdysone (Fig. 27). For this work it was necessary to synthesize tritium — labeled ecdysone of high specific activity, preferably using a minimum number of radiochemical transformations. Grignard alkylation of a bisnorcholenal derivative (see Fig. 26) introduced the required unsaturation at C-23 for later catalytic tritiation. Selenium dioxide hydroxylation at C-14 then allowed smooth equilibration to the A/B-cis ring junction characteristic of all the molting hormones. Careful acid hydrolysis produced the free pentahydroxy unsaturated derivative of α-ecdysone which was catalytically tritiated over platinum, to afford 23,24-³H-α-ecdysone having a specific activity of 50 Ci/mmole (110).

Evidence for conversion (see Fig. 27) of this hormone into ecdysterone (β-ecdysone) by actively molting shrimp *Crangon nigricauda*, premolt fiddler crabs *Uca pugilator* and fifth instar larvae of the blowfly *Calliphora vicina* hinged on microchemical cleavage of the vicinal 20,22-glycol system which distinguishes ecdysterone from α-ecdysone. The possible intermediacy of α-ecdysone was first suggested by Horn *et al.* (58) and further work in progress may allow definition of earlier oxidative stages in ecdysterone biosynthesis. Such observation of facile hydroxylation *in vivo* raises the possibilities that deoxyanalogs of the molt-ing hormones may owe their observed biological activities (109) to their suitability for enzymatic hydroxylation or at higher doses to feedback inhibition of true molting hormone biosynthesis.

More recent studies by Thomson *et al.* (117) have shown that ³H-labeled 25-deoxyecdysone (110) is hydroxylated *in vivo* to inokosterone, β-ecdysone and ponasterone-A. Since the latter is not normally present in insects, 25-deoxyecdysone is probably not a normal precursor of β-ecdysone but may be involved in phytoecdysone biosynthesis.

G. Metabolism of Ecdysones

To date, the structures of molting hormone metabolites are unknown, but the isolation of 20,26-dihydroxy-α-ecdysone by Thompson and co-workers, (65, 66) may define the first such metabolite. Here a presently philosophical question of which compounds are the real hormones is involved. In this connection, Oberlander (118) has suggested that differ-ent events in metamorphosis may be controlled by distinct molting hor-mones.

Recent work by the Melbourne group (118a) has shown that previously suggested (58) side chain scission of β-ecdysone does occur in *Cal-*

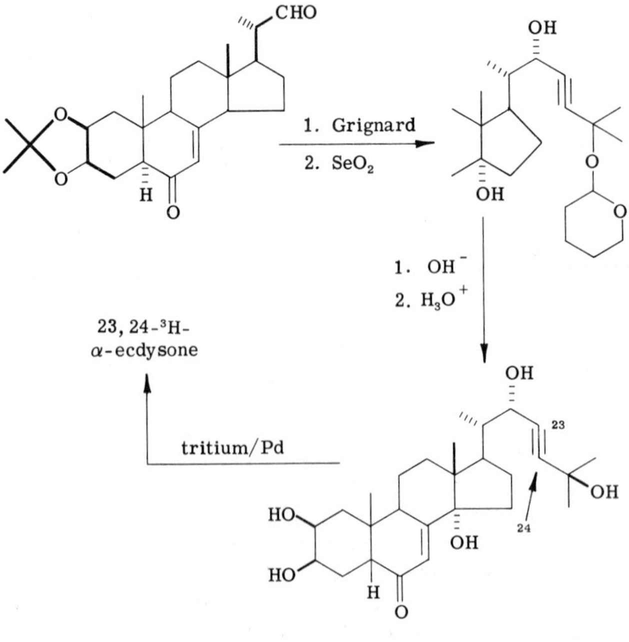

Fig. 26. Synthesis of tritiated α-ecdysone.

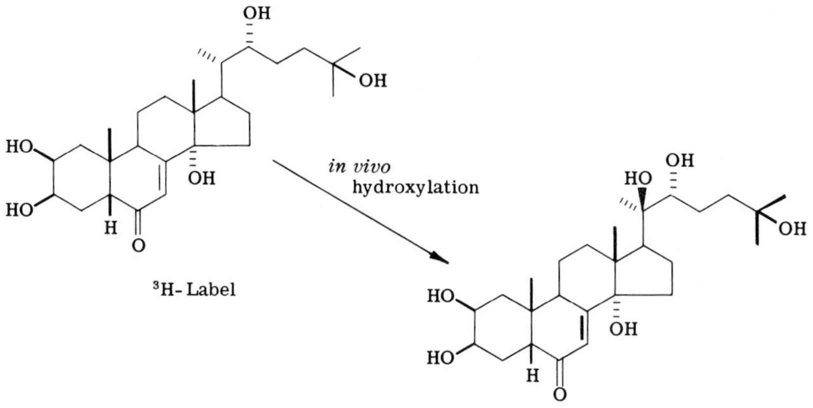

Fig. 27.   Conversion of α-ecdysone to β-ecdysone.

liphora stygia, giving rise to 4-hydroxy-4-methyl-pentanoic acid, but is not a major catabolic pathway.

Somewhat surprising in view of the high incorporation (116) of α-ecdysone into ecdysterone are various reports (119) that ecdysones are rapidly inactivated *in vivo*, when determination of residual molting hormone titer is made by bioassay. However, differences in the species examined and their differing stages of development undoubtedly contribute to the problem. Evidence for an inactivating enzyme system in the fat body of *Calliphora* has recently been presented by Karlson and Bode (120) and may provide an avenue to chemical study of ecdysone metabolites. Inhibition of ecdysone metabolism and biosynthesis may provide new methods for control of invertebrate populations but this requires considerably more detailed knowledge of the pathways of these complex processes.

References

1. Harvard Business School Reports (1966). "Selective Insect Control." Nimrod Press, Boston, Massachusetts.
2. Engelmann, F. (1968). *Ann. Rev. Entomol.* 13, 1.
3. Riddiford, L. M., and Williams, C. M. (1967). *Proc. Natl. Acad. Sci. U.S.* 57, 595.
4. Röller, H., and Dahm, K. H. (1968). *Recent Progr. Hormone Res.* 24, 666.
5. Wigglesworth, V. B. (1936). *Quart. J. Microscop. Sci.* 79, 91.
6. Sehnal, F., and Meyer, A. S. (1968). *Science* 159, 981.
7. Lawrence, P. A. (1968). *J. Cell. Sci.* 3, 391.
8. Ilan, J. (1968). *J. Biol. Chem.* 243, 5859.
9. Williams, C. M. (1956). *Nature* 178, 212; see also Williams, chapter 6.

10. Röller, H., Dahm, K. H., Sweeley, C. C., and Trost, B. M. (1967). *Angew. Chem.* 79, 190; see also (1967). *Angew. Chem. Intern. Ed. English* 6, 179.
11. Röller, H., and Dahm, K. H. (1968). *Recent Progr. Hormone Res.* 24, 651.
12. Bowers, W. S., Thompson, M. J., and Uebel, E. C. (1965). *Life Sci.* 4, 2323.
13. Dahm, K. H., Trost, B. M., and Röller, H. (1967). *J. Am. Chem. Soc.* 89, 5292.
14. Meyer, A. S., Schneiderman, H. A., and Hanzmann, E. (1968). *Federation Proc.* 27, 393; see also (1968). *Proc. Natl. Acad. Sci. U.S.* 60, 853.
15. Dahm, K. H., Röller, H., and Trost, B. M. (1968). *Life Sci.* 7, 129.
16. Braun, B. M., Jacobson, M., Schwarz, M., Sonnet, P. E., Wakabayashi, N., and Waters, R. M. (1968). *J. Econ. Entomol.* 61, 866.
17. Corey, E. J., Katzenellenbogen, J. A., Gilman, N. W., Roman, S. A., and Erickson, B. W. (1968). *J. Am. Chem. Soc.* 90, 5618.
18. Johnson, W. S., Li, T., Faulkner, D. J., and Campbell, S. F. (1968). *J. Am. Chem. Soc.* 90, 6225.
19. Zurflüh, R., Wall, E. N., Siddall, J. B., and Edwards, J. A. (1968). *J. Am. Chem. Soc.* 90, 6224.
20. Brady, S. F., Ilton, M., and Johnson, W. S. (1968). *J. Am. Chem. Soc.* 90, 2882.
21. Siddall, J. B., and Zurflüh, R. (1970). In preparation.
22. The author is greatly indebted for these assays performed by (a) Dr. H. Röller; (b) Dr. G. B. Staal; (c) Dr. V. B. Wigglesworth.
23. Siddall, J. B., Biskup, M., and Fried, J. H. (1969). *J. Am. Chem. Soc.* 91, 1853.
24. Corey, E. J., and Katzenellenbogen, J. A. (1969). *J. Am. Chem. Soc.* 91, 1851.
25. van Tamelen, E. E., and Sharpless, K. B. (1967). *Tetrahedron Letters* p. 2655.
26. van Tamelen, E. E. (1968). *Accounts Chem. Res.* 1, 111.
27. Clayton, R. B., see chapter 10.
28. Wigglesworth, V. B. (1969). *J. Insect Physiol.* 15, 73; see also Wigglesworth, V. B. (1969). *Nature* 221, 190.
29. Schneiderman, H. A., and Gilbert, L. I. (1964). *Science* 143, 325.
30. Sláma, K., and Williams, C. M. (1965). *Proc. Natl. Acad. Sci. U.S.* 54, 411.
31. Williams, C. M., see chapter 6, section XVII.
32. Bowers, W. S., Fales, H. M., Thompson, M. J., and Uebel, E. C. (1966). *Science* 154, 1020).
33. Černý, V., Dolejš, L., Labler, L., Sorm, F., and Sláma, K. (1967). *Tetrahedron Letters* p. 1053.
34. Suchý, M., Sláma, K., and Šorm, F. (1968). *Science* 162, 582.
35. Sláma, K., Suchý, M., and Šorm, F. (1968). *Biol. Bull.* 134, 154.
36. Sláma, K., Romaňuk, M., and Šorm, F. (1969). *Biol. Bull.* 136, 91.
37. Romaňuk, M., Sláma, K., and Šorm, F. (1967). *Proc. Natl. Acad. Sci. U.S.* 57, 349.
38. Law, J. H., Yuan, C., and Williams, C. M. (1966). *Proc. Natl. Acad. Sci. U.S.* 55, 576.
39. Masner, P., Sláma, K., and Landa, V. (1968). *J. Embryol. Exptl. Morphol.* 20, 25.
40. Masner, P., Sláma, K., and Landa, V. (1968). *Nature* 219, 395.
41. Conference on Insect-Plant Interactions. (1968). *Bio Science* 18, 791.
42. Mori, K., and Matsui, M. (1968). *Tetrahedron* 24, 3127.
43. Ayyar, K. S., and Rao, G. S. K. (1968). *Can. J. Chem.* 46, 1467.
44. Pawson, B. A., Cheung, H. C., Gurbaxani, S., and Saucy, G. (1968). *Chem. Commun.* p. 1057; see also *Chem. Commun.* (1969). pp. 715, 1016.
45. Birch, A. J., Macdonald, P. L., and Powell, V. H. (1969). *Tetrahedron Letters* p. 351.
46 Williams, C. M. (1966). *Science* 152, 677.
47. Spielman, A., and Williams, C. M. (1966). *Science* 154, 1043.
48. Vinson, J. W., and Williams, C. M. (1967). *Proc. Natl. Acad. Sci. U.S.* 58, 294.
49. Spielman, A., and Skaff, V. (1967). *J. Insect Physiol.* 13, 1087.
50. Bryant, P. J., and Sang, J. H. (1968). *Nature* 220, 393.
51. Williams, C. M. (1967). *Sci. Am.* 217, 13.
52. White, D., and Lamb, V. (1968). *J. Insect Physiol.* 14, 395.
53. Lessing, L. (1968). *Fortune* July, p. 88.
54. Bowers, W. S. (1968). *Science* 161, 895.
55. Bowers, W. S. (1969). *Science* 164, 323.

56. Butenandt, A., and Karlson, P. (1954). Z. *Naturforsch.* **9b**, 389.
57. Hampshire, F., and Horn, D. H. S. (1966). *Chem. Commun.* p. 37.
58. Horn, D. H. S., Middleton, E. J., Wunderlich, J. A., and Hampshire, F. (1966). *Chem. Commun.* p. 339.
59. Karlson, P. (1956). *Vitamins Hormones* **14**, 227.
60. Hoffmeister, H. (1966). *Angew. Chem.* **78**, 269.
61. Hoffmeister, H. (1966). Z. *Naturforsch.* **21b**, 335; see also (1967). Z. *Naturforsch.* **22b**, 66.
62. Hoffmeister, H., and Grützmacher, H. F. (1966). *Tetrahedron Letters* p. 4017.
63. Hocks, P., Watzke, E., Schulz, G., and Karlson, P. (1967). *Naturwissenschaften* **54**, 44.
64. Hocks, P., and Wiechert, R. (1966). *Tetrahedron Letters* p. 2989.
65. Kaplanis, J. N., Thompson, M. J., Yamamoto, R. T., Robbins, W. E., and Loulodes, S. J. (1966). *Steroids* **8**, 605.
66. Thompson, M. J., Kaplanis, J. N., Robbins, W. E., and Yamamoto, R. T. (1967). *Chem. Commun.* p. 650.
67. Galbraith, M. N., Horn, D. H. S., Middleton, E. J., and Hackney, R. J. (1968). *Chem. Commun.* p. 83.
68. Faux, A., Horn, D. H. S., Middleton, E. J., Fales, H. M., and Lowe, M. E. (1969). *Chem. Commun.* p. 175.
69. Lowe, M. E., Horn, D. H. S., and Galbraith, M. N. (1968). *Experienta* **24**, 518.
70. Krishnakumaran, A., and Schneiderman, H. A. (1968). *Nature* **220**, 601.
71. Kurata, H. (1968). *Nippon Suisan Gakkaishi* **34**, No. 10.
72. Huber, R., and Hoppe, W. (1965). *Chem. Ber.* **98**, 2403.
73. Siddall, J. B., Horn, D. H. S., and Middleton, E. J. (1967). *Chem. Commun.* p. 899.
74. Hüppi, G., and Siddall, J. B. (1968). *Tetrahedron Letters* p. 1113.
75. Nakanishi, K., Koreeda, M., Sasaki, S., Chang, M. L., and Hsu, H. Y. (1966). *Chem. Commun.* p. 915.
76. Nakanishi, K., Koreeda, M., Chang, M. L., and Hsu, H. Y. (1968). *Tetrahedron Letters* p. 1105.
77. Takemoto, T., Ogawa, S., and Nishimoto, N. (1967). *Yakugaku Zasshi* **87**, 325.
78. Galbraith, M. N., and Horn, D. H. S. (1966). *Chem. Commun.* p. 905.
79. Kaplanis, J. N., Thompson, M. J., Robbins, W. E., and Bryce, B. M. (1967). *Science* **157**, 1436.
80. Jizba, J., Herout, V., and Sorm, F. (1967). *Tetrahedron Letters* p. 1689.
81. Rimpler, H., and Schulz, G. (1967). *Tetrahedron Letters* p. 2033.
82. Jizba, J., Herout, V., and Šorm, F. (1967). *Tetrahedron Letters* p. 5139.
83. Takemoto, T., Hikino, Y., Hikino, H., Ogawa, S., and Nishimoto, N. (1968). *Tetrahedron Letters* p. 3053.
84. Takemoto, T., Arihara, S., Hikino, Y., and Hikino, H. (1968). *Tetrahedron Letters* p. 375.
85. Rimpler, H. (1969). *Tetrahedron Letters* p. 329.
86. Takemoto, T., Nomoto, K., and Hikino, H. (1968). *Tetrahedron Letters* p. 4953.
87. Galbraith, M. N., Horn, D. H. S., Porter, Q. N., and Hackney, R. J. (1968). *Chem. Commun.* p. 971.
88. Takemoto, T., Hikino, Y., Arai, T., and Hikino, H. (1968). *Tetrahedron Letters* p. 4061.
89. Imai, S., Fujioka, S., Murata, E., Sasakawa, Y., and Nakanishi, K. (1968). *Tetrahedron Letters* p. 3887.
90. Takemoto, T., Hikino, Y., Okuyama, T., Arihara, S., and Hikino, M. (1968). *Tetrahedron Letters* p. 6095.
91. Imai, S., Hori, M., Fujioka, S., Murata, E., Goto, M., and Nakanishi, K. (1968). *Tetrahedron Letters* p. 3883.
92. Galbraith, M. N., Horn, D. H. S., Middleton, E. J., and Hackney, R. J. (1969). *Chem. Commun.* p. 402.
93. Imai, S., Fujioka, S., Murata, E., Otsuka, K., and Nakanishi, K. (1969). *Chem. Commun.* p. 82.
93a. Hikino, H., Nomoto, K., and Takemoto, T. (1969). *Tetrahedron Letters* p. 1417.
94. Kerb, U., Wiechert, R., Furlenmeier, A., and Fürst, A. (1968). *Tetrahedron Letters* p. 4277.

95.   Takemoto, T., Hikino, Y., Arihara, S., Hikino, H., Ogawa, S., and Nishimoto, N. (1968). *Tetrahedron Letters* p. 2475.

96.   Takemoto, T., Hikino, Y., Nomoto, K., and Hikino, H. (1967). *Tetrahedron Letters* p. 3191.

97.   Takemoto, T., Nomoto, K., Hikino, Y., and Hikino, H. (1968). *Tetrahedron Letters* p. 4929.

98.   Siddall, J. B., Marshall, J. P., Bowers, A., Cross, A. D., Edwards, J. A., and Fried, J. H. (1966). *J. Am. Chem. Soc.* **88**, 379.

99.   Siddall, J. B., Cross, A. D., and Fried, J. H. (1966). *J. Am. Chem. Soc.* **88**, 862.

100.   Kerb, U., Hocks, P., Wiechert, R., Furlenmeier, A., Fürst, A., Langemann, A., and Waldvogel, G. (1966). *Tetrahedron Letters* p. 1387.

101.   Horn, D. H. S. (1970). "Naturally Occurring Insecticides." Dekker, New York.

102.   Mori, H., Shibata, K., Tsuneda, K., and Sawai, M. (1968). *Chem. Pharm. Bull (Tokyo)* **16**, 563, 1593.

103.   Furutachi, N., Nakadaira, Y., and Nakanishi, K. (1968). *Chem. Commun.* p. 1625. *cf.* Syntex Corp., Fr. P. 1, 524, 924 (1968).

104.   Harrison, I. T., Siddall, J. B., and Fried, J. H. (1966). *Tetrahedron Letters* p. 3457.

105.   Mori, H., Shibata, K., Tsuneda, K., and Sawai, M. (1968). *Chem. Pharm. Bull. (Tokyo)* **16**, 2416.

106.   Hüppi, G., and Siddall, J. B. (1967). *J. Am. Chem. Soc.* **89**, 6790.

107.   Kerb, U., Wiechert, R., Furlenmeier, A., and Fürst, A. (1968). *Tetrahedron Letters* p. 4277.

108.   The author is most grateful for these assays performed by Dr. J. N. Kaplanis and Dr. D. H. S. Horn.

109.   Hocks, P., Jäger, A., Kerb, U., Wiechert, R., Furlenmeier, A., Fürst, A., Langemann, A., and Waldvogel, G. (1966). *Angew. Chem. Intern. Ed. English* **5**, 673.

110.   Hafferl, W., Wren, D., Marshall, J. P., Calzada, M. C., and Siddall, J. B. (1970). *J. Labelled Compounds* (in preparation).

111.   Bern, H. A. (1967). *Am. Zoologist* **7**, 815.

112.   Karlson, P., and Hoffmeister, H. (1963). *Z. Physiol. Chem.* **331**, 298.

113.   Karlson, P. (1967). *Pure Appl. Chem.* **14**, 75.

114.   Robbins, W. E., Thompson, M. J., Kaplanis, J. N., and Shortino, T. J. (1964). *Steroids* **4**, 635.

115.   Sauer, H. H., Bennett, R. D., and Heftmann, E. (1968). *Phytochemistry* **7**, 2027.

116.   King, D. S., and Siddall, J. B. (1969). *Nature* **221**, 955.

117.   Thomson, J. A., Siddall, J. B., Galbraith, M. N., Horn, D. H. S., and Middleton, E. J. (1969). *Chem. Commun.* p. 669.

118.   Oberlander, H. (1969). *J. Insect Physiol.* **15**, 297.

118a. Galbraith, M. N., Horn, D. H. S., Middleton, E. J., Thomson, J. A., Siddall, J. B., and Hafferl, W. (1969). *Chem. Commun.* p. 1134.

119.   Ohtaki, T., Milkman, R. D., and Williams, C. M. (1968). *Biol. Bull.* **135**, 322.

120.   Karlson, P., and Bode, C. (1969). *J. Insect Physiol.* **15**, 111.

Author Index

Numbers in parentheses are reference numbers and indicate that an author's work is referred to although his name is not cited in the text. Numbers in italics show the page on which the complete reference is listed.

A

Aaronson, S., *261*
Abdul-Wahab, A. S., 47, 50, *66*
Achiwa, K., 33, *40*
Ackman, R. G., 274(219), *280*
Adams, D. K., 144, *154*
Addink, A. D. F., 264(175), *279*
Adler, J., 98, *101*
Agaki, S., 268(181), *279*
Agranoff, B. W., 237(39), 238(41), *276*
Akhtar, M., 255(118, 123, 129, 132, 133), *278*
Alexander, A. J., 163, 189, *210*
Allee, W. C., 2, *19*
Alsop, D. W., 163, 168, 180, 209, 210, *212, 213*
Alsop, R., 173, *212*
Altmann, S. A., 134, *153*
Amdur, B. H., 228, 232, 237(29), *275*
Ames, S. R., 242(61), *276*
Amourig, L., 147, *153*
Anderson, D. G., 240(48), *276*
Anderson, J. M., 230, *233*
Anderson, R. J., 249(99), *277*
Andreasen, A. A., *261*
Andrewartha, H. G., 85, *99*
Aneshansley, D., 168, 186, 187, 189, 205, *210*
Aoki, J., 204, *213*
Aplin, R. T., 86, 87, 99, 171, 172, 195, 202, *210*, 272(208), *280*
Ap Rees, T., 53, *67*
Archer, B. L., 239(43), *276*
Ardao, M. I., 205, *213*
Arigoni, D., 237(26), 247(79), 248(79), 253(106), 254(79), 258(136), *275, 277, 278*

Arihara, S., 109, *131*, 294(84, 90, 95), 295(84, 90), *305, 306*
Armstrong, D., 183, *214*
Arndt, A., 15, *18*
Arsenault, G. P., 33, 39, 147, *153*, 265(177), *279*
Atema, J., 135, 144, *154*
Atz, J. W., 223, *233*
Avigan, J., 255(125, 126), *278*
Ayyar, K. S., 290(43), *304*

B

Bachhawat, B. K., 237(28), *275*
Backus, R. H., 222, *232*
Bader, S., 258(136), *278*
Baisted, D. J., 258(143), *278*
Baker, G., 58, *66*
Baker, J. E., 92, 99, *100*
Banbury, G. H., 34, 39, *40*
Bandoni, R. J., 29, *40*
Barbier, M., *261*
Bardach, J. E., 135, 144, *154*, 221, 230, *232*
Barker, R. J., 242(59), *276*
Barkley, D. S., 5, 13, *18, 19*, 147, *154*
Barksdale, A. W., 33, 39, 40, 41, 147, *153*, 265(177, 178), *279*
Barnard, D., 239(43), *276*
Barton, D. H. R., 255(130), *278*
Barton, L. V., 46, *66*
Bautz, E., 48, *66*
Bayer, I., 54, 59, *69*
Beadle, L. C., 220, *232*
Beck, S. D., 91, 92, *100*
Becker, Y., 50, *66*
Bedard, W. D., *244*

S

Subject Index

A

Abies balsamea, 126, 290
Absinthin, 46
Acanthomyops, 143, 144, 241
Acanthomyops claviger, 139–142, 201, 243
Acetic acid, 165, 179, 180
Acetylinic compounds, 58
3-Acetyl-6-methoxybenzaldehyde, 46, 73
Acheta, 242
Achlya, 30–33, 267
  hormone A in, 32–33
  hormone C in, 33
  sexual hormones in, 31–33
  sexual progression in, 31–33
Achlya bisexualis, 147, 265
Achyranthis radix, 294
Acilius sulcatus, 266
Aconite, 57
Acrasin *see* 3′,5′-Adenosine monophosphate, cyclic
Actinomycetes, 76
3′,5′-Adenosine monophosphate, cyclic, 5, 12–15, 17, 146, 147
Adenostoma fasciculatum, 46
Aedes aegypti, 200
Agave, 56
Agglutination, yeast, 29
Aggregation, 5, 8, 13, 134, 137, 197, 199, 207, 221, 224
Aggregation inhibition, 15
Aggregation substance, 146, 149
Agria affinis, 87, 88
Agrobacterium tumaefaciens, crown-gall formation, 79
Agropyrene, 47
Agropyron repens, 47
Ailanthus, 46
Ajugasterone, 294, 295
Alarm reaction, 135, 137, *see also* Pheromones, alarm substance
Aleurodidae, 173

Algae, 24, 79
Aldehyde(s), 206
Aliphatic acid(s), 179, 206
Aliphatic aldehydes, 179
Alkaloid(s), 46, 56–57, 195
  senecio, 171, 172
Alkanes, 141–143, 148
Allelopathy, 44–65, 73–79
  autotoxicity, 50
  bacterial effects of microorganisms, 48
  definition of, 44
  effects of dead plant residues, 47
    of fire and heat on, 49
    on growth of desert plants, 73–75
    on plant succession, 49
    of rain wash from leaves, 46
    of root exudates, 47
    of volatilization from leaves, 46
  intensification by drought, 45
  by secondary plant substances, *see* Secondary plant substances
Allium chamaemoly, 52
Allomone(s), 62, 133, 134
Allomyces sp., 29, 30, 147, 267
Amaranthaceae, 111, 264
Amarasterone A, 295
Amarasterone B, 295
Amebae, 2, 147
Amphibia, 193
Amygdalin, 47, 57
Anastamosis, multicellular, 9, 10
Androstan, 301
C_{17}-Androstene, 193
Angiosperm(s), 110, 111
Anisomorpha sp., 168, 199
Anisomorpha buprestoides, 165, 177, 179, 181, 183, 199, 201, 243
Anisomorphal, 161, 241, 243
Annelida, 135
Antheraea pernyi, 107, 108
Antheraea polyphemus, 105, 106, 118, 120
Antheridiol, 147, 265
Anthranilic acid, 202